THE "DRUNKEN" SYNAPSE

Studies of Alcohol-Related Disorders

THE "DRUNKEN" SYNAPSE
Studies of Alcohol-Related Disorders

Edited by

Yuan Liu and
Walter A. Hunt
National Institute on Alcohol Abuse and Alcoholism
National Institutes of Health
Bethesda, Maryland

Springer Science+Business Media, LLC

Proceedings of a National Institute on Alcohol Abuse and Alcoholism Symposium on The "Drunken" Synapse: Studies of Alcohol-Related Disorders, held in conjunction with the 27th Annual Meeting of the Society for Neuroscience on October 25, 1997, in New Orleans, Louisiana

ISBN 978-0-306-46111-8 ISBN 978-1-4615-4739-6 (eBook)
DOI 10.1007/978-1-4615-4739-6

Originally published by Kluwer Academic / Plenum Publishers in 1999

10 9 8 7 6 5 4 3 2 1

A C.I.P. record for this book is available from the Library of Congress

PREFACE

Over the past two years, the National Institute on Alcohol Abuse and Alcoholism (NIAAA) has begun a series of symposia to highlight the need for more integrative research to understand how ethanol alters behavior. Much of the research to date has dealt either at the molecular level or has been whole animal studies. More studies are needed to build our base of knowledge between these two extremes by focusing more on cellular and network levels of organization.

To begin this focus on the intermediate steps in this scheme, the NIAAA presented a satellite symposium entitled "Approaches for Studying Neural Circuits: Application to Alcohol Research" held at the Annual Meeting of the Society for Neuroscience in Washington, DC, on November 16, 1996. This symposium brought together a group of scientists who presented their work on techniques used to study neural circuits. The proceedings of that symposium were published (Y. Liu (Ed.) Approaches for Studying Neural Circuits: Application to Alcohol Research. *Alcohol Clin Exp Res* 1998 Feb; 22:1–66).

The following year the NIAAA convened a symposium on the latest research on the actions of ethanol on the synapse. Entitled "The 'Drunken' Synapse: Studies of Alcohol-Related Disorders" and held in conjunction with the Annual Meeting of the Society for Neuroscience on October 25, 1997, in New Orleans, LA, the symposium brought together a distinguished cast of scientists who study synaptic function from various perspectives. This book represents the proceedings of this symposium and will provide not only scholarly accounts of presentations of the speakers, but also the discussion that ensued in response to these presentations. The symposium was organized around three sessions: synaptic transmission, synaptic modulation, and synaptic plasticity. The overview provides a synopsis of the chapters to follow. More specific details can be found in the individual chapters. In addition, we edited all the discussions between the audience and the speakers, grouped them under related subtitles, and organized them into three chapters placed at the end of each section. We highly encourage the readers to go through these chapters as well—the in-depth discussions during the symposium provided a wealth of information, and it is reflected in these chapters.

The NIAAA presented another symposium in its series at the Annual Meeting of the Society for Neuroscience on November 7, 1998, in Los Angeles, CA, to discuss the application of gene knockout techniques to alcohol research. This symposium specifically addressed how these techniques can be used to explore the roles of various gene products in biological changes induced by ethanol and its behavioral effects. Our hope is that these symposia will generate more interest in these exciting areas of alcohol and neuroscience research.

We would like to thank the National Institute on Alcohol Abuse and Alcoholism on sponsoring and supporting this symposium. We would also like to thank Ms. Brenda Hewitt for designing the cover of this book.

Yuan Liu, Ph.D.
Walter A. Hunt, Ph.D.

CONTENTS

1

OVERVIEW OF THE SYMPOSIUM

Walter A. Hunt and Yuan Liu

Neurosciences and Behavioral Research Branch
National Institute on Alcohol Abuse and Alcoholism
Bethesda, Maryland 20892-7003

1. INTRODUCTION

Ethanol is the most abused drug in the country. According to the Ninth Special Report to the U.S. Congress on Alcohol and Health, Americans consume 2.24 gallons of ethanol per year. Although this is the lowest level of consumption since 1964, ethanol still underlies a multitude of ills in this country. Alcohol contributes to many highway deaths, homicides, suicides, and accidents, and causes numerous medical problems in long-term abusers, inducing liver cirrhosis, pancreatitis, brain damage, and of course dependence. Almost 14 million people are classified as either alcohol abusers or alcohol dependent, using standard psychiatric instruments such as DSM-IV. Economic costs of alcohol abuse to the nation are enormous at $148 billion due to lost productivity and health care expenditures.

Most of the problems caused by ethanol consumption relate to its actions on the brain. Ethanol being a drug with many effects on neurons, researchers have been challenged to find the relevant targets on which ethanol acts. Around the turn of the century, anesthetics including alcohols were believed to act on the membranes of neurons. The Meyer-Overton Principle was formulated to explain their effects based on lipid solubility by stating that the more lipid soluble the anesthetic, the more potent it was (Meyer, 1899; Overton, 1896). Although numerous alcohols followed this principle within limits (McCreery & Hunt, 1978), actions just on the lipids themselves were insufficient to explain the various, sometimes apparently specific, effects of ethanol. Because ethanol is a simple molecule and is also amphiphilic, it unlikely has a specific receptor, as do most other drugs. Thus, research over the last decade has focused on common actions of ethanol on various functional entities in membranes responsible for neuronal excitability and neurotransmitter release. These entities are complex proteins that traverse the neuronal membrane and dangle in the intracellular and extracellular plasma. Collectively, these proteins constitute the receptors and ion channels that regulate the excitability of neurons and transmit impulses from one neuron to another. An important site on the neuron where many of these important actions of ethanol take place is at the synapse.

The "Drunken" Synapse, edited by Liu and Hunt.
Kluwer Academic / Plenum Publishers, New York, 1999.

1.1. History of Synaptic Research

Gordon Shepherd from Yale University opened the symposium with an overview describing the history of the discovery of the synapse. He outlined how the synapse was found, in part as a result of a feud over one hundred years ago between two famous neuroanatomists, Camillo Golgi and Santiago Ramon y Cajal. Golgi used his new staining technique to identify the first axon collaterals but saw these collaterals as connecting with each other into a fine network. On the other hand, Cajal noticed "breaks" between neurons and believed that neurons were not directly connected. In the 1890s, Sir Charles Sherrington coined the term "synapse", from a Greek word meaning "connection" or "junction", to describe the gap that Cajal saw. Chemical transmission between neurons had not yet been discovered and raised the question about how impulses could jump such a gap if not electrically.

For the first half of the 20th century, a debate raged as to whether transmission was electrical or chemical, a debate described by Shepherd as the "soup versus sparks" debate. By the 1950s, with synapses visualized with the electron microscope and its electrical properties characterized with microelectrodes, the notion that chemical mediators were responsible for transmission between neurons was increasingly accepted. Research in the 1970s consolidated the thinking that the synapse released mediators in a quantal fashion from stored vesicles. The vesicular membrane fused with the neuronal membrane, with the transmitters being released by diffusion into the synaptic cleft.

Finally, Shepherd emphasized the importance of neural circuits. When neurons connect with one another, they are often part of defined circuits that traverse the nervous system from one part to another. These circuits mediate a variety of functions including motivation and reward systems, which are relevant to the problem of alcoholism. He made a connection to his own research with the olfactory bulb to suggest that odor could contribute to the sensory input underlying eventual rewarding properties of ethanol. The olfactory system at the molecular level involves receptors, some of which are metabotropic, coupled to G-proteins. These are systems also affected by ethanol, as presented in chapters of this book.

As we proceed into the rest of this book, Shepherd reminds us that the synapse not only serves to relay impulses between neurons but can also alter the manner in which impulses are generated. This may be a clue into the complexities of action of ethanol on neurons.

2. SYNAPTIC TRANSMISSION

A prevailing belief in alcohol research is that the synapse is the most sensitive part of the neuron to ethanol. This belief is reflected in the pattern of research with ethanol over the last 25 years. Traditional notions state that synaptic transmission is the predominant activity at the synapse and involves the release of a transmitter, its reuptake and storage in vesicles, and its actions on pre- and postsynaptic receptors. In recent years, mechanisms by which transmitter release could be modulated have become increasingly important in understanding synaptic transmission. In this section, the discussion will revolve around research on the kinetics of neurotransmitter release and on synaptic receptors, which have been postulated as molecular targets for ethanol.

2.1. Mechanisms of Transmitter Release

To lead the session on synaptic transmission, Richard Tsien from Stanford University presented cutting-edge research on mechanisms of transmitter release and provided some insights on two aspects that had not yet been investigated in alcohol research—kinetics and regulation of presynaptic vesicle recycling and saturation of postsynaptic glutamatergic receptors.

Transmitter release has been considered an "all or none" event when synaptic vesicles release their entire store of transmitters after fusing with the presynaptic membrane. The overall recycle time of synaptic vesicles was estimated as about 40 seconds. Through a series of elegantly designed experiments, Tsien and colleagues provided evidence that this "classical" view may need to be modified (Klingauf et al., 1998). By using the fluorescent dye FM1-43, which can reversibly stain vesicle membranes, Tsien's group followed the time course of exocytosis and endocytosis closely. They first preloaded the dye into presynaptic terminals at hippocampal synapses in culture, then stimulated the cells with a high K^+, normal Ca^{++} solution to induce exocytosis and consequential endocytosis. In the classical view, if a vesicle fuses completely with the presynaptic membrane, it will lose its entire transmitter contents, as well as the dye. When monitoring with quantitative fluorescent microscopy, this event will be captured as a continuous decay of the fluorescence intensity—destaining of the vesicle. However, this was not the sole picture that Tsien's group saw. Before bleaching completely, more than one-third of the stained terminals displayed what Tsien called a "kink", a second destaining phase 20–60 seconds after the initial destaining. Tsien's interpretation of this interesting phenomenon is that some of the vesicles are performing a "kiss and run" show. Instead of collapsing with the terminal membrane completely, these vesicles pinch-off early, retaining some unreleased transmitters inside the vesicle and part of the dye in the vesicle membrane. In other words, a rapid endocytosis happened after exocytosis, then followed by a second-phase release of the freshly retrieved vesicle. To test the rate of this rapid endocytosis, they repeated the experiments with another fluorescent dye, FM2-10, which has much lower affinities to the lipid membrane than FM1-43. As predicted, terminals loaded with FM2-10 had a larger first destaining phase and a smaller second phase. The time constant of the rapid endocytosis process estimated by these experiments is about 6 seconds. Furthermore, it appears that this process of rapid endocytosis is under some regulatory control, by both calcium and staurosporine, a non-selective kinase inhibitor that only reduces dye destaining but not transmitter releasing. If this phenomenon of variable rates of vesicular recycling has physiological significance, this could be one more potential action site of ethanol in modifying synaptic transmission.

During the second part of his presentation, Tsien tackled another challenging question in synaptic research of the central nervous system—what causes the unitary synaptic current at the central fast glutamatergic synapse to be so variable? Is it due to the variability of presynaptic release or the saturation of the postsynaptic receptor? To answer this question, they again used the FM1-43 dye to visualize individual presynaptic boutons. In order to isolate individual synapses, a very small local application of the high K^+, normal Ca^{++} solution to one bouton resulted in release from the same bouton without affecting others 2 μm away. The synaptic events in the postsynaptic cell were recorded with whole-cell patch-clamp in the soma. Consistent fast time-course of the excitatory postsynaptic currents (EPSCs) was seen, however, the EPSC amplitudes varied significantly among many trials of a single bouton (Liu and Tsien, 1995). Two interpretations could explain this phenomenon: first, variations of presynaptic glutamate release without saturating postsynaptic receptors; and second, functional changes of saturated postsynaptic receptors. By comparing the local high K^+-evoked synaptic events with local "puff" of glutamate-induced synaptic events, they found the answer is the former. The dose-response curves of the two clearly indicated that postsynaptic receptors were not reaching saturation by evoked-release. These results indicate that the cause of the variability of unitary synaptic current at the hippocampal synapse probably is due to variations of transmitter release.

Through these two sets of experiments, Tsien showed us states-of-the-art methodologies of studying synaptic transmission dynamics and separating presynaptic from postsynaptic effects at the level of a single synapse. These have been major challenges for synaptic research in the central nervous system. Adapting these approaches into alcohol research will provide powerful tools to study the effect of ethanol at a single synaptic site.

2.2. L-Type Calcium and Calcium-Activated Potassium Channels

Movement of ions through the neuronal membrane involves pores or channels, allowing ions to pass through that would not ordinarily do so to any great extent. Ion channels come in two types, voltage-gated and ligand-gated channels. Voltage-gated channels open and close in response to changes in the electrical potential across the membrane. On the other hand, ligand-gated channels function in response to the binding of a particular neurotransmitter that induces a conformational change to open the channel.

Steven Treistman from the University of Massachusetts presented a series of experiments that study two different voltage-gated ion channels in neurohypophysial nerve terminals. This preparation was chosen for its relevance to ethanol-induced diuresis and its ease for measuring in tandem both antidiuretic hormone (AVP) release and electrophysiological properties of the terminals. In addition, it is a model system for studying tolerance from the molecular to the behavior level. Based on Treistman's studies, inhibition of AVP release by ethanol is caused at least by actions on two types of presynaptic channels, the L-type calcium channel and the large conductance, calcium-activated potassium (BK) channel. The beauty of Treistman's experiments is his use of several different preparations of the same channels and obtaining similar results across preparations.

Calcium is a primary regulator of neurotransmitter and hormone release, including AVP. Treistman found that the long-lasting calcium currents, mediated by L-type calcium channels are quite sensitive to ethanol. This action appears to be on the gating properties of the channel protein rather than on the permeability characteristics. Further characterization using single-channel recording techniques suggests that one ethanol molecule interacts with one channel protein molecule to inhibit the L-type calcium channel.

Studies of the BK channels indicate that ethanol is stimulatory. BK channels are important because they regulate spike width and bursting pacemaker activity. The enhanced action of this channel would further hyperpolarize the membrane, which would then inhibit the release of AVP. As with the L-type calcium channel, single channel analysis suggests that ethanol acts on the gating properties of the BK channel protein. The selectivity and permeability of the channel are unaffected by ethanol. Since calcium activates BK channels, ethanol might act by interfering or augmenting the sensor for calcium. Because of channel heterogeneity of natural membranes, further analysis was conducted using cloned mslo BK channels expressed in *Xenopus* oocytes. Similar results were found in this preparation as with the natural membranes. Ethanol enhanced channel activity by modifying gating properties rather than channel conductance. In addition, ethanol did not alter the voltage-sensitivity of the gating process, but the sensitivity of the channel to calcium was augmented. Finally, the lipid environment in natural membranes, whose composition is unknown, modulates BK channels. Using planar bilayers in which t-tubule BK channels were reconstituted, ethanol's site and mechanisms of action were preserved and comparable to the other preparations. These findings will allow for study of more specific questions regarding the role of lipids in the actions of ethanol.

2.3. $GABA_A$ and Nicotinic ACh Receptors

The remaining chapters in this section explore actions of ethanol on several ligand-gated postsynaptic ion channels. Toshio Narahashi from Northwestern University focused on the "A" type, γ-aminobutyric acid $(GABA)_A$ and nicotinic acetylcholine (ACh) receptors. The $GABA_A$ receptor regulates the flow of chloride ions into neurons, exerting an inhibitory effect on electrical activity by hyperpolarizing the synaptic membrane. On the other hand, the nicotinic ACh receptor when activated by ACh promotes an excitatory response at the synapse.

Narahashi discusses the controversial results reported in the literature on the effect of ethanol on the $GABA_A$ receptor. One group of investigators finds that ethanol augments the ability of GABA to increase chloride fluxes into neurons, whereas another group finds no effect. To resolve this discrepancy, Narahashi addressed the importance of the state of channel activation and the controversial area of subunit composition of the $GABA_A$ receptor. In human embryonic kidney (HEK) cells, ethanol had different effects on cells expressing two different GABA receptor subunit compositions, $\alpha 1\beta 2\gamma 2$ or $\alpha 6\beta 2\gamma 2$. Ethanol enhanced the decay of the chloride current without affecting its amplitude in cells expressing $\alpha 6\beta 2\gamma 2$ subunits but not in those with a $\alpha 1\beta 2\gamma 2$ subunit composition. Using single channel recording techniques, ethanol increased the frequency and duration of channel openings as well as bursts without changing amplitude. Collectively, these results suggest that ethanol enhances desensitization of the $GABA_A$ receptor, which depends on subunit composition.

In other studies, Narahashi describes studies indicating a potentiation of nicotinic ACh receptors by ethanol in undifferentiated PC12 cells. Similar to the $GABA_A$ receptor, ethanol enhances the desensitization of the receptor by accelerating the rate of current decay in whole-cell recordings. In addition, ethanol decreased the rate of dissociation of ACh from the receptor. This action leads to more receptors in the desensitized state. Further evidence is cited where ethanol induced bursts of channel openings. Since this effect was less in differentiated PC12, where β2 mRNAs are upregulated, subunit composition may contribute to the effects observed.

The ACh receptor contains a variety of α and β subunits including the $\alpha 3\beta 4$ and $\alpha 3\beta 2$ combinations. Narahashi tested the sensitivity of tsA201 cells, a derivative of HEK cells, expressing either $\alpha 3\beta 4$ or $\alpha 3\beta 2$ combinations of ACh receptors. Ethanol increases the desensitization of $\alpha 3\beta 4$ receptors without having much effect of the $\alpha 3\beta 2$ receptors. In $\alpha 3\beta 4$ expressed cells, ethanol did not alter the rate of current desensitization but enhanced the current amplitude. No effect was observed in cells expressing $\alpha 3\beta 2$ receptors. In all, the results obtained in these experiments suggest that actions of ethanol on ACh receptors also depend on the subunit composition.

2.4. $5\text{-}HT_3$ Receptors

Type 3 serotonin ($5\text{-}HT_3$) receptors are synaptic neurotransmitter receptors/channels that have not been extensively studied in alcohol research. They are the only 5-HT receptors that are ligand-gated channels. The others are coupled to second messenger systems. The $5\text{-}HT_3$ receptor has one known functional unit that has several splice variants. Its structure is similar to the nicotinic ACh receptor-like class of ligand-gated ion channels. This receptor may play a role in the actions of ethanol after both acute and chronic administration by modulating the activity of other neurotransmitters such as dopamine.

David Lovinger from Vanderbilt University has reported that ethanol potentiates $5\text{-}HT_3$-induced currents by increasing the affinity of the receptor to 5-HT. When testing different compounds, more lipid-soluble ones were more potent and had higher efficacy in enhancing the action of 5-HT. Using whole-cell patch clamp recording and rapid agonist application, the increased potency of 5-HT in the presence of ethanol resulted from an elevated rate of receptor-channel activation, along with a reduced rate of desensitization and deactivation state. Computational modeling revealed that the potentiation of the 5-HT_3 receptor involved changes in the receptor/channel activation and deactivation rate constants as well as those for the desensitization and resensitization rate constants. In addition, ethanol did not raise the maximum conductance of the channel. Finally, Lovinger argues for a possible common region for ethanol's actions on several receptors. 5-HT_3, $GABA_A$, glycine, and ACh receptors all have common amino acid sequences that may correspond to a site of action for ethanol on these receptors.

In summary, results obtained from alcohol research on both pre- and postsynaptic molecular targets presented in chapters of this section draw several common conclusions: ethanol only alters the channel opening probability and desensitization state but does not influence channel conductance and selectivity. Furthermore, most of the interactions between ethanol and its targets depend on the subunit composition of the target protein.

3. SYNAPTIC MODULATION

A major advance in the understanding of synaptic transmission is the finding that transmitter release can be modulated by numerous and diverse means. Some of these modulators are second messengers coupled to G-proteins, some are retrograde messengers, and some involve protein phosphorylation and dephosphorylation. Among many known synaptic transmitter receptors, GABA receptors have received considerable attention as possible molecular targets for ethanol. We have already learned that ethanol enhances chloride currents induced by GABA. The exact mechanism by which this is done is not known. Some evidence in the literature suggests that factors other than a direct interaction of ethanol with the GABA receptor may play a role in ethanol-induced increases in GABA currents.

3.1. Depolarization-Induced Suppression of Inhibition (DSI)

Bradley Alger from the University of Maryland presents a mechanism not yet explored in alcohol research. This mechanism involves DSI and may act to soften excess reductions in GABAergic inhibition. DSI has been found in hippocampal pyramidal and cerebellar Purkinje cells. Briefly, DSI is a suppression of $GABA_A$ receptor-mediated inhibitory postsynaptic currents by a short depolarization of the principle cell. According to his data, it seems that the consequence of DSI initiation is to reduce inhibitory postsynaptic currents by suppressing GABA release from interneurons. There is no change in the postsynaptic $GABA_A$ receptor sensitivity. Thus, DSI is a novel form of "retrograde" signaling, whereby the nominally postsynaptic target cell, the pyramidal cell, actually influences its own state of excitation, by regulating the inhibitory inputs it receives.

In his chapter, Alger first described the induction of DSI in a hippocampal slice model, then he discussed the role of various factors, such as intracellular calcium and the metabotropic glutamate receptors, in the expressions of DSI. The hippocampal DSI can be induced by various formats of stimulation, such as intracellular voltage steps in the pyramidal cells. However, DSI does not appear to require action potentials in the postsynaptic cells for its induction. On the other hand, it does require an increase in intracellular calcium, entering through the N-type voltage-gated calcium channel in postsynaptic pyramidal neurons. Under some experimental conditions, where calcium influx occurs primarily through L-type channels, DSI can also be induced. Using various pharmacological tools in sophisticated electrophysiological experiments, Alger provided abundant evidence to support his hypothesis on the mechanism of the expression of DSI. According to his data, it seems that the consequence of DSI initiation is to reduce inhibitory postsynaptic currents by suppressing GABA release from interneurons. This action may occur as a result of calcium-dependent glutamate release from the pyramidal cell, which activates group I metabotropic glutamate receptors on neighboring GABAergic interneurons, which in turn inhibit GABA release for a short period of time. As an endogenous mechanism for down regulation in the strength of $GABA_A$ inhibition, DSI may conceivably play a role in processes that are influenced by reduction in inhibition, such as the induction of LTP and LTD, as well as the onset of certain epileptic seizures.

Alger suggests that DSI could underlie some of the effects of ethanol because of its actions on the L-type calcium channels. Since activation of L-type channels can induce DSI, inhibition of these channels by ethanol would reduce DSI, augment GABA release, and possibly contribute to ethanol-induced depression. On the other hand, after chronic ethanol exposure, with the upregulation of L-type calcium channels, DSI could be augmented, GABA release suppressed, and ultimately contribute to learning deficits. DSI, thus, provides a new area of research in understanding the effect of ethanol on GABAergic synaptic connections.

3.2. Models for Studying Neurotransmitter Receptor Subunits

A continuing controversy in alcohol research is whether the subunit composition of various transmitter receptors determines or contributes to the sensitivity of the GABA receptor to ethanol. Some of the divergent results reported could depend in part on the preparation used to study different subunit compositions. These preparations employ recombinant receptors expressed in different systems, such as *Xenopus* oocytes and mammalian cell lines. An effective new technique that allows one to characterize the effect of ethanol on single cells, then extract and determine the receptor subtypes, could provide an alternative to recombinant systems. Developing and utilizing this technique, Hermes Yeh from the University of Connecticut examined this issue in several different preparations to determine how subunit composition of the $GABA_A$ receptor could modulate the action of ethanol on this receptor. He found strikingly different results across preparations.

Ethanol is known to augment some GABA receptors and not others, presumably because of the molecular structures of the receptors. In addition, controversy surrounds the importance of the $\gamma 2_L$ subunit of the receptor. Yeh could not find a requirement of the $\gamma 2_L$ for ethanol's actions using several preparations containing native $GABA_A$ receptors. For example, in retinal bipolar cells, ethanol augmented only the $GABA_C$ receptor. Using retinal ganglion cells, two populations of GABA receptors were found, one sensitive to ethanol and the other sensitive to diazepam. Finally, Yeh cleverly used rat cerebellar Purkinje cells at different stages of maturation, taking advantage of the observation that the relative abundance of the $\gamma 2$ splice variants changes over time. The $\gamma 2_S$ variant is expressed at birth and remains steady during development, whereas the $\gamma 2_L$ variant does not appear until 7 days postnatally. Ethanol potentiated the actions of GABA on its receptor in the absence of the $\gamma 2_L$ variant. These results further support Yeh's notion that when studying the effects of ethanol on synaptic membrane proteins, the differences in native cells and expression systems have to be carefully taken into account.

3.3. Adenosine Receptors

In the second section, the actions of ethanol directly on "classical" neurotransmitter receptors and ion channels were emphasized. However, the effects of ethanol on other molecular targets may also play a role in mediating its behavioral effects. One such substance is adenosine. Thomas Dunwiddie from the University of Colorado gave a thorough review of the evidence concerning how ethanol acts on the adenosine system and how it might modulate synaptic transmission.

Adenosine exerts an inhibitory modulation of synaptic activity through receptors coupled to a family of related G-proteins. Four receptors work through different pathways. The A_1 receptor inhibits adenylyl cyclase and hyperpolarizes neurons through actions on a pertussis toxin-sensitive G-protein that activates a potassium channel, thereby inhibiting transmitter release particularly of glutamate. The A_{2A} and A_{2B} receptors also activate adenylyl cyclase, whereas the A_3 receptors activate phospholipase C. These latter receptors are less understood than the A_1 receptor.

Adenosine does not appear to be a typical transmitter because it is not released in a calcium-dependent manner. Also, it usually is found in low concentrations in neurons except during metabolic stress, such as ischemia and stress. The modulatory role of adenosine depends on the status of the transporters that move adenosine into and out of the cell. During transmitter release when adenosine triphosphate (ATP) is co-released, adenosine is formed from the break down of ATP. Sufficient adenosine usually exists in the extracellular space to exert a tonic inhibitory action on neurons.

Dunwiddie reviews several possible mechanisms by which ethanol could alter adenosinergic activity. The possibilities that sufficient adenosine is formed from the metabolism of ethanol to acetate, and that ethanol acts on adenosine receptors, are not supported by the literature. He favors a mechanism by which ethanol inhibits adenosine transport into cells thereby increasing its concentration at adenosine receptors. Activation of A_2 receptors leads to stimulation a G-protein and the formation of cyclic-3',5'-adenosine monophosphate. On the other hand, chronic ethanol exposure desensitizes the A_2 receptor to adenosine as well as to a number of other neurotransmitters, a process called heterologous desensitization. This occurs as a result of altered regulation of the adenosine transporter by protein kinase A (PKA). PKA phosphorylates the transporter, making it sensitive to ethanol. With chronic ethanol exposure, the transporter reverts mostly to the dephosphorylated state, making it insensitive to ethanol. These studies have been performed predominately in cultured cells.

Experiments to examine the effect of ethanol and adenosine on neurons *in vivo* have been equivocal. Dunwiddie describes experiments with variable results but suggests that adenosine-mediated effects of ethanol may depend on high concentrations of a particular adenosine transporter, the *es* transporter, *i.e.*, brain region dependent. Further studies are needed to resolve this issue.

3.4. Metabotropic Regulation of Ethanol Sensitivity

Another source of modulation of synaptic function is metabotropic regulation. Metabotropic regulation involves G-protein-coupled transmitter receptors, such as the $GABA_B$ receptor and the metabotropic glutamate receptors. The $GABA_B$ receptor has been understudied compared to the $GABA_A$ receptor. The chapter by George Siggins from the Scripps Research Institute suggests a greater role of the $GABA_B$ receptor in the regulation of other neurotransmitter receptors than has been previously appreciated. $GABA_B$ receptors are found both pre- and postsynaptically in various areas of the brain, including the hippocampus and nucleus accumbens, the two areas Siggins studied.

Enhancement of $GABA_A$ receptor activity by ethanol in the hippocampus has been difficult to demonstrate. Siggins reevaluated these findings by isolating responses specifically mediated by the subtypes of GABA and glutamate receptors. Through various pharmacological manipulations, Siggins found that $GABA_B$ receptors suppressed the effect of ethanol on the $GABA_A$ receptor. When generating a pure monophasic $GABA_A$-inhibitory postsynaptic potential (IPSP) in the presence of a $GABA_B$ inhibitor, ethanol induced a small elevation in the peak amplitude of the $GABA_A$-IPSP, with a pronounced elongation of the response. These results suggest that in the hippocampus, $GABA_B$ receptors, possibly presynaptic, modulate the action of ethanol on $GABA_A$ receptors.

Various laboratories have found that ethanol inhibits *N*-methyl-D-aspartate (NMDA) receptor function in several brain areas. Siggins and his colleagues found ethanol blunted NMDA-induced excitatory postsynaptic potentials (EPSPs) in the nucleus accumbens, an area involved in ethanol reinforcement. The potency of ethanol was much greater in this area of the brain compared to other areas, with ethanol having an IC_{50} of 13 mM. Normally, these experiments were conducted in the presence of L-α-amino-3-hydroxy-5-methyl-4-isoxazole proprionate (AMPA) and $GABA_A$ antagonists to minimize any contribution of these receptors in the effect of ethanol. Siggins found, however, that not only ethanol blocked the NMDA-induced EPSPs, but a $GABA_B$ agonist did so as well. Moreover, a $GABA_B$ antagonist suppressed both responses, suggesting that $GABA_B$ receptors regulate ethanol's effect on the NMDA receptor in the nucleus accumbens.

Unlike the hippocampus, the nucleus accumbens has $GABA_A$ receptors sensitive to ethanol, but the effect was mostly easily observed when $GABA_B$ receptors were simultaneously antagonized. In addition, glutamate could stimulate an effect of ethanol on the $GABA_A$ receptor. However, ionotropic glutamate antagonists could not block this augmented effect, but inhibitors of metabotropic glutamate receptors could antagonize this effect. This modulation may occur through G-proteins and protein kinases and modulate the sensitivity of the receptor system to ethanol. Using an activator of protein kinase C (PKC), the number of cells with GABA currents responsive to ethanol significantly increased, an effect blocked by a PKC inhibitor. Taken together, the studies outlined in this section testify to the complexity by which ethanol's actions on neurotransmitter function are modulated.

4. SYNAPTIC PLASTICITY

Two well-known concepts in alcohol research are tolerance and dependence. Tolerance refers to the diminished response of a given dose of a drug after repeated administration. Dependence occurs when some drugs, such as ethanol, are repeatedly administered. When the drug is abruptly withdrawn, a withdrawal syndrome can result. Also, acute and chronic ethanol exposure can cause deficits in learning and memory. These changes in responsiveness of the brain to ethanol exposure have been widely studied and have been thought to result from neuroadaptation. One approach to investigate neuroadaptation is to study forms of synaptic plasticity. Synaptic plasticity involves changes in the efficiency of neurotransmission in response to a stimulus and is often used as a model for learning and memory. Charles Stevens from the Salk Institute opened this session with a discussion of experimental approaches at the single synaptic level that might begin to clarify the processes associated with synaptic plasticity.

4.1. Role of the Readily Releasable Pool of Neurotransmitter in Synaptic Plasticity

For decades, neurotransmitters have been known to be released from a readily releasable pool of synaptic vesicles. These studies were based on neurochemical studies from tissue containing multiple synapses. Stevens has performed a series of elegant experiments that study this process at the single synapse level, giving a clearer picture of the specific events involved. He built on the concept of the readily releasable pool by showing that this pool is located in active zones in the presynaptic terminal. These active zones contain a cluster of vesicles, some of which are "docked" to the presynaptic membrane, that are ready to release their contents into the synaptic cleft.

Stevens further discussed the factors influencing the probability of transmitter release. His research group first estimated the pool size of releasable quanta of transmitter at a single synapse both electrophysiologically and morphologically. Then they determined the time needed to refill them. By locally stimulating only a few synapses with a hypertonic solution, they measured the rate of postsynaptic miniature potentials produced from each synapse. They also measured the rate of endocytotic uptake of a membrane dye FM1–43 with quantitative fluorescent microscopy (Steven & Tsujimoto, 1995; Murthy & Stevens, 1998). From these studies, they concluded that each active zone contained about 5–10 vesicles in the readily releasable pool (Dobrunz & Stevens, 1997), with a refill time of 10 seconds (Rosenmund & Stevens, 1996). This conclusion was further confirmed by electron microscopic studies on the same synapses (Murthy et al., 1997). The results were also similar when transmitter release was evoked by a presynaptic action potential, indicating that the local hypertonic stimulation actually depleted the same releasable pool as by a nerve impulse (Rosenmund & Stevens, 1996). They then investigated the relationship between neurotransmission and the readily releasable pool and found that the greater the size of this pool, the greater the probability that transmitter release would occur (Dobrunz & Stevens, 1997; Murthy et al., 1997). Thus, the probability of transmitter release depends on the size of the readily releasable pool. Finally, they studied the role of this pool in synaptic plasticity by measuring the ratio of the pool size before and after establishing one form of synaptic plasticity, long-term depression (LTD). Their data indicate that when synaptic plasticity was induced at autaptic and reciprocal synapses formed by cultured hippocampal cells, the size of the readily releasable pool also changed (Goda & Stevens 1998). The same phenomenon was observed in the hippocampal slice preparation (Dobrunz & Steven, unpublished observation). In conclusion, synaptic plasticity involves increases and decreases in the size of the readily releasable pool rather than changing other aspects of a synapse. Such an observation provides a possible variable in the actions of ethanol on synaptic plasticity.

4.2. Role of the NMDA Receptors in Ethanol-Induced Inhibition of Long-Term Potentiation (LTP)

Two forms of synaptic plasticity that have been studied extensively are LTP and, more recently, LTD. Researchers have investigated the effect of acute and chronic ethanol exposure of LTP and to a limited extent LTD. LTP is of interest because it is presumed to be an electrophysiological antecedent of learning and memory, behaviors that are compromised by exposure to ethanol. To induce LTP, a pathway is exposed to high frequency stimulation, after which the response ultimately obtained is greater than that found after the initial pre-tetanic stimulus. On the other hand, LTD develops after low frequency stimulation, where the response obtained is reduced compared to the initial stimulus.

The exact mechanisms by which LTP and LTD occur are controversial. For example, there are issues of whether LTP is initiated pre- or postsynaptically or is mediated by NMDA or non-NMDA mechanisms. These issues are also present in alcohol research in attempting to identify how ethanol interferes with synaptic plasticity. In the next three chapters, the authors place increasing layers of complexity of possible mechanisms underlying the influence of ethanol on various forms of synaptic plasticity.

A leading candidate mechanism underlying LTP involves NMDA receptors. Michael Browning from the University of Colorado focused on these receptors by first acquiring sufficient dose-response effects of ethanol on the NMDA receptor to begin to relate them to potential molecular mechanisms. Using hippocampal slices and measuring the potentiation of extracellular field excitatory postsynaptic potentials (fEPSPs), Browning found that ethanol did not inhibit LTP until a concentration of 50 mM was reached. Complete inhibition occurred at 100 mM. These effects were completely reversible. In addition, ethanol's action appeared to occur on the induction of LTP, not on the maintenance phase.

Browning then attempted to relate the inhibition of LTP to inhibition of NMDA receptors. Unlike many investigators who studied the effect of ethanol on NMDA receptors on cell culture or recombinant preparations, hippocampal slices were used where blocking AMPA, $GABA_A$, and $GABA_B$ receptors could isolate the NMDA-mediated fEPSPs. Ethanol had only a modest, magnesium-dependent, inhibitory effect on NMDA receptor-mediated synaptic activities even at 100 mM concentrations.

Finally, Browning assessed whether the modest effect of ethanol on NMDA fEPSPs was sufficient to block LTP. Using two antagonists of the NMDA receptor, their effectiveness to block LTP induced by high frequency stimulation was compared to that of ethanol. Although the dose of the NMDA antagonists reduced the slope of the fEPSP and inhibited induction of LTP, they were less effective than 100 mM ethanol. These results indicate that NMDA receptor inhibition by ethanol is insufficient to account for the entire effect of ethanol on LTP. Browning suggests that other receptors such as the $GABA_A$ receptor may interact with the NMDA receptor to mediate ethanol's inhibitory effect on LTP.

4.3. Ethanol Effects on Non-"Classical" Forms of Synaptic Plasticity

As discussed earlier in this section, several forms of synaptic transmission exist in addition to NMDA receptors-mediated LTP, the "classical" type of synaptic plasticity. Richard Morrisett from the University of Texas describes experiments addressing the effects of ethanol on "classical" and other forms of synaptic plasticity. He began by discussing NMDA receptor-mediated potentiation and depression. With LTP, induced by high frequency stimulation, NMDA receptors are activated, with postsynaptic calcium concentrations rising and second and retrograde messengers induced. As a result, kainate/AMPA receptors become more sensitive to glutamate, previously silent synapses become active, and glutamate release increases. All this leads to increased synaptic strength. On the other hand, with LTD, induced by low frequency stimulation, calcium influx is less than with initiation of LTP. Phosphatases are activated, thereby reducing the phosphorylation state of the kainate/AMPA receptor and decreasing synaptic strength.

Recording population spikes from the dentate gyrus in the hippocampus, Morrisett measured LTP in response to θ-like conditioning stimuli. Ethanol in a concentration of 75 mM reduced NMDA receptor-dependent potentiation in this paradigm. However, when synaptic transmission was activated by low frequency stimulation, NMDA receptor-dependent LTD was not affected. Morrisett suggests that the different responses to ethanol could relate to differences for the two forms of synaptic plasticity in the necessary NMDA receptor subtypes or the degree of activation of NMDA receptors.

Morrisett then turned his attention to the effect of chronic exposure on NMDA-mediated processes. In these experiments, he used hippocampal explants that could be chronically exposed to ethanol *in vitro*. Using recordings of extracellular field potentials, he found that when ethanol was withdrawn, NMDA receptor-mediated field potentials were enhanced. Thus, this enhanced NMDA receptor-mediated activity contributes to the generation of seizure activity associated with ethanol withdrawal.

Synaptic plasticity can also be induced by means of non-NMDA receptor-mediated mechanisms. One of these mechanisms involves voltage-gated calcium channels. When stimulated excessively, these channels can mediate enhanced calcium influx and induce synaptic plasticity. Ethanol is known to act on these calcium channels. Acute ethanol exposure decreases channel function, whereas chronic exposure increases it. Morrisett further analyzed these actions of ethanol using whole-cell patch clamp recordings of miniature synaptic currents. Depolarizing the cell potentiated both the frequency and amplitude of the currents resulting through activation of L-type calcium channels. Ethanol in a concentration of 75 mM blocked this potentiation, suggesting another site at which ethanol might interfere with information processing.

4.4. Hippocampal Plasticity and the Importance of Subcortical Inputs

The discussion of synaptic plasticity to this point has been based on experiments performed *in vitro* using hippocampal slice preparations. The stimulus for developing synaptic plasticity has been provided locally without consideration for the possible involvement of other areas of the brain contributing both to the induction of hippocampal synaptic plasticity and the sites at which ethanol acts to disrupt it. Scott Steffensen from the Scripps Research Institute examines various aspects of synaptic plasticity using anesthetized and freely-moving animals. He concentrated on several subfields of the hippocampus, the CA1, CA3, and dentate gyrus. In the ethanol experiments, the findings not only include a direct action of ethanol on the hippocampus, but also on subcortical regions projecting to the hippocampus. These studies add the final layer of complexity provided by this section of the book on synaptic plasticity.

Using various electrophysiological parameters, Steffensen found that when administered locally or systemically, ethanol had similar effects on all subfields of the hippocampus. Ethanol decreased the amplitudes of evoked population spikes of the principal cells and concomitantly increased the number of discharges of GABAergic interneurons. Steffensen suggests that the reduced excitability of the principal cells is a result of enhanced inhibition mediated by GABA.

Interestingly, the actions of ethanol on some forms of synaptic plasticity are region-specific. Of particular importance here was the finding that the dentate gyrus responded differently to ethanol depending on whether it was administered locally or systemically. Systemic ethanol administration increased recurrent inhibition and decreased LTP in the dentate gyrus, but local ethanol exposure had no such effect. These results suggest a region outside the hippocampus mediates the effect of ethanol on LTP.

In the last set of experiments, Steffensen explores this possible role of other regions of the brain in mediating the effects of ethanol on LTP in the hippocampus. He concentrated on two areas known to project to the hippocampus, the septum and the ventral tegmental area (VTA). The septum provides the major cholinergic afferents to the hippocampus. Lesioning the septal-hippocampal pathway blocks ethanol-induced enhancement of recurrent inhibition and resulting in suppression of LTP in the dentate gyrus. When the VTA is stimulated, evoked responses in the dentate gyrus can be modulated and do so without altering excitatory monosynaptic transmission. Systemic ethanol administration reduces this modulatory effect of VTA stimulation. Local injection of ethanol, glutamate, or bicuculline into the VTA enhances recurrent inhibition in the dentate gyrus, the same effect as found after systemic ethanol administration. Dopaminergic antagonists to the septal area attenuate both ethanol-induced enhancement of recurrent inhibition and suppression of LTP in the dentate gyrus.

In studies of non-dopaminergic neurons in the VTA *in vivo,* Steffensen found them to be especially sensitive to ethanol inhibition. His hypothesis is that when the activity of these neurons is blocked by ethanol, dopaminergic neurons may become disinhibited and contributes to actions of ethanol in the hippocampus. Collectively, the data presented by Steffensen suggest that ethanol acts on synaptic plasticity in the hippocampus not only locally but also as a result of its actions in subcortical areas such as the lateral septum and VTA. In addition, because of the role of the hippocampus in learning and the role of the VTA in the reinforcing effects of ethanol, the learning component of these reinforcing effects may involve a neural circuit from the VTA to the lateral septum to the hippocampus. Thus, the VTA may be a common origin for two neural circuits mediating the reinforcing effects of ethanol.

5. FUTURE DIRECTIONS

One conclusion that can be easily drawn from this symposium is the complex nature in which ethanol exerts its many effects on the central nervous system. The fact that ethanol acts on multiple sites is not surprising and follows from its simple molecular structure and amphiphilic nature. However, ethanol does not alter everything with which it comes in contact. There appears to be some yet to be identified three-dimensional configuration that increases the probability of interaction between ethanol and the molecules, even though they have different functions. The nature of the interaction between ethanol and synaptic membrane proteins, and possible involvement of a lipid or hydrophobic environment as proposed by Treistman, is not known. Also, unknown is the relative importance of actions of ethanol pre- and postsynaptically.

This symposium presented effects of ethanol on both sides of the synaptic cleft. It is not clear which effects are direct and which are indirect. For example, do changes in release reflect a direct disruption by ethanol or do they reflect an action on the various afferent inputs to the neuron? A research opportunity exists for studying directly the effects of ethanol on the machinery involved in neurotransmitter release. Many ethanol studies have been reported measuring transmitter release, but nothing has been published on the release mechanism itself. This could relate to the issues raised by Tsien and Stevens about conditions that alter the probability and efficiency of release from single synapses. For example, vesicle-associated proteins could interact with ethanol, resulting in altered release. This is an area of research currently not studied in alcohol research.

The subunit composition of neurotransmitter receptors that renders them sensitive to ethanol has been controversial, in part on the preparations being used. Differences in results after ethanol exposure obtained from recombinant, transiently or permanently transfected cell lines, and native cells, as found in Yeh's studies, suggest a need for finding ways to effectively standardize results obtained across preparations and extrapolate them to *in vivo* preparations.

Returning to the theme of complexity, another general facet of understanding how ethanol acts on the brain is appreciating the integrative aspects of synaptic events. From basic neurobiology, we know that the excitability of a neuron is largely determined by the sum of its inputs. Numerous neurotransmitters may contribute to these influences. Much of alcohol research has focused on one particular molecular target, such as NMDA or GABA receptors, either on the presynaptic or postsynaptic site, as reflected in most of the chapters in this book. However, a few of them examine understudied receptors, such as $5HT_3$, adenosine, and nicotinic ACh receptors, as presented by Lovinger, Dunwiddie, and Narahashi, respectively.

Alcohol researchers have begun to peel back each layer of the process. For example, Siggins shows the importance of neurotransmitter interactions with findings that actions of ethanol on pre- and postsynaptic GABA receptor subtypes can modify the effect of ethanol on NMDA receptors in the hippocampus. Moreover, Steffensen showed the importance of input from other areas of the brain in synaptic plasticity in the hippocampus. This raises the critical point of the multifaceted way in which different receptors in various parts of the neuron, inhibitory inputs from interneurons, and afferents from other brain areas can regulate the responsiveness of a cell to ethanol. Furthermore, dendritic influences, as discussed by Morrisett, suggest integrated synaptic input on the neuron on which ethanol can act to change the neuron's responsiveness to ethanol. What all of this leads to is the importance of studying the effects of ethanol at the single synapse level, as elegantly shown by Tsien and Stevens. In this way, it will be possible to better understand how the action of ethanol on each cell can contribute to the whole response in a particular brain region.

Alcohol researchers are only scratching the surface on how ethanol alters synaptic plasticity. Initially, inhibition by ethanol of LTP appeared to be mediated by NMDA receptors. However, more recent studies, such as those of Browning, suggest that NMDA receptors are not the whole story and probably involve other pathways as well. Since all the studies of ethanol on synaptic plasticity have been in the hippocampus, more research is needed on other forms of synaptic plasticity, such as LTD and non-NMDA mediated LTP, and in other areas of the brain, such as the cerebellum, mesolimbic systems, and frontal cortex. Another possible direction is Alger's new form of synaptic plasticity. The model he described shows promise in understanding how ethanol modifies synaptic plasticity through modulation of synaptic integration.

Ultimately, a higher level of integration will include understanding neural circuits and how ethanol can affect them. Much of alcohol research in this area has concentrated on individual areas of a circuit. Further research at the neural circuit level is crucial and can now be pursued by applying several recent technical advances (Liu, 1998). Extracellular, single-unit recording in anaesthetized animals has long been used in neurophysiological studies. Recent modifications of single-unit recording in freely behaving animal models have converted this classic technique to a powerful new tool to study ethanol-induced impairments in motor and cognitive behaviors (Givens et al., 1998). Another recently developed technique, the multi-electrode single-unit recording in freely behaving animals, is even more powerful in neural circuit studies (Woodward et al., 1998). With this approach, patterns of electrical activity of individual neurons from different areas of a distinct neural circuit can be recorded simultaneously during a specific behavioral paradigm. Neurochemically, *in vivo* microdialysis, iontophoresis, or fast-scan cyclic voltammetry, when combined with electrophysiological and behavioral approaches, provide means of simultaneously recording neurophysiological and neurochemical activities in real-time with ongoing behaviors. When these techniques were combined with local delivery of ethanol and neurotransmitter receptor agonists or antagonists to individual neurons, transmitter release can be related to ethanol-induced events from the same neuron in freely behaving animals (Rebec, 1998; Ludvig et al., 1998).

These above-mentioned cutting-edge approaches have great promise in studying neural circuits. Once they are incorporated into alcohol research, more mechanistic questions of neural circuits mediating ethanol-related brain dysfunction can be addressed from several different levels. Other appropriate approaches or combined methods to determine the role of specific neural circuits in the behavioral effects of ethanol are needed.

Demonstrating how the results from studies presented in this book can explain the behavioral effects of ethanol has been a difficult challenge. Once the targets of ethanol at different levels are better understood, perhaps behavior studies with specific pharmacological agents or with mutants, gene knockouts, and transgenic animals will bridge the gap of our knowledge on the cause of alcoholism between molecular and behavioral levels.

As the new century approaches, the great progress in understanding the synapse, as reviewed by Shepherd, provides an expectation of new and exciting paradigms on the horizon that will offer a more complete appreciation of how the brain works. The knowledge obtained will better help identify the precise and relevant mechanisms underlying the etiology of alcoholism.

REFERENCES

Dobrunz LE, Stevens CF (1997) Heterogeneity of release probability, facilitation, and depletion at central synapses. Neuron 18 (6):995–1008

Givens B, Williams J, Gill TM (1998) Cognitive correlates of single neuron activity in task-performing animals: application to ethanol research. Alcohol Clin Exp Res 22 (1) 23–31

Goda Y, Stevens CF (1998) Readily releasable pool size changes associated with long term depression. PNAS (USA) 95 (3): 1283–1288

Klingauf J, Kavalali ET, Tsien RW (1998) Kinetics and regulation of fast endocytosis at hippocampal synapses. Nature 39:581–585

Liu G, Tsien RW (1995) Properties of synaptic transmission at single hippocampal synaptic boutons. Nature 375:404–408

Liu Y (1998) Approaches for Studying Neural Circuits: Application to Alcohol Research. Alcohol Clin Exp Res 22 (1):1–2

Ludvig N, Fox SE, Kubie JL, Altura BM, Altura B (1998) The combined single cell recording and intracerebral microdialysis method in freely behaving animals. Alcohol Clin Exp Res 22 (1):41–50

McCreery MJ, Hunt WA (1978) Physico-chemical correlates of alcohol intoxication. Neuropharmacol 17:451–461.

Meyer H (1899) Welche Eigenschaft der Anasthetica bedingt ihre narkitische Wirkung? Naunyn-Schmiedebergs Archiv Exp Pathol Pharmakol 42:109–118.

Murthy VN, Sejnowski TJ, Stevens CF (1997) Heterogeneous release properties of visualized individual hippocampal synapses. Neuron 18 (4):599–612

Murthy VN, Stevens CF (1998) Synaptic vesicles retain their identity through the endocytic cycle. Nature 392:497–501

Overton E (1896) Über die osmotischen Eigenschaften der Zelle in ihrer Betdeutung für die Toxikologie und Pharmakologie. Z physik Chem. 22:189–209.

Rebec GV (1998) Behaviorally relevant assessments of neurochemical function: single-unit recording, iontophoresis and voltammetry in awake unrestricted rats. Alcohol Clin Exp Res 22 (1): 32–40

Rosenmund C, Stevens CF (1996) Definition of the readily releasable pool of vesicles at hippocampal synapses. Neuron 16(6):1197–1207

Stevens CF, Tsujimoto T (1995) Estimates for the pool size of releasable quanta at a single central synapse and for the time required to refill the pool. PNAS 92 (3):846–849

Woodward DJ, Janak PH, Chang JY (1998) Ethanol action on neural networks studied with multineuron recording in freely moving animals. Alcohol Clin Exp Res 22 (1):10–22

2

A PERSPECTIVE ON THE SYNAPSE

Gordon M. Shepherd

Section of Neurobiology
Yale Medical School
333 Cedar Street
New Haven, Connecticut 06510

1. INTRODUCTION

Since the synapse is a fundamental building block for nervous organization, it is important to gain some perspective on how we define that unit. It's particularly appropriate to do that in this year, because it was just 100 years ago that the term was introduced by Sherrington (summarized in Shepherd and Erulkar, 1997). I will then consider briefly some of our own studies, to illustrate how work in a particular model system can have relevance to understanding the circuit functions of synapses.

2. CENTENARY OF THE SYNAPSE

The immediate background for the time at which Sherrington introduced the term "synapse" was the battle that had arisen in the 1890s between Camillo Golgi on the one hand and Santiago Ramon y Cajal on the other over a basic concept of how the nervous system is organized at the cellular level (summarized in Shepherd, 1991; Jones, 1994).

Golgi invented the stain that enabled one to see an individual nerve cell in its entirety. This advance, which he made in 1873 but didn't reach the rest of the world until 1886, revolutionized the brain sciences and initiated the cellular study of brain function and structure. Amongst many fundamental discoveries was the first identification of axon collaterals. However, because even with his stain it was difficult to see how the finest twigs terminate, he believed that the axon collaterals merge into a network of fine anastomoses with each other. He therefore became an adherent of the network theory of nervous organization, the idea that the axon collaterals form a continuous network, much as the capillaries form a continuous network within the vascular system.

Cajal, on the other hand, with his use of the Golgi stain, came up with a strikingly different picture. From his first publications he showed the neurons of the cerebellum more distinctly stained, with the axons and axon collaterals ending bluntly. From this observation he deduced that each cell is an anatomical unit in which the fibers end without anastomosing. But how do nerve cells interact if they are not connected with each other? He gained his essential insight by focusing his attention on the part of the cerebellum in which what he called the "descending fringes" of the basket cells end in relation to the cell bodies of the Purkinje cells. That apposition he called a contact or site of articulation, and it was there he surmised that the two cell types interact. He generalized this view to other regions where the axons end by contact rather than by continuity on their target neurons.

This was the background for Sherrington's studies during the 1890s on the reflex organization of the spinal cord. When it came time to interpret how nerve cells mediate reflexes in the spinal cord, Sherrington was impressed with the fact that there seemed to be a sharp contrast between the properties of axon conduction and the properties of the spinal reflexes. The impulses in the axons had all or nothing properties, with distinct thresholds, and were followed by refractory periods. By contrast, the reflexes, as recorded by the muscle contractions induced by sensory nerve activation, had graded properties, with no distinct threshold and no refractory periods.

Sherrington was asked by Foster to revise the section on the nervous system in Foster's textbook of physiology. When it came to the point of generalizing from his studies of how nervous transmission occurred through the spinal cord, he had to describe the articulation between axons and nerve cells, and felt the need for a new term. He was well acquainted with Cajal's work and ideas, having hosted Cajal when the latter visited London in 1894 to deliver the Croonian Lectures. Basing his interpretation largely on Cajal's depictions of the articulations between afferent nerve endings and motor neurons, he introduced a new term (Sherrington, 1897): "So far as our present knowledge goes, we are led to think that the tip of the twig of the arborescence of an axon is not continuous but merely in contact. Such a special connection might be called a synapse." Later, in his great book "The Integrative Action of the Nervous System" (Sherrington, 1906), he speculated on what the functional properties of the synapse might be: "Such a surface might restrain diffusion, back up osmotic pressure, restrict movement of ions, electric charges, double electric layer, altering shape, surface tension, intervene as the membrane between dilute solutions of electrolytes."

Now the point is not that any of these mechanisms is the correct one, but rather that he was drawing on everything then known about the functional properties of cells to speculate that one or more of these properties could be involved in the interactions between cells at these sites of contact. This is a useful perspective to have as we ourselves struggle with describing the increasing range of interactions that takes place between nerve cells in general and between nerve cells at their specific points of contact, the synapses.

At that early stage in describing the possible mechanisms at the synapse Sherrington didn't mention anywhere the possibility that a chemical released by one cell acts on another one. That idea came from another line of work, chiefly from a group in Cambridge under John Langley. The first specific statement is found in an early paper by one of his students, Thomas Elliott (1904), who suggested in an analysis of smooth muscle and other glands that "...adrenaline might be released by nerve fibers to act on receptive substances in the postsynaptic target."

From around 1900 to 1950 a great debate took place, which is sometimes called the "soup versus sparks" debate. This debate between biochemists and pharmacologists, on the one hand, and electrophysiologists, on the other, dealt with whether the action at a synapse is primarily or exclusively electrical or chemical.

It was the coming of age of neuroscience, to be able to integrate those two great basic approaches to nervous function, biochemical and electrical. Here I would like to fast-forward to the time of the 1950s when the evidence for the synapse and its functional properties finally was obtained anatomically by the electron microscope and physiologically by microelectrodes.

The introduction of the intracellular electrode by Ling and Gerard (1949) was every bit as big a step forward for the development of neuroscience as the recent introduction of the patch electrode has been for current studies of the nervous system. This innovation moved the analysis of brain mechanisms from the organ level to the single-cell level, and thereby permitted the melding of the electrophysiological and neuropharmacological approaches.

In the early 1950s, Bernard Katz and his collaborators (Fatt and Katz, 1951) used the new recording technique to reveal the end-plate potential from the skeletal muscle activated by a single impulse in a skeletal muscle nerve. They showed that the end-plate potential arises after a delay following an impulse invading the nerve terminals, and that it in turn gives rise to the action potential of the muscle. Fatt and Katz (1952) also found that very small potentials, called miniature end-plate potentials or miniature EPSPs, underlie the large-amplitude end plate response. There had been evidence for the large-amplitude endplate potential ever since the late 1930s, but the miniature end-plate potentials were new. In fact, they are becoming more and more central to the concept of a fundamental physiological unit of the synapse.

At about the same time, John Eccles and his collaborators recorded the excitatory postsynaptic potentials (EPSPs) in spinal motor neurons that depolarize the cell membrane to give rise to action potentials (Brock et al., 1952). And very soon came the evidence for hyperpolarizations of the cell membrane to produce inhibitory postsynaptic potentials (IPSPs) that oppose the depolarizing actions of EPSPs (Fatt and Katz, 1953; Eccles et al., 1954).

In addition to this evidence for the action of single types of synapses, Eccles began to identify the synaptic circuits that are involved in mediating the reflex control of the motor neuron. The first circuit identified was the connections of axon collaterals onto Renshaw cells that feed back recurrent and lateral inhibition to control the output of the motor neurons (Eccles et al., 1954). Pharmacological agents were needed to analyze the excitatory and inhibitory actions at the different synapses in these circuits. This was the beginning of what now is our standard approach to applying pharmacological tools, together with physiology and the evidence from anatomy, to the analysis of synaptic circuits.

These first fundamental advances in the physiology of synapses were made before knowledge of the structure of the synapse. Shortly thereafter, in the middle 1950s, Sanford Palay and George Palade on the one hand, and Eduardo de Robertis and H. S. Bennett on the other, applied the electron microscope to describing the junctions between cells at the level of the cell membranes and organelles. Individual contacts could be seen characterized by increased densification under the postsynaptic membrane; in the presynaptic terminal were collections of synaptic vesicles (Palade and Palay, 1954; Palay, 1956; de Robertis and Bennett, 1954; de Robertis, 1958). Since that time the combination of membrane density and local vesicle cluster has been the consensus definition of a synapse. Two features have been remarkable about this morphological definition. One is how well it has applied to many of the contacts present throughout the nervous system. The other is how variable the morphology can be: this variability can include contacts with densities but few or no vesicles; vesicles with little or no membrane density; large contacts, small contacts; large vesicles and small vesicles; round vesicles and flattened vesicles.

The possibility that a physiological quantum is equivalent to the action of a single synaptic vesicle was obvious from the start, and soon was incorporated into the concept of the synapse as a functional unit. Combined morphological and functional studies ultimately converged on a concept of the synapse best summarized in the work of Heuser, Reese and Landis (1974), in which vesicular release was identified as involving fusion with the membrane, followed by recycling through reconstitution of the vesicles, and reloading of them with transmitter for release again. In the last several years, this mechanism is beginning to have a number of different kinds of alternatives, such as vesicular fusion by very brief fusion events, the so-called "kiss and run" type of fusion release, rather than the whole cycle.

A range of mechanisms is now emerging from the analysis of every step along the way, from the formation of vesicles, loading of vesicles, movement to the presynaptic membrane, release, diffusion, action on a postsynaptic receptor, and the extraordinary range of postsynaptic target mechanisms. This range of actions was, in fact, already becoming apparent to Bernard Katz, who in his book on "Nerve, Muscle and Synapse"(1966), wrote "The more one finds out about properties of different synapses, the less grows one's inclination to make general statements about their mode of action."

As if that isn't enough, we have to remember that in analyzing the interactions between nerve cells we have to take account of an increasing number of what can be called nonsynaptic interactions. There are electrical synapses/gap junctions that mediate metabolic exchange and electrical interactions between cells at specific sites. There are electrical field potential interactions. There are presynaptic autoreceptors in great quantity, which may or may not be acting at what one can call a classical synapse. There is nonvesicular and calcium-independent transmitter release. Slow actions of neuropeptides and neurohormones grade over into long-lasting effects that are difficult to characterize as specifically "synaptic". Diffuse transmitter actions occur at a distance: gaseous messengers provide for local interactions in all directions, not just from pre- to postsynaptic components; and finally, the functions of neuroglia are very rich field for the supporting and mediating the interactions between nerve cells (summarized in Shepherd and Erulkar, 1997).

3. RELEVANCE TO ALCOHOLISM

The relevance of these topics for understanding the synaptic basis not only of normal function but also of disorders such as alcoholism is illustrated by a recent report to the Congress on the effects of alcohol on health (NIAAA, 1997). It summarizes the evidence for the effects of alcohol at GABAergic neurons, at glutamatergic synapses, on adenosine transporters, and so forth. And this is not including effects on ion channels.

Although it is natural to focus on the microstructure and microfunction of the synapse in understanding its fundamental nature, the functions of a synapse are always expressed in behavior by synaptic circuits, and so we are led always back to the circuits. That same report summarizes some of the basic circuits involved in any effects of alcohol or other drugs of abuse on the motivation and reward systems at the core of the brain running through the median forebrain bundle, connecting different parts of the brain. In addition to GABAergic systems, the opioid systems, both long-range endorphin systems and short-range enkephalin systems, are extremely important.

I would like to end by summarizing briefly some of our recent studies that are relevant to the topic of the day. They involve a model system, the olfactory bulb, receiving input from the olfactory receptors in the nose and having connections indirectly or directly to many of the basal forebrain systems that are involved in the development of alcoholism. We might wonder why the olfactory pathway would play any role in these systems. Imagine a wine expert, an oenologist, taking a sniff of the bouquet of a glass of wine. He's going to take a drink, which is okay, because he's going to spit it out. But much of the problem that the NIAAA is dealing with is people who don't know when they shouldn't be taking that next drink, and the next one and the next one.

The main sensory input driving this behavior is mediated through the nose. Everyone is aware of the olfactory contribution to wine tasting, but I don't know whether the contribution to alcoholism has ever been adequately addressed. Perhaps it has, but it seems to me that it must play a role, so let me remind you of some things we now know about this system. There is, in fact, an increasing amount of work at all stages of the olfactory system, from the periphery to the central, and at all levels, from molecular to systems (summarized in Shepherd and Greer, 1998).

Based on the findings of Buck and Axel (1991) regarding a putative olfactory receptor protein, we have carried out computer modeling studies to obtain insight into the nature of odor-receptor interactions. These studies (summarized in Shepherd et al, 1996) suggest that there is a binding pocket in the G-protein coupled seven-transmembrane domain olfactory receptors. This appears to be similar to the binding pocket for epinephrine in a β-adrenergic receptor, except that we postulate that olfactory receptors interact with broader affinities for different odor molecules. This would explain the fact that olfactory cells appear to respond broadly to variable numbers of odor ligands, though they may express only one type of receptor molecule, or at most a very few.

Using carefully selected and analyzed homologous series of different odorous chemical compounds, Kensaku Mori and his collaborators in Japan have shown that a given cell has what they term a "molecular receptive range" that is characteristic for that given cell, with peak responses for a given member of a series and surrounding smaller responses (Mori and Yoshihara, 1995). From these experiments one can begin to identify the olfactory circuit, from receptor cells through mitral cells to the olfactory cortex, in which the glomerulus, receiving the input from a subset of olfactory neurons, defines the response spectrum of a given mitral cell that is connected to it. Different glomeruli have different response spectra that are overlapping but different, with lateral inhibition mediated by granule cells onto mitral cells (Rall and Shepherd, 1968; Yokoi et al., 1995) then coming into play to sharpen the responses of a given cell.

It has also been shown from the work of Nakanishi and his colleagues (Kaba et al., 1994) in the accessory olfactory bulb that these same inhibitory synapses are the sites of metabotropic receptors that are involved in an olfactory kind of memory. One can postulate that this might be a more general type of memory mechanism also found in the main olfactory bulb. Learned responses to different odors—and here alcohol might have an odor that is learned—could contribute to the substrate for identifying and remembering a particular odor on which one then becomes dependent. It is a possibility that deserves testing.

Finally, we have reported recently, using dual-patch recordings from the mitral cell body and the distal dendrite, that with increasing excitatory synaptic input to the distal dendritic tuft the site of impulse initiation changes from being in the soma and axon hillock first and dendrites second to the dendrites first and the soma and axon hillock second (Chen et al, 1997). This enlarges the possibility of multiple sites of impulse initiation in soma-dendritic trees, building on the recent work of Stuart and Sakmann and others. We also show that the site of impulse initiation can be controlled by the level of synaptic inhibition through GABAergic receptors in the secondary dendrites.

These results show that the significance of the synapse lies not just in its role of controlling the impulse output through the axon initial segment, but also in an additional role of controlling the site of action potential initiation throughout the entire extent of the neuron, from initial axonal segment through the soma to the furthest reaches of the dendritic tree. When synapses get "drunk", they change not just the response to the input at those synaptic sites, but also potentially alter the entire integrative framework of that neuron in generating its impulse output. The implications for the functional organization of that neuron are thus profound. The following chapters will shed light on some of the properties of synapses that are most critical with regard to the functional organization of the normal neuron, and in addition will provide insights into some of the effects on those functions that may accompany alcoholism and related disorders.

ACKNOWLEDGMENTS

Our work has been supported by the National Institute for Deafness and other Communicative Disorders (National Institutes of Health), by the National Institute for Neurological Diseases and Stroke (National Institutes of Health), and by the National Aeronautics and Space Agency, National Institute of Mental Health, and the National Institute for Deafness and other Communicative Disorders and the National Institute of Alcohol Abuse and Alcoholism (National Institutes of Health) through the Human Brain Project.

REFERENCES

Brock LG, Coombs JS, Eccles JC (1952) The recording of potentials from motorneurones with an intracellular electrode. J Physiol (Lond) 117: 431–460.

Buck LD, Axel R (1991) A novel multigene family may encode odorant receptors: a molecular basis for odorant recognition. Cell 65:175–187.

Chen WR, Midtgaard J, Shepherd GM (1997) Forward and backward propagation of dendritic impulses and their synaptic control in mitralcells. Science 278: 463–467.

De Robertis E (1958) Submicroscopic morphology of the synapse. Int Rev Cytol 8: 61–96.

De Robertis E, Bennett HS (1954) Submicroscopic vesicular component in the synapse. Fed Proc 13: 35.

Eccles JC, Fatt P, Koketsu K (1954) Cholinergic and inhibitory synapses in a pathway from motor-axon collaterals to motoneurones. J Physiol (Lond) 126: 524–562.

Elliott TR (1904) On the action of adrenaline. J Physiol (Lond) 31, XX-XXIP.

Fatt P, Katz B (1951) An analysis of the end-plate potential recorded with an intra-cellular electrode. J Physiol (Lond) 115: 320–370.

Fatt P, Katz B (1952) Spontaneous subthreshold activity at motornerve endings. J Physiol (Lond) 117: 109–128.

Fatt P, Katz B (1953) The effect of inhibitory nerve impulses on a crustacean muscle fibre. J Physiol (Lond) 121: 374–389.

Heuser JE, Reese TS, Landis DMD (1974) Functional changes in frog neuromuscular junctions studied with freeze-fracture. J Neurocytol 3: 109–131.

Jones EG (1994) The neuron doctrine 1891. J Hist Neurosci 3: 3–20.

Kaba H, Hayashi Y, Higuchi T, Nakanishi S (1994) Induction of an olfactory memory by the activation of a metabotropic glutamate receptor. Science 265: 262–264.

Katz B (1966) Nerve, Muscle, and Synapse. New York: McGraw-Hill.

Ling G, Gerard RW (1949) The normal membrane potential of frog sartorius muscle. J Cell Comp Physiol 34: 383–396.

Mori K, Yoshihara Y (1995) Molecular recognition and olfactory processing in the mammalian olfactory system. Progr Neurobiol 45:585–619.

The Ninth Special Report to the U.S. Congress on Alcohol and Health, (1997) NIAAA, Bethesda, MD.

Palade GE, Palay SL (1954) Electron microscope observations of interneuronal and neuromuscular synapses. Anat Rec 118: 335–336.

Palay SL (1956) Synapses in the central nervous system. J Biophys Biochem Cytol 2: 193–202.

Rall W, Shepherd GM (1968) Theoretical reconstruction of field potentials and dendro-dendritic synaptic interactions in olfactory bulb. J Neurophysiol 31: 884–915.

Shepherd GM (1991) Foundations of the Neuron Doctrine. New York: Oxford University Press.
Shepherd GM, Erulkar SD (1997) Centenary of the synapse: from Sherrington to the molecular biology of the synapse and beyond. Trends Neurosci 20: 385–392.
Shepherd GM, Greer CA (1998) Olfactory bulb. In G. M. Shepherd (Eds.), The Synaptic Organization of the Brain (pp. 159–203). New York: Oxford University Press.
Shepherd GM, Singer MS, Greer CA (1996) Olfactory receptors: a large gene family with broad affinities and multiple functions. Neuroscientist 2: 262–271.
Sherrington C (1906) The Integrative Action of the Nervous System. New Haven: Yale University Press.
Sherrington CS (1897) Nervous System. In M. Foster (Ed.) Textbook of Physiology.
Yokoi M, Mori K, Nakanishi S (1995) Refinement of odor molecule tuning by dendrodendritic synaptic inhibition in the olfactory bulb. Proc Natl Acad Sci (USA) 92:3371–3375.

Section I

SYNAPTIC TRANSMISSION

3

MOLECULAR TARGETS UNDERLYING ETHANOL-MEDIATED REDUCTION OF HORMONE RELEASE FROM NEUROHYPOPHYSIAL NERVE TERMINALS

Steven N. Treistman, Benson Chu, and Alejandro M. Dopico

Department of Pharmacology and the Neuroscience Program
University of Massachusetts Medical Center
55 Lake Avenue North
Worcester, Massachusetts 01655

1. INTRODUCTION

In developing a model system in which to study the molecular basis for the acute and chronic actions of ethanol in the nervous system, our basic philosophy has been: 1) to work with a relevant molecular target (i.e., a mediator of a behavioral or physiological consequence of ethanol ingestion), which is 2) amenable to analysis at the molecular level, and in which 3) we can identify the biophysical parameters responsible for acute modulation by the drug. It will also be possible to follow alterations in the function and ethanol response of this target during chronic exposure of the animal to ethanol, and the development of various forms of tolerance in those systems or behaviors subserved by the target molecule.

For years, it has been known that plasma vasopressin (AVP, also known as anti-diuretic hormone) levels in animals, including humans, are depressed after acute exposure to ethanol (Dopico et al., 1995). We have recently made significant progress in understanding the basis for the acute inhibition of AVP release from the neurohypophysis by ethanol, identifying two populations of membrane channels that are inhibited and potentiated, respectively, by this drug, resulting in reduced release of AVP and oxytocin. The ability to measure release and electrophysiology in tandem from both the intact posterior pituitary and isolated terminals makes mechanistic studies possible. These conditions make ethanol inhibition of diuresis ideal for the study of tolerance.

The "Drunken" Synapse, edited by Liu and Hunt.
Kluwer Academic / Plenum Publishers, New York, 1999.

1.1. Anatomy of the Vasopressin and Oxytocin Neurosecretory System

The neurohypophysis (also known as the neural lobe or posterior pituitary), consists of numerous nerve terminals that originate mainly from magnocellular neurons in hypothalamic nuclei. Other cellular elements in the neurohypophysis include fibroblasts (relatively scarce), macrophages, mast cells, and an astrocyte-derived population of supportive cells named pituicytes (Lederis, 1965; Bergland and Torack, 1969). Among the different hypothalamic nuclei from which the neurohypophysial terminals originate, the most clearly defined are the supraoptic (SON) and the paraventricular (PVN), composed of 30,000–50,000 large magnocellular neurons, as well as numerous smaller parvicellular neurons (Scheithauer et al., 1992; Kozlowki, 1990). The projections from SON and PVN give rise to the supraopticohypophysial and the paraventriculohypophysial tract, respectively, whereas the projections found in the posterior wall are usually referred to as the tuberohypophysial tract. The latter is thought to originate in a number of different hypothalamic regions, including the PVN and the central, posterior, mammillary, and tuberal regions (Stopa et al., 1993; Reichin, 1992).

Two peptide hormones, arginine-vasopressin (AVP) and oxytocin (OT), are synthesized in magnocellular neurons in the SON and PVN. Vasopressin and oxytocin are synthesized together with their respective neurophysins in the rough endoplasmic reticulum, stored in granules in the Golgi complex, and transported along the axons into the neurohypophysis where the content of the granules is released by exocytosis (Brownstein et al., 1982; Pickering et al., 1986; Schmale et al., 1987). It has been estimated that approximately 2,000 terminals emanate from each cell body in the hypothalamus, and each nerve ending contains thousands of peptide-containing granules (Nordmann, 1977).

2. RESULTS

2.1. Inhibition of Calcium Channels and Inhibition of AVP Release

The role of inhibition of calcium (Ca^{++}) channels by ethanol in the inhibition of release of AVP from neurohypophysial terminals has been documented by a combination of biochemical and electrophysiological techniques. Release of AVP from the intact neurohypophysis, and from nerve terminals isolated from the rat neurohypophysis is very sensitive to ethanol (Wang et al., 1991a,b). However, ethanol does not affect the release of AVP from terminals that had been permeabilized with digitonin, suggesting that voltage-gated calcium channels might be the targets of ethanol's actions. Patch clamping of these terminals indicated that both inactivating and long-lasting calcium currents were reduced in ethanol, but that the long-lasting currents were more sensitive. Ethanol-induced decreases in plasma AVP levels can be at least partly explained by ethanol's inhibition of calcium currents in the nerve terminals. Most recently, we have shown that the open time duration of the long-lasting or L-type Ca^{++} channel is very sensitive to ethanol, and that this effect could contribute to the reduction in AVP release (Figure 1). For this channel, as well as the potassium channel discussed below, the primary action of ethanol is on the gating properties of the channel protein, with little effect on other properties, such as voltage-dependency, ion selectivity, or channel conductance. These studies are described more fully in published manuscripts (Wang et al., 1991a,b; 1994).

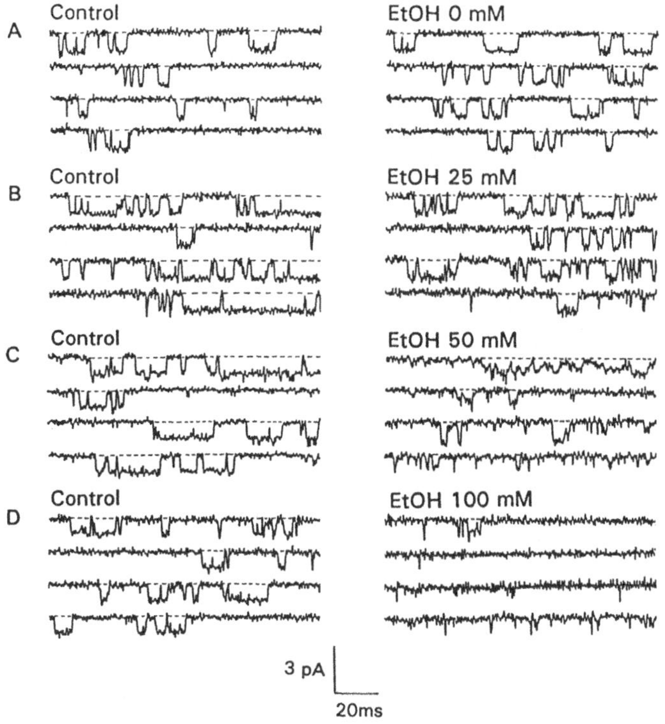

Figure 1. Effects of ethanol on neurohypophysial terminal L-type Ca^{++} channel currents, obtained during a step to +10 mV (Vh=-50 mV), in the presence of 5 μM Bay K 8644. A-D: Representative cell-attached single channel current traces recorded from a nerve terminal before (left panel) and after (right panel) exposure to concentrations of ethanol as noted. (Taken from Wang et al., J Neurosci 14:5453–5460 (1994) with permission from publisher).

2.2. Bimolecular Interaction between Ethanol and L-Type Channels

This aspect of the work used sophisticated single channel techniques and analysis to unequivocally identify the channel type being inhibited by ethanol, and to fully characterize the inhibition (Wang et al., 1994). In addition, our single channel results were consistent with a bimolecular interaction between the drug and the channel. This work represented, to our knowledge, the most complete characterization of the interaction of ethanol with a voltage-gated channel that had been done at that time. However, in this chapter, I will be focussing on the other primary target of ethanol action in the terminal, the Ca^{++}-activated K^+ channel, whose activity is potentiated by ethanol, at similar concentrations to those that inhibit the activity of the voltage-gated Ca^{++} channel.

2.3. Ethanol Potentiates Terminal Large-Conductance Calcium-Activated Potassium (BK) Channels

Ca^{++}-activated potassium (K^+) current is prominent in the terminals, and since it is an important determinant of spike width and bursting pacemaker activity, which in turn, are determinants of peptide release, we decided to use single channel techniques to examine whether a part of ethanol's inhibition of AVP release was related to augmentation of activity in this channel (called BK, for big K^+ conductance). These studies were done with inside-out patches, minimizing the contribution of intracellular components, and allowing us to control Ca^{++} concentrations at the intracellular side of the channel. The channels under study typically show high selectivity for K^+, and show a reversal potential close to 0 mV in excised, inside-out patches in symmetric 145 mM K^+ concentrations. They have a unitary conductance of 218 pS, and no rectification was observed from -60 mV to +60 mV of membrane potential. Channel activity increased at more positive membrane potentials (10–15 mV for an e-fold change in NP_o at low P_o values (N = the number of active channels in the patch, and P_o = the probability of a channel being open at a given time) and/or upon elevating the $[Ca^{++}]$ at the intracellular surface of the patch. All these characteristics are typical of BK channels, first reported in cultured bovine chromaffin cells, and since described in a wide variety of preparations, including, by us, rat neurohypophysial terminals (Wang et al., 1992).

Ethanol (10–100 mM) applied to the cytosolic surface of the patch rapidly and reversibly activates BK channels (i.e., increased NP_o) (Figure 2A), in a concentration-dependent manner (Figure 2B). The ethanol-induced increase in BK channel activity occurred with no change in the single channel conductance, the reversal potential, or the shape of the I/V relationship. These findings suggest that ethanol activation does not modify the high selectivity and permeation characteristics of this channel. Thus, the pore forming region of the channel is functionally unaffected by ethanol. Since activation of BK channels by ethanol occurred without modification of the single channel conductance (γ), and an ethanol-induced increase in the number of functional channel proteins present in the patch membrane (N) is unlikely in excised (*i.e.*, cell-free) patches, activation is due to a modification of channel gating properties by ethanol. Since the activity of BK channels critically depends on voltage, we examined the relationship between this key activator of the channel and ethanol. If the voltage activation is described by a Boltzmann relationship, a plot of the natural log of NP_o as a function of voltage should be linear at low values of P_o. The reciprocal of the slope (a measure of the voltage sensitivity) is the potential needed to produce an e-fold change in P_o, at low P_o. Voltage dependency of activation was studied in the presence and absence of a fixed concentration of 50 mM ethanol. Ethanol did not affect the voltage sensitivity of the channels, indicating that the voltage-sensor of the channel is unaffected by ethanol. Detailed analysis of BK is difficult in natural membranes because of apparent channel heterogeneity. To circumvent this, we expressed the cloned mslo BK channel in oocytes, to examine whether ethanol functionally interacts with the channel Ca^{++} sensor.

2.4. Ethanol Differentially Modulates the Voltage and Calcium-Sensitivity of the BK Channels

The study of ethanol's action on cloned channels, compared to natural BK channels, has the advantage of providing a homogeneous population of channels. Dual sensitivities

Figure 2A. 50 mM ethanol reversibly increases BK channel activity in isolated rat neurohypophysial terminals. Representative single channel recordings obtained from excised, inside-out patches before (upper panel), during (middle panel), and 6 minutes after (lower panel) exposure of the cytosolic side of the patch to ethanol. The arrows at the top of each panel indicate the baseline; BK channel openings are shown as upward deflections. Four selected traces are shown for each condition; P_o values were calculated from 90–180 sec of recording under each condition. The solution facing the intracellular and extracellular sides of the patch (symmetric conditions) is described in Mol Pharmacol 49:40–48 (1996). The membrane potential was set to +40 mV. P_o: probability that a particular BK channel is open. (Taken from Dopico et al., Mol Pharmacol 49:40–48 (1996), with permission from publisher).

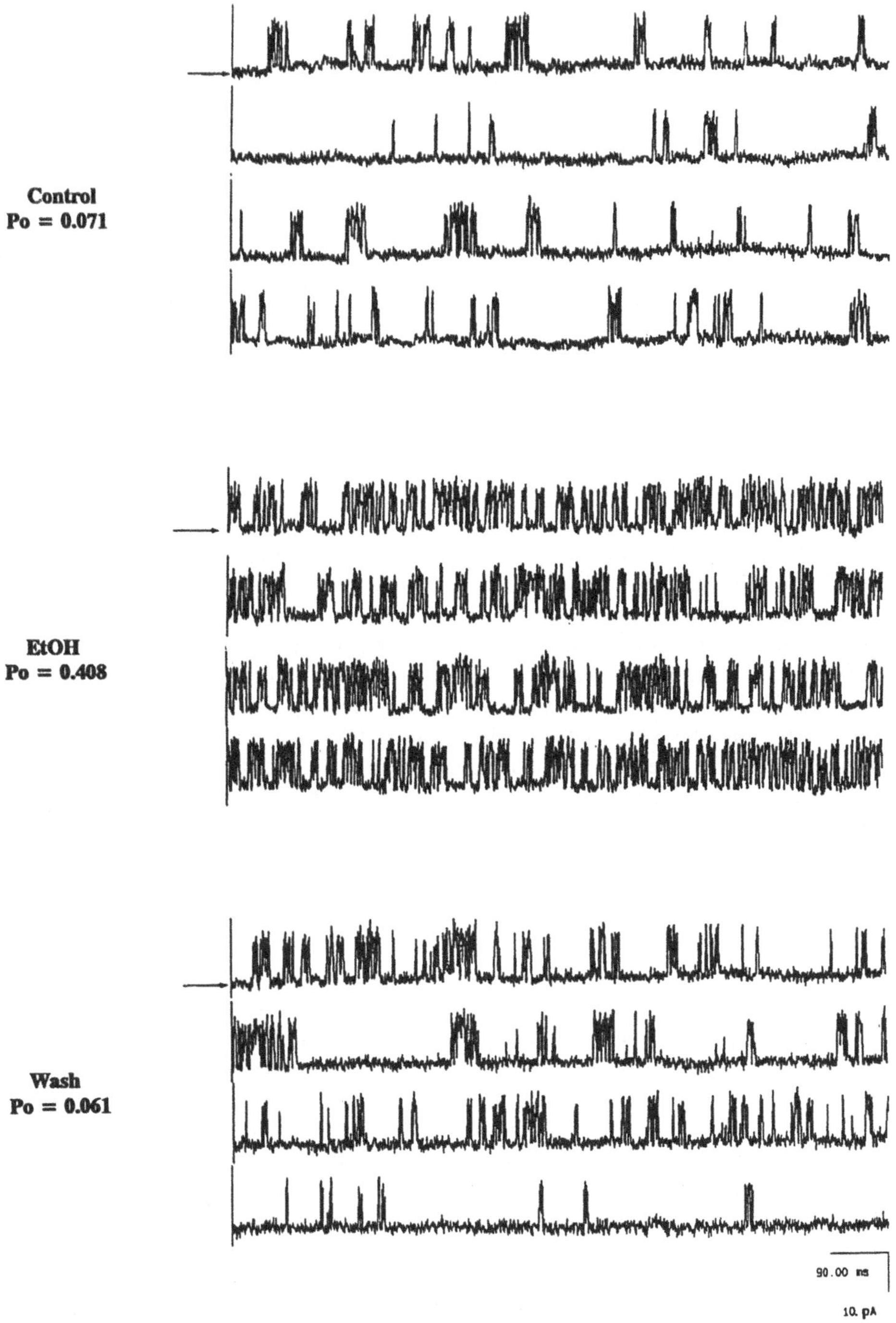
Control
Po = 0.071
EtOH
Po = 0.408
Wash
Po = 0.061
90.00 ms
10. pA

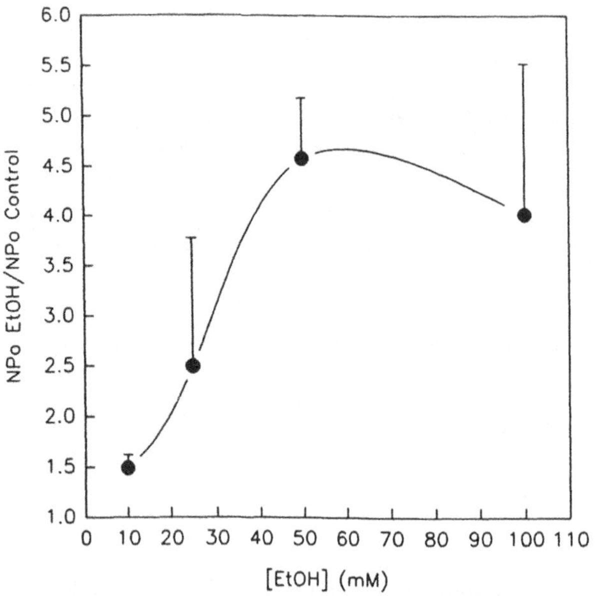

Figure 2B. Increase of BK channel NP_o as a function of ethanol concentration. Results are expressed as the ratio of NP_o values obtained in the presence and absence (before presentation) of the drug, determined in the same patch. Each point of the graph is the mean ± SEM of at least 4 determinations; each determination was obtained in a different patch, and each patch was excised from a different terminal. (Taken from Dopico et al., Mol Pharmacol 49:40–48 (1996), with permission from publisher).

to activating Ca^{++} and transmembrane depolarizing voltage play a key role in the modulation of BK channel gating, without modifying single channel conductance or ion selectivity (McManus, 1991; Behrens et al., 1989; Magleby and Pallotta, 1983). We first demonstrated that ethanol increases channel activity in the cloned mslo BK channel, by modifying channel gating, without altering either single channel conductance or the selectivity of the channel for K^+. We next tested whether the drug enhances mslo channel activity by increasing the sensitivity of the channel to voltage and/or intracellular Ca^{++} concentration. For a fixed intracellular concentration, $[Ca^{++}]_i$, if the voltage activation is described by a Boltzmann relationship, a plot of the natural log of NP_o (or P_o if N=1) as a function of voltage should be linear at low values of P_o (Singer and Walsh, 1987). Low values of P_o were achieved by working at low $[Ca^{++}]_i$ (10–316 nM). When the natural log of NP_o was plotted as a function of voltage, at low P_o. the reciprocal of the slope (a measure of voltage sensitivity), is the potential needed to produce an e-fold change in NP_o (Singer and Walsh, 1987). The voltage sensitivity of mslo channels found here was quantitatively homogenous among different channels from different patches: 18.1 ± 3.6 mV per e-fold change in NP_o (n=5). The reciprocal of the slope in the NP_o-V relationship gives an effective valence (z) of 1.43 ± 0.28, calculated from: 1/slope=RT/zF, where R, T, and F have their usual meaning. These values are within the range of values reported in the literature for this and other BK clones encoded by slo genes, and native BK channels (Dopico et al., 1996; DiChiara and Reinhart, 1995; Butler et al., 1993; Toro et al., 1990). More importantly, 50 mM ethanol did not modify the slope of this relationship. Typically, control and ethanol values were obtained in the same patch, and the lack of effect of ethanol on the slope was similar in five patches obtained from different oocytes (Figure 3A). The lack of change in the slope after the exposure to ethanol indicates that the drug does not modify the voltage-sensitivity of the cloned channel (similar to what was seen when recording from the terminal membrane), although the probability of the channel being open at a given potential is markedly increased. In other words, ethanol displaces the equilibrium between conducting and nonconducting states of the mslo channel without changing the voltage dependence of the gating process.

The ethanol-induced parallel shift of the voltage activation curve towards negative potentials is consistent with the possibility that the drug increases the Ca^{++}-sensitivity of mslo channels (DiChiara and Reinhart, 1995; Barrett et al., 1982). To directly explore this possibility, we evaluated ethanol action on channel activity at a range of free $[Ca^{++}]_i$ at a fixed potential. For BK channels, given a fixed voltage, a plot of the natural log of $(N)P_o$ as a function of $[Ca^{++}]_i$ should be linear at low values of activity (Barrett et al., 1982). Figure 3B shows a representative plot of NP_o as a function of free $[Ca^{++}]_i$, at +40 mV in the presence and absence of 50 mM ethanol. The plot shows that ethanol increases the apparent Ca^{++}-sensitivity of mslo channels, since channels in the presence of the drug require less cytosolic free Ca^{++} to attain a given level of activity. However, in spite of this shift to the left, ethanol decreased the slope of the NP_o-$[Ca^{++}]_i$ relationship: 1.75 vs. 1.04 in the absence and presence of 50 mM ethanol, respectively. Therefore, for a given ethanol concentration, the increase in mslo channel NP_o is an inverse function of $[Ca^{++}]_i$. In the case depicted in Figure 3B, ethanol increased channel activity to 292.9% of control values at 10nM free $[Ca^{++}]_i$, whereas the drug increased channel NP_o to only 122.8% of control values at 316.2 nM free $[Ca^{++}]_i$. The decrease in the slope of the NP_o-free $[Ca^{++}]_i$ relationship may be interpreted as ethanol altering the capability of the Ca^{++}-sensing site(s) to respond to increases in $[Ca^{++}]_i$. Thus, it is important, when considering the overall effects of ethanol on cellular excitability, to take into account that the efficacy of ethanol in increasing channel BK activity is a function of the cytosolic Ca^{++} concentration. This influence will result in varying degrees of ethanol potentiation of the BK channel as the activity level of the target neuron increases. The data obtained using cloned channels may be seen in more complete form in our publication (Dopico et al., 1998).

2.5. Planar Bilayer Studies

For many years, alcohol-mediated alterations of the membrane lipid phase dominated theories of the mechanisms of action of this drug. Recently, a direct interaction between ethanol and the target protein has been more commonly accepted. However, a full understanding of ethanol's actions require that we also examine the role of lipids in the perturbation of protein activity. The activity of BK channels is modulated by several lipid species. A variety of fatty acids have been reported to activate BK channels in pulmonary and mesenteric artery smooth muscle cells (Dopico et al., 1994; Kirber et al., 1992). Manipulations of the cholesterol content of rabbit aorta smooth muscle membranes markedly modify membrane fluidity and, in parallel, the gating kinetics of BK channels (Bolotina et al., 1989): an increase in membrane cholesterol leads to an overall inhibition of BK channel function. In addition, an increase in the cholesterol/ phospholipid ratio of lipid bilayers has been reported to reduce BK channel activity by specifically favoring the appearance of long-closures in the channel dwell time distribution (Chang et al., 1995). This result is of particular relevance to our data, since we found that ethanol activates BK channels by essentially doing the opposite: it suppresses long-closures in the channel dwell-time distribution (Dopico et al., 1998; Dopico et al., 1996; Chu et al., 1998). In all these studies, even those performed in cell-free membrane patches, the use of native channels of unknown sequence, the potential presence of unknown regulatory proteins, as well as the complex and unknown lipid composition and architecture of the natural membrane, impose a serious limitation on our ability to evaluate lipid modulation of both ion channel function, and ethanol modulation of function. A system where the channel protein is identified and its structure known, and where the lipid composition of the bilayer can be controlled, alleviates this difficulty.

A major laboratory effort was undertaken to determine the feasibility of the planar lipid technique for the proposed studies. Our efforts have focussed on ethanol's actions at the single channel level, on muscle t-tubule BK channels reconstituted into planar bilayer membranes of known composition. T-tubule BK channels are a convenient model for studying alcohol action because they reliably fuse into planar bilayers where they have been well characterized (Moczydlowski and Latorre, 1983; Moczydlowski et al., 1985). The results which we obtained demonstrate the feasibility of the work, including evidence that native lipids surrounding the channel effectively exchange with the artificial lipids of the bilayer (see below), and that part (if not all) ethanol's sites and mechanisms of action are preserved in this greatly simplified preparation.

Ethanol applied to the "intracellular" side of the bilayer increased the activity of BK channels in a concentration-dependent manner. Although a plateau in the concentration-response curve was apparent above 100 mM ethanol concentrations, no clear saturation of the ethanol effect was observed up to 200 mM ethanol, at which point the bilayer became unstable. We examined whether the increase in BK currents by ethanol was solely due to an increase in channel activity, or was accompanied by an increase in the channel unitary conductance. As previously reported for native and cloned BK channels expressed in natural membranes (Dopico et al., 1998; Dopico et al., 1996), ethanol did not significantly affect BK channel conduction for K^+ in this minimal system. Rather, as we had previously found for native and cloned channels, the primary effect of ethanol exposure was on channel gating. Ethanol increased the relative proportion of long openings, without changing their duration, which resulted in a mild increase in the channel mean open time. In addition, the drug markedly reduced the mean closed time, with this being the predominant determinant of ethanol-induced channel activation. These findings parallel those observed for the action of ethanol on neurohypophysial BK channels studied *in situ* (Dopico et al., 1996), and cloned (α subunit, mslo) BK channels expressed in *Xenopus* oocytes (Dopico et al., 1998), suggesting that the activation of all BK channels by ethanol share site(s) and mechanism(s) of action.

It is important that significant exchange occur between native lipids contained in the membrane vesicles, and the lipids comprising the planar bilayer. Several pieces of evidence from the literature and from this study indicate that exchange does occur. For exam-

Figure 3. A) Representative plot of NP_o as a function of voltage from mslo channels in the presence (hollow circles) and absence (filled circles) of 50 mM ethanol, obtained in the same I/O patch. Given a fixed intracellular $[Ca^{++}]$ (in this case $[Ca^{++}]_{free}$ ~100 nM), when the voltage activation is described by a Boltzmann relationship, a plot of the natural log of NP_o as a function of voltage should be linear at low values of P_o. The reciprocal of the slope, a measure of voltage sensitivity, is the potential needed to produce an e-fold change in NP_o: 21.5 mV (r = 0.995) and 22.1 mV (r = 0.993) in the presence and absence of ethanol, respectively. This representative result was confirmed in 4 other patches from different oocytes, with the voltage sensitivity of mslo channels being 17.52 ± 2.18 vs. 18.07 ± 3.61 mV/e-fold change in NP_o, in the presence and absence of 50 mM ethanol, respectively (not statistically significant, P>0.9). Thus, ethanol activates mslo channels without modifying the voltage dependence of activation. B) Representative plot of mslo channel NP_o as a function of $[Ca^{++}]_i$, at a fixed potential (+40 mV) in the presence (hollow circles) and absence (filled circles) of 50 mM ethanol, obtained in the same inside-out (I/O) patch. A plot of the natural log of NP_o as a function of free $[Ca^{++}]_i$ should be linear at low values of P_o. The slope, a measure of the response in channel activity upon recognition of Ca^{++}-sensing sites to increases in $[Ca^{++}]_i$ is markedly decreased by ethanol: 1.748 (r = 0.997) vs. 1.044 (r = 0.977) in the presence and absence of the drug, respectively. Similar data to this representative case were obtained in 6 other patches from different oocytes, with slopes of 1.824 ± 0.210 vs. 0.769 ± 0.225 in the absence and presence of 50 mM ethanol, respectively (n = 7; P<0.02, paired two-tailed t-test). (Taken from Dopico et al., J Pharmacol Exp Ther 284:258–268, 1998, with permission of publisher).

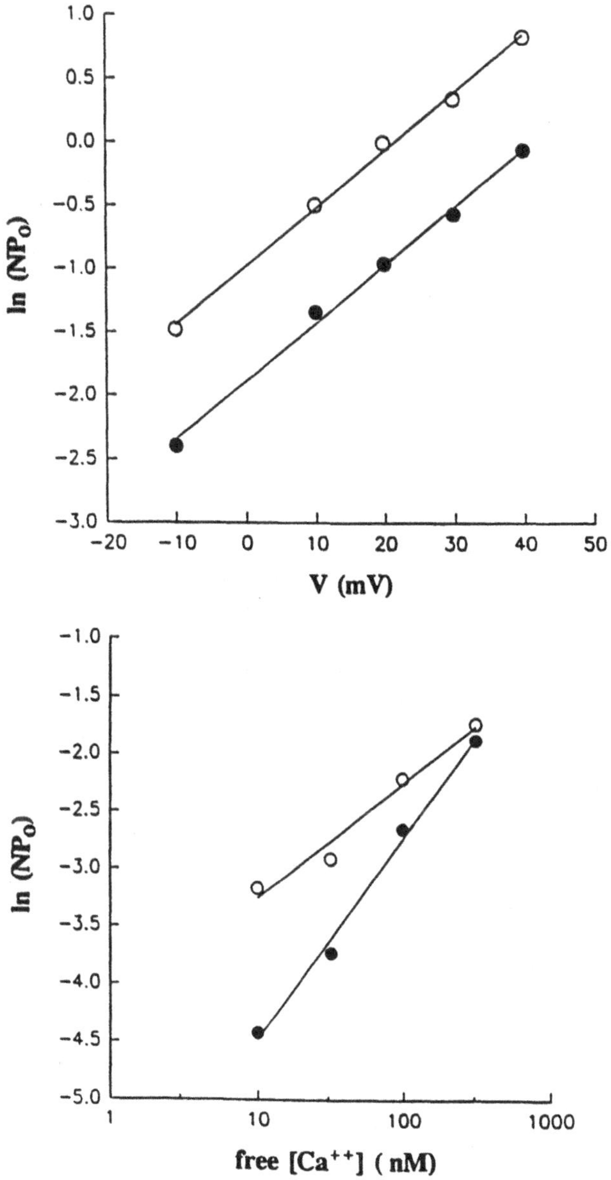
1.0
0.5
0.0
-0.5
-1.0
-1.5
-2.0
-2.5
-3.0
ln (NP$_O$)
-20
-10
0
10
20
30
40
50
V (mV)
-1.0
-1.5
-2.0
-2.5
-3.0
-3.5
-4.0
-4.5
-5.0
ln (NP$_O$)
1
10
100
1000
free [Ca^{++}] (nM)

ple, the activity of nystatin, a peptide that requires ergosterol to form channels, is lost when membrane vesicles containing nystatin and ergosterol are incorporated into ergosterol-free membranes, presumably due to diffusion of ergosterol away from the channel complex (Woodbury and Miller, 1990). Electron spin resonance studies of reconstituted nicotinic acetylcholine receptors indicate that the lipid at the protein/lipid boundary is relatively motionally restricted, but, nevertheless, can exchange with the bulk lipid. This exchange rate is rapid, on the order of 10^7 sec^{-1}, and is slowed by high protein/lipid ratios (Barrantes, 1993; Ellena et al., 1983). We might expect this exchange to be faster in our system since the protein/lipid ratio is likely far lower than that in biological membranes. More importantly, our own data indicate that BK channels were modulated by the amount of fixed charge present in the bilayer, with channels in neutral phosphatidylethanolamine (PE) bilayers exhibiting lower open channel probabilities and conductances than channels in negatively charged 3:1 phosphatidylethanolamine/phosphatidylserine (PE/PS) bilayers. These data are qualitatively identical to those of Moczydlowski et al. (1985) and indicate that replacement of native with bilayer lipid is extensive, possibly complete. Thus, these findings strongly support the conclusion that the bilayer lipid substitutes for the native lipid immediately surrounding incorporated channels, allowing us to assess the actions of ethanol on channel protein function in a lipid bilayer of minimal complexity.

2.6. Tolerance

Finally, our criteria outlined at the beginning of this chapter stated that we would like to study a system in which those molecular targets of acute ethanol action can be demonstrated to play a role in behavioral tolerance to the effects of ethanol. Our preliminary data suggest that this criterion is also fulfilled by the preparation that we are using. We have examined the effects of acute ethanol on vasopressin release from isolated neurohypophysial terminals from rats that have been on an alcohol-containing diet for three weeks. Our results to date indicate that acute ethanol has significantly less effect on AVP release from these terminals than it does on terminals taken from calorie-yoked rats, which were ethanol-naive. Since we know so much about the acute actions of ethanol on the calcium and Ca^{++}-activated potassium channels, which are involved in vasopressin release, we are now poised to ask how channel function and ethanol response are altered after chronic exposure, in a manner which can hopefully be correlated with the altered effects of acute ethanol challenge on release, after chronic exposure.

ACKNOWLEDGMENT

We would like to acknowledge the support of the National Institute on Alcohol Abuse and Alcoholism (NIH) for funding the research from our laboratory that is discussed in this chapter.

REFERENCES

Barrantes FJ (1993) Structural-functional correlates of the nicotinic acetylcholine receptor and its lipid microenvironment. FASEB J 7:1460–1467.

Barrett JN, Magleby KL, Pallota BS (1982) Properties of single calcium-activated potassium channels in cultured rat muscle. J Physiol 331:211–230.

Behrens MI, Oberhauser A, Bezanilla F, Latorre R (1989) Batrachotoxin-modified sodium channels from squid optic nerve in planar bilayers. Ion conduction and gating properties. J Gen Physiol 93:23–41.

Bergland RM, Torack RM (1969) An electron microscopic study of the human infundibulum. Z Zellforsch 99:1–12.

Bolotina V, Omelyanenko V, Heyes B, Ryan U, Bregestovski P (1989) Variations of membrane cholesterol alter the kinetics of Ca^{2+}-dependent K^+ channels and membrane fluidity in vascular smooth muscle cells. Pflügers Arch 415: 262–268.

Brownstein MJ, Russell JT, H Gainer (1982) Biosynthesis of posterior pituitary hormones. In: Frontiers in Neuroendocrinology Vol 7 (Ganong WF, Martini L, eds), pp. 31–43. New York, NY: Raven Press.

Butler A, Tsunoda S, McCobb DP, Wei A, Salkoff L (1993) mSlo, a complex mouse gene encoding "Maxi" calcium-activated potassium channels. Science261:221–224.

Chang HM, Reitstetter R, Mason RP, Gruener R (1995) Attenuation of channel kinetics and conductance by cholesterol: An interpretation using structural stress as a unifying concept. J Membr Biol 143:51–63.

Chu B, Dopico AM, Lemos JR and Treistman SN (1998) Ethanol potentiation of calcium- activated potassium channels reconstituted into planar lipid bilayers. Mol. Pharmacol. 54:397–406.

DiChiara TJ, Reinhart PH (1995) Distinct effects of Ca^{2+} and voltage on the activation and deactivation of cloned Ca^{2+}-activated K^+ channels. J Physiol 489:403–418.

Dopico AM, Kirber MT, Singer JV, Walsh JV (1994) Membrane stretch directly activates large conductance Ca^{++}-activated K^+ channels in smooth muscle cells freshly dissociated form rabbit mesenteric artery. Am J Hypert 7:82–89.

Dopico AM, Lemos JR, Treistman SN (1995) Alcohol and the release of vasopressin and oxytocin. In: Alcohol and Hormones (Watson RR, ed) pp. 209–226. Boca Raton, FL: CRC Press.

Dopico AM, Lemos JR, Treistman SN (1996) Ethanol increases the activity of large conductance, Ca^{2+}-activated K^+ channels in isolated neurohypophysial terminals. Mol Pharmacol 49:40–48.

Dopico AM, Anantharam V, Treistman, SN (1998) Ethanol increases the activity of Ca^{++}-dependent K^+ (mslo) channels: Functional interaction with cytosolic Ca^{++}. J Pharmacol Exp Ther 284:258–268.

Ellena JF, Blazing MA, McNamee MG (1983) Lipid-protein interactions in reconstituted membranes containing acetylcholine receptor. Biochem 22:5523–5535.

Kirber MT, Ordway RW, Clapp LH, Walsh, JV, Jr, Singer JJ (1992) Both membrane stretch and fatty acids directly activate large conductance $Ca2^+$- activated K^+ channels in vascular smooth muscle cells. FEBS Lett 297:24–28.

Kozlowski GP (1990) Alcohol-neuroendocrine interactions: vasopressin and oxytocin. In: Biochemistry and Physiology of Substances Abuse Vol 2 (Watson RR, ed) pp. 257–277. Boca Raton, FL: CRC Press.

Lederis K (1965) An electron microscopic study of the human neurohypophysis. Z Zellforsch 65:847–868.

Magleby KL, Pallotta, BS (1983) Calcium dependence of open and shut interval distributions from calcium-activated potassium channels in cultured rat muscle. J Physiol 344;585–604.

McManus O (1991) Calcium-activated potassium channels: Regulation by calcium. J Bioenerg Biomembr 23:537–560.

Moczydlowski E, Alvarez O, Vergara C, Latorre R (1985) Effect of phospholipid surface charge on the conductance and gating of a Ca^{2+}-activated K^+ channel in planar lipid bilayers. J Membr Biol 83:273–282.

Moczydlowski E, Latorre R (1983) Gating kinetics of Ca^{++}-activated K^+ channels from rat muscle incorporated into planar lipid bilayers. J Gen Physiol 82:511–542.

Nordmann J (1977) Ultrastructural morphometry of the rat neurohypophysis. J Anat 123:213–218.

Pickering BT, Swann RW, Gonzalez CB (1986) Biosynthesis and processing of neurohypophysial hormones. In: Neuropeptides and Behavior Vol 2, pp. 1–22. Oxford: Pergamon Press.

Reichin S (1992) Neuroendocrinology. In: William's Textbook of Endocrinology, 8th ed (Wilson JD, Foster DW, eds), Ch 5 pp. 135–220 Philadelphia, PA: Saunders Co.

Scheithauer BW, Horvath E, Kovacs K (1992) Ultrastructure of the neurohypophysis. Microsc Tech 20:177–186.

Schmale H, Fehr S, Richter D (1987) Vasopressin biosynthesis: From gene to peptide hormone. Kidney Int 32 (Suppl 21): 8–13.

Singer JJ, Walsh, JV (1987) Characterization of calcium-activated potassium channels in single smooth muscle cells using the patch-clamp technique. Pflugers Arch 408:98–111.

Stopa EG, Kuo LeBlanc V, Hill DH, Anthony ELP (1993) A general overview of the anatomy of the neurohypophysis. In: Ann NY Acad Sci, Vol 689: The neurohypophysis: a window on brain function. (North WG et al. eds). pp. 6–15. New York, NY.

Toro L, Ramos-Franco J, Stefani E (1990) GTP-dependent regulation of myometrial KCA channels incorporated in lipid bilayers. J Gen Physiol 96:373–394.

Wang G, Thorn P, Lemos JR (1992) A novel large-conductance Ca^{++}-activated potassium channel and current in nerve terminals of the rat neurohypophysis. J Physiol 457:47–74.

Wang X, Dayanithi G, Lemos JR, Nordmann JJ, Treistman SN (1991a) Calcium currents and peptide release from neurohypophysial terminals are inhibited by ethanol. J Pharmacol Exp Ther 259:705–711.

Wang X, Lemos JR, Dayanithi G, Nordmann JJ, Treistman SN (1991b) Ethanol reduces vasopressin release by inhibiting calcium currents in nerve terminals. Brain Res 551:338–341.
Wang X, Wang G, Lemos JR, Treistman SN (1994) Ethanol directly modulates gating of a dihydropyridine-sensitive Ca^{2+} channel in neurohypophysial terminals. J Neurosci 14:5453–5460.
Woodbury DJ, Miller C (1990) Nystatin-induced liposome fusion: a versatile approach to ion channel reconstitution into planar bilayers. Biophys J 58:833–839.

ALCOHOL AND GENERAL ANESTHETIC MODULATION OF $GABA_A$ AND NEURONAL NICOTINIC ACETYLCHOLINE RECEPTORS

Toshio Narahashi,[1] Gary L. Aistrup,[1] Jon M. Lindstrom,[2] William Marszalec,[1] Haruhiko Motomura,[1] Keiichi Nagata,[1] Hideharu Tatebayashi,[1] Fan Wang,[2] and Jay Z. Yeh[1]

[1]Department of Molecular Pharmacology and Biological Chemistry
Northwestern University Medical School
303 East Chicago Avenue
Chicago, Illinois 60611-3008
[2]Department of Neuroscience
University of Pennsylvania School of Medicine
217 Stemmler Hall, 36th and Hamilton Walk
Philadelphia, Pennsylvania 19104-6074

1. INTRODUCTION

A number of studies have been performed during the past ten years or so in an attempt to elucidate the cellular and molecular mechanisms of action of alcohols and general anesthetics. One of the many questions asked is which receptor(s) and channel(s) is (are) the important target sites of alcohols. The history of the study of alcohol-channel interactions goes back to 1964 when Armstrong and Binstock (1964) and Moore et al. (1964) found that ethanol and higher alcohols suppressed both sodium and potassium currents in squid giant axons. The potency of ethanol was very low, and at 650 mM and 1300 mM it suppressed the sodium current only by 18% and 41%, respectively (Moore et al., 1964). The general anesthetic halothane is also known to inhibit sodium and potassium currents in squid giant axons only at concentrations much higher than those that cause clinical anesthesia (Franks and Lieb, 1994).

It was not until late 1980s that the study of alcohol and anesthetic actions on various types of receptors/channels flourished owing to the development of patch clamp techniques (Hamill et al., 1981). Almost all types of neurotransmitter receptors/channels have been the subjects of investigation of alcohol and anesthetic actions. Practically all of them have been found to be modulated by alcohols. Some of them, such as voltage-gated cal-

The "Drunken" Synapse, edited by Liu and Hunt.
Kluwer Academic / Plenum Publishers, New York, 1999.

cium channels and NMDA-, AMPA- and kainate-activated channels, are inhibited by alcohols, while some others such as $GABA_A$ receptors, glycine receptors, and 5-HT_3 receptors are augmented (reviewed by Crews et al., 1996; Diamond and Gordon, 1997; Peoples et al., 1996). However, data for some of these receptors and channels are quite controversial. For example, the $GABA_A$ receptors were reported to be augmented by ethanol by some investigators (Nishio and Narahashi, 1990; Nakahiro et al., 1991; Aguayo, 1990; Aguayo and Pancetti, 1994; Reynolds and Prasad, 1991; Harris et al., 1995b, 1997; Whitten et al., 1996; Yeh and Kolb, 1997), while other investigators found no effect at all (White et al., 1990; Osmanovic and Shefner, 1990; Harris et al., 1995b, 1997). Furthermore, the potencies of ethanol on various receptors and channels were in most cases low, requiring 30 to 100 mM ethanol to observe sizable effects. It should be noted that the maximum legal blood levels of ethanol for driving are approximately 20 mM in most states of the U.S. The ethanol concentrations of 30 to 100 mM would bring the human to the state of complete drunkenness or even coma.

General anesthetics are also known to affect a variety of receptors and channels including voltage-gated calcium channels, $GABA_A$ receptors, glycine receptors, glutamate receptors, and muscle-type acetylcholine (ACh) receptors (reviewed by Franks and Lieb, 1994, 1996).

We have recently found that the neuronal nicotinic (ACh) receptor is very sensitive to ethanol, being modulated at concentrations as low as 1 mM or less (Nagata et al., 1996). This is in sharp contrast with the muscle nicotinic ACh receptor, which is much less sensitive to ethanol (Forman et al., 1989). This chapter highlights our recent studies of the mechanisms of action of alcohols on the $GABA_A$ receptor channel and the neuronal nicotinic ACh receptor channel. Some comparisons are also made between alcohols and general anesthetics as they share certain aspects of action.

2. $GABA_A$ RECEPTOR CHANNELS

Whereas inhibitory synapses have been suggested as a potential target site of general anesthetics (Nicoll, 1972), it was not until 1989 that the potentiating action of inhalational general anesthetics on GABA-induced chloride currents was directly demonstrated by patch clamp techniques (Nakahiro et al., 1989). This anesthetic modulation of the $GABA_A$ receptor has since been confirmed and elaborated (Reviewed by Franks and Lieb, 1994; Harris et al., 1995a). On the contrary, the effects of ethanol on the $GABA_A$ receptor are very controversial. For example, ethanol augmentation of GABA-induced currents was demonstrated in newborn rat dorsal root ganglion (DRG) neurons (Nishio and Narahashi, 1990; Nakahiro et al., 1991), in hippocampal and cortical neurons cultured from fetal mice (Aguayo, 1990; Aguayo and Pancetti, 1994), in embryonic chick cerebral cortical neurons (Reynolds and Prasad, 1991), in the $\alpha 1\beta 1\gamma 2L$ and $\alpha 1\beta 2\gamma 2L$ GABA receptor subunits expressed in mouse L(tk$^-$) cells (Harris et al., 1995b, 1997; Whitten et al., 1996), and in bipolar cells and ganglion cells of the rat retina (Yeh and Kolb, 1997). However, ethanol potentiation of GABA-induced currents was not observed in adult rat DRG neurons (White et al., 1990), in locus coeruleus neurons (Osmanovic and Shefner, 1990), and in mouse L(tk$^-$) cells expressing subunit combinations that lacked the γ 2L subunit (Harris et al., 1995b, 1997). This controversy may be due to different species of animals and to different ages of animals which may contain different combinations of $GABA_A$ receptor subunits. Our recent studies address two of these issues. One is channel state dependence of alcohol action and the other is subunit dependence.

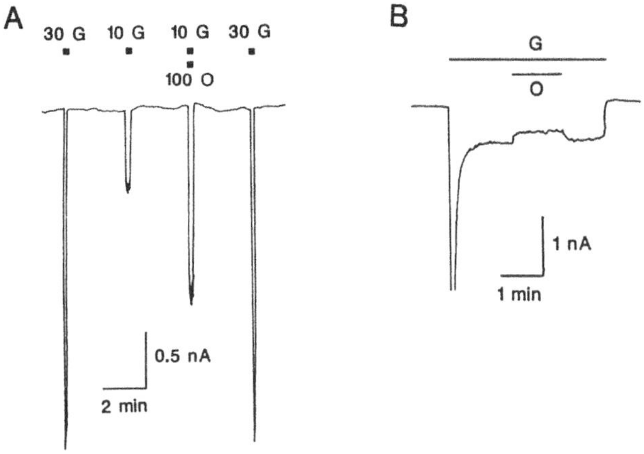

Figure 1. Dual effect of n-octanol on GABA-induced currents in rat DRG neurons. (A) n-Octanol (O) at 100 μM enhances the non-desensitized current induced by 10 μM GABA (G). (B) n-Octanol at 100 μM suppresses the desensitized current induced by 300 μM GABA. (From Nakahiro et al., 1991, with permission).

2.1. $GABA_A$ Receptor Modulation by General Anesthetics and Alcohols Is Channel State Dependent

Alcohols had a dual effect on GABA-induced currents. While the non-desensitized current induced in rat DRG neurons by a low concentration (10 μM) of GABA was greatly augmented by 100 μM n-octanol, the desensitized current induced by a high concentration (300 μM) of GABA was suppressed by 100 μM n-octanol (Figure 1).

Halothane also exhibited a dual effect. At a concentration of 0.86 mM (~2 MACs), halothane markedly augmented the chloride current induced by a low, non-desensitizing concentration (3 μM) of GABA in rat DRG neurons (Nakahiro et al., 1989). Isoflurane and enflurane had essentially the same stimulating effect. However, when these anesthetics were applied after the current had been desensitized by a high concentration (300 μM) of GABA, suppression rather than augmentation was observed. Thus when enflurane was applied during the decaying phase of GABA-induced current, an initial augmentation was followed by suppression (Nakahiro et al., 1989).

These results clearly show the dual modulation of GABA-induced currents by general anesthetics and alcohols depending on the state of the receptor channel, the augmentation of non-desensitized currents and the suppression of desensitized currents.

2.2. Is Modulation of the $GABA_A$ Receptor by General Anesthetics and Alcohols Subunit Dependent?

The $GABA_A$ receptor comprises a pentameric receptor protein with an integral chloride channel (Nayeem et al., 1994), and is endowed with several allosteric binding sites for various agents such as benzodiazepines, barbiturates and picrotoxin (Olsen and Tobin, 1990; Sieghart, 1992; Macdonald and Olsen, 1994). There are at least 16 subunits including 6α, 4β, 3γ, 1δ, 2ρ, and the combinations of these subunits are known to differ depending on the area in the brain (Burt and Kamatchi, 1991).

In human embryonic kidney (HEK) cells expressing three combinations of $GABA_A$ receptor subunits, currents induced by 3 μM GABA responded to 0.9 mM halothane differently. The currents in the α1β2γ2s combination were augmented, those in the α1β2 combination were augmented and then suppressed, and those in the α6β2γ2s combination were suppressed (Tanguy et al., 1995). However, the different responses to halothane were due to different states of receptor activation, since the receptors with the three subunit combinations had different affinities for GABA. At GABA concentrations lower than the EC_{50} values, halothane augmented GABA-induced currents in all three subunit combinations, and at GABA concentrations higher than the EC_{50} values, halothane suppressed the currents in all three subunit combinations (Tanguy et al., 1995). Therefore, halothane modulation of the GABA receptor is not subunit dependent but state dependent (see also Nakahiro et al., 1989). Propofol potentiation of GABA responses was also found to be independent of 18 combinations of α, β, and γ subunits (Sanna et al., 1995). However, the ρ1 subunit is an exception being inhibited by both inhalational anesthetics and ethanol (Mihic and Harris, 1996). The importance of transmembrane domains TM2 and TM3 of the GABA ρ1 receptor and the glycine α1 receptor in the action of ethanol and enflurane has recently been determined using chimeric receptor constructs (Mihic et al., 1997).

By contrast, alcohol modulation of GABA-induced currents was subunit dependent as well as state dependent (Marszalec et al., 1994). In both α1β2γ2s and α6β2γ2s combinations, 100 μM n-octanol augmented the non-desensitized currents induced by low concentrations of GABA, whereas it suppressed the desensitized currents induced by higher concentrations of GABA. Thus n-octanol modulation of GABA-induced currents is state dependent in agreement with the results obtained using DRG neurons (Nakahiro et al., 1991).

Although ethanol had no effect on the amplitude of GABA-induced currents in the α1β2γ2s and α6β2γ2s combinations, it accelerated the current desensitization in a subunit-dependent manner (Marszalec et al., 1994). In HEK cells expressing the α6β2γ2s subunits, 100 mM ethanol greatly accelerated the current decay without much changing the amplitude (Figure. 2). However, no such acceleration of current decay was observed by ethanol in the α1β2γ2s combination.

These results indicate that there are some differences between halothane and ethanol in their subunit dependence of modulation of GABA-induced currents. Halothane modulation is not subunit dependent but state dependent, whereas ethanol modulation is both subunit and state dependent.

2.3. Single $GABA_A$ Receptor Channel Modulation by Ethanol, n-Octanol, and Other GABAergic Agents

The potencies of various alcohols to augment GABA-induced currents are known to be linearly related to their carbon chain lengths, which are in turn related to their oil/water partition coefficients (Nakahiro et al., 1991, 1996). This raises a question as to whether these alcohols exert the GABA receptor modulating actions through the same basic

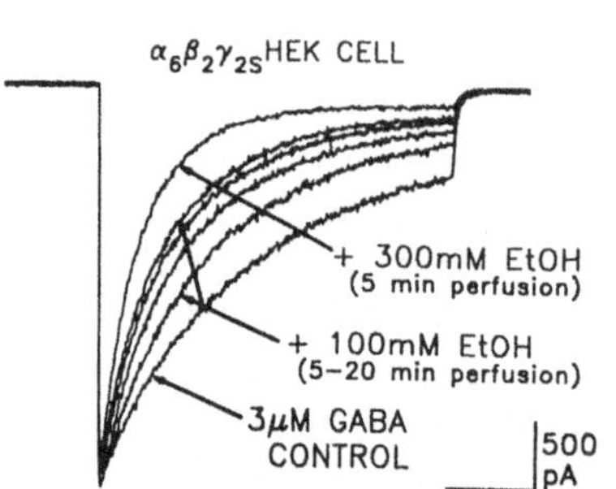

Figure 2. Ethanol at 100 mM and 300 mM accelerates desensitization of GABA-induced currents in a HEK cell expressing the α6β2γ2s subunit combination. (From Marszalec et al., 1994, with permission).

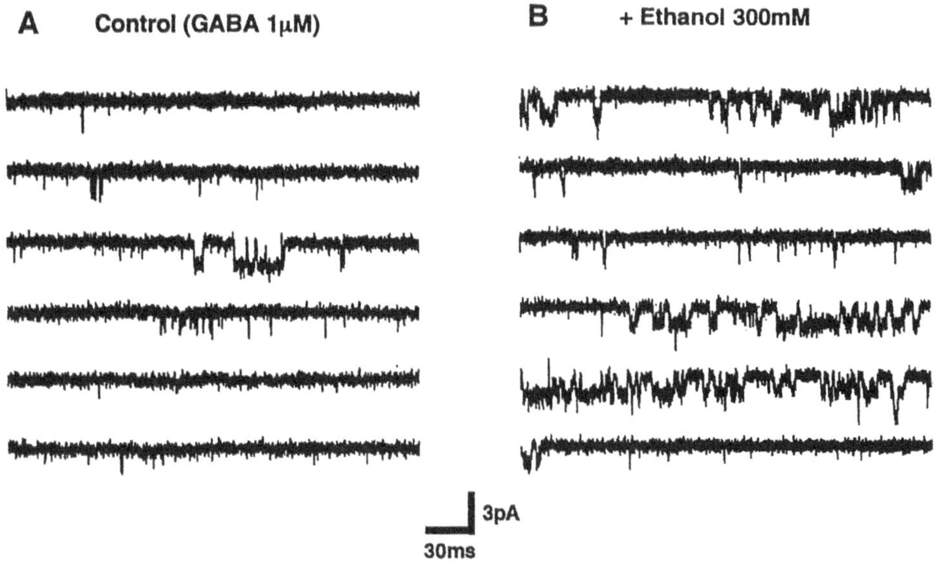

Figure 3. Single-channel currents recorded from an outside-out membrane patch excised from a rat dorsal root ganglion neuron. (A) Control currents induced by pressure ejection of 1 μM GABA from a micropipette. (B) Currents induced by 1 μM GABA plus 300 mM ethanol. Ethanol altered temporal properties of single channels without changing single-channel conductance. (From Tatebayashi et al., 1998, with permission).

mechanism. This issue was answered by single-channel recording experiments which clearly showed that ethanol and n-octanol modified the activity of the GABA$_A$ receptors in the identical manner aside from the difference in potency (Tatebayashi et al., 1998).

An example of an experiment is shown in Figure 3. Single-channel currents were recorded from an outside-out membrane patch isolated from a rat DRG neuron in the presence of 1 μM GABA in the external solution. Ethanol at 300 mM increased the frequency of channel openings without changing the current amplitude. Ethanol at 30 and 100 mM had a similar but less efficacious effect. n-Octanol at concentrations of 30 to 300 μM also had a similar effect. When the data were compiled and analyzed, it became clear that ethanol and n-octanol had identical effects as exhibited by a great increase in the frequencies of channel openings and bursts, a moderate increase in the mean open time and mean burst duration, and a great decrease in the mean close time. Thus, it was concluded that ethanol and n-octanol act on the GABA$_A$ receptor single channel in the identical manner with the exception of the difference in potency. A corollary of this result is that n-octanol, and possibly some other longer-chain alcohols, could be used conveniently, due to their high potencies as surrogates of ethanol for its action on GABA$_A$ receptors. It should also be noted that the modulation of GABA$_A$ receptor single channels by ethanol and n-octanol are different from that caused by other GABAergic agents including halothane, barbiturates, benzodiazepines, steroids and terbium (Table 1).

3. NEURONAL NICOTINIC ACETYLCHOLINE RECEPTORS

3.1. Potent Modulation of Neuronal Nicotinic ACh Receptors by Ethanol

We have recently found that ethanol is a potent modulator of the neuronal nicotinic ACh receptor (Nagata et al., 1996). The experiments were performed using undifferentiated

Table 1. Comparison of modifications of single-channel parameters caused by alcohols and other agents that stimulated the GABA system[a]

	Ethanol	n-Octanol	Halothane[18]	Barbiturates	Benzodiazepines	Steroids	Terbium[7]
Frequency of openings	↑↑	↑↑	↑↑	↑[3]	↑[2,3,4]	↑[5]	O
Mean open time	↑	↑	↑	↑[1,3]	O[3]	↑[5,6]	↑
Mean close time	↓↓	↓↓	O	O[1]		↓[5]	↓
Frequency of bursts	↑↑	↑↑	↑	O[2]	↑[2]		O
Mean burst duration	↑	↑	↑↑	↑[2]	O[2,4]	↑[5]	↑

[a]From Tatebayshi et al. (1998).
[1]Macdonald et al. (1989).
[2]Twyman et al. (1989).
[3]Study and Barker (1981).
[4]Vicini et al. (1987).
[5]Twyman and Macdonald (1992).
[6]Barker et al. (1987).
[7]Ma et al. (1994).

PC12 cells without prior treatment with nerve growth factor (NGF). Although the effect of ethanol on the amplitude of ACh-induced current was variable, ethanol invariably accelerated the rate of current decay in a concentration-dependent manner with an EC_{50} of 90 μM.

To elucidate the mechanism underlying the ethanol-induced desensitization of the ACh receptor, two types of experiments were performed; one was brief application of ACh and the other was single-channel analysis (Nagata et al., 1996). The current generated by 10-msec application of 1 mM ACh decayed with a single exponential time course (Figure 4A). The decay phase represents the rate at which ACh is dissociated from the receptor, and was found to be slowed in the presence of ethanol (Figure 4B). This suggests that ethanol increases the affinity of the ACh receptor for ACh leading ultimately to the desensitized state in the continuous presence of ACh. Single-channel openings in the presence of 30 μM ACh and 1 mM ethanol occurred as bursts and clusters separated by long closure intervals (Nagata et al., 1996) indicating the desensitization of nicotinic ACh receptors (see Colquhoun and Ogden, 1988).

3.2. Ethanol Modulation of the ACh Receptor Is Subunit and State Dependent

Our initial study of ethanol effects on the ACh receptor was performed using undifferentiated PC12 cells (Nagata et al., 1996), in which the major portion of ACh-induced currents is carried by ACh receptors containing various combinations of α3, α5, β2, β3 and β4 subunits, including α3β4 and α3β2 combinations (Henderson et al., 1994; McGehee and Role, 1995; Rogers et al., 1992). After differentiation by NGF, β2 mRNAs were reported to be up-regulated (Rogers et al., 1992). Our experiments showed that ethanol was much less potent on the NGF-differentiated PC12 cells than on the undifferentiated PC12 cells, suggesting that ethanol modulation of ACh-induced currents is dependent on subunits.

Experiments were performed to test this hypothesis using tsA201 cells, a derivative of human embryonic kidney cells, in which the human α3β4 or α3β2 subunits were expressed. In the α3β4 subunit combination, ethanol at 100 μM and 100 mM accelerated the decay rate of 300 μM ACh-induced current. The changes in peak current amplitude were variable among different cells. When a much lower concentration (30 μM) of ACh that caused only slight desensitization was used in the α3β4 combination, ethanol had a negligible effect on

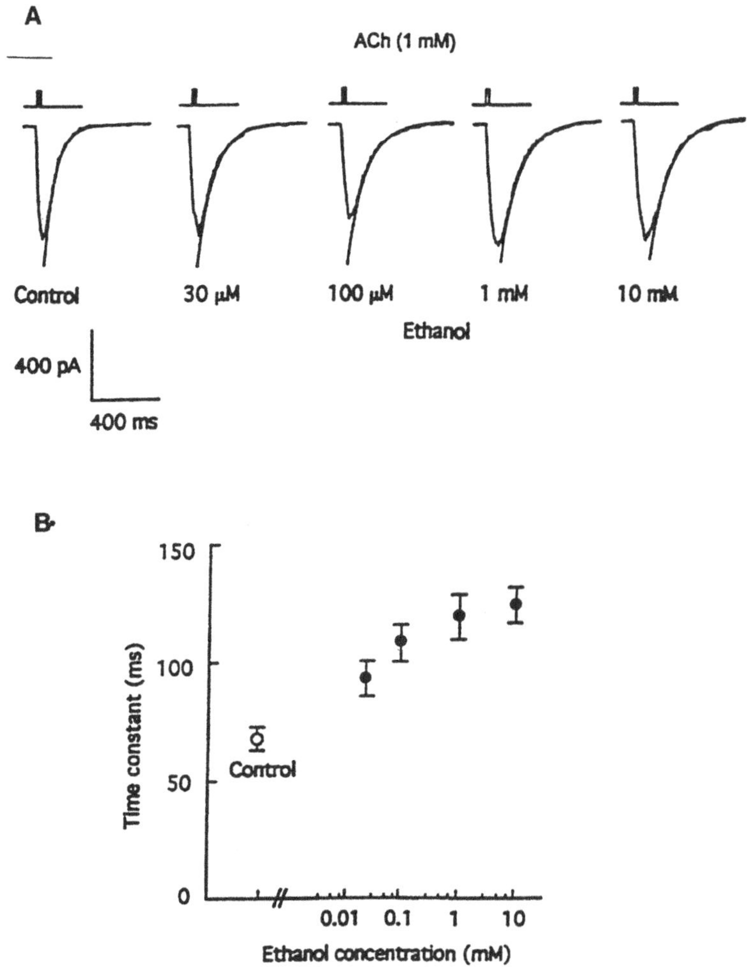

Figure 4. Ethanol slows the decay of current generated by a brief application of ACh in PC12 cells. (A) The decay phase of the current in the presence of the various concentrations of ethanol was fitted to a single exponential function using the least squares method. (B) The time constant of current decay increased with increasing concentrations of ethanol. Mean ± SD (n = 3). (From Nagata et al., 1996, with permission).

the current decay. In the α3β2 combination, however, ethanol even at a high concentration of 100 mM exhibited little or no effect on currents induced by 100 µM ACh.

Our most recent experiments suggest that ethanol even at very low concentrations may leach a chemical or chemicals from certain types of plastic syringes and tubing used in the perfusion system. These chemicals obviously had no effect on the GABA$_A$ and glutamate receptors, but might have affected the neuronal nicotinic ACh receptors. After changing the plastic perfusion system to the one using Teflon and glass, some of the changes in ACh receptors previously observed using the plastic perfusion system in the presence of low concentrations of ethanol disappeared and different types of changes in currents were disclosed. In experiments using the tsA201 cells expressing the α3β4 receptor combination in which the Teflon/glass system was used, ethanol at concentrations ranging from 30 mM to 300 mM did not accelerate the rate of current desensitization but

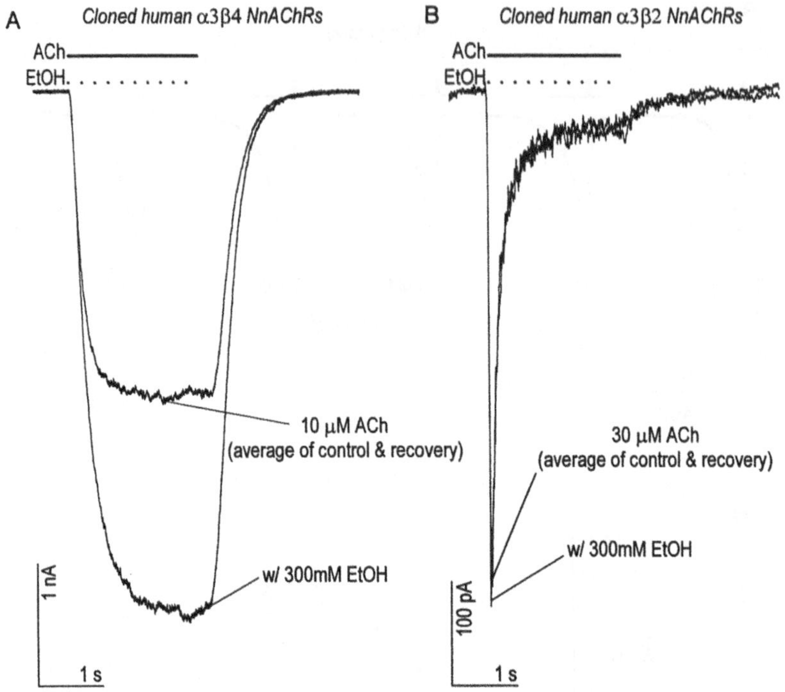

Figure 5. Effects of co-application of a high concentration (300 mM) of ethanol on ACh-induced currents in tsA201 cells in which human α3β4 (A) and α3β2 (B) neuronal nicotinic ACh receptor subunits are expressed. The current is reversibly potentiated by ethanol in the α3β4 combination, and is hardly affected in the α3β2 combination. Teflon and glass were used in the perfusion system.

potentiated the current amplitude reversibly (Figure 5A). The ethanol sensitivity varied considerably among cells. The currents evoked by 3 μM ACh in the most sensitive cells (~5% of the total tested) were significantly potentiated 4% by 1 mM ethanol and 9% by 3 mM ethanol. However, in the α3β2 combination, there was no effect of ethanol even at 300 mM in the Teflon and glass system in agreement with the aforementioned data using the plastic system (Figure 5B). It is concluded that the effect of ethanol in modulating the ACh-induced current is subunit dependent.

It should be noted that certain types of plastic have been shown to contain bis(2,2,6,6-tetramethyl-4-piperidinyl sebacate) (BTMPS) which inhibits ACh-induced currents in *Xenopus* oocytes expressing various combinations of neuronal nicotinic ACh receptors (Papke et al., 1994). However, since the inhibitory action of BTMPS was independent of subunit combinations and reversed only slowly after washout, the reversible and subunit-dependent ethanol modulation of ACh-induced currents observed in our previous experiments using the plastic perfusion system was unlikely to be caused by BTMPS. However, it remains to be seen to what extent our initial observations of the potent ethanol modulation of the ACh receptor of undifferentiated PC12 cells were due to a yet-to-be identified chemical or chemicals that might have been leached out of the plastic perfusion system.

In experiments using *Xenopus* oocytes expressing rat neuronal nicotinic receptor subtypes, ethanol at a concentration of 100 or 300 mM potentiated the ACh-induced current in the rat α3β4 subunit combination (135–305% of control), but the effects at 1 mM ethanol indicated mixed results of potentiation and inhibition (Covernton and Connolly, 1997). The effect of ethanol at 100 or 300 mM on the α3β2 combination was less than that on the α3β4 combination and was only potentiation (86–136% of control) (Covernton and Connolly, 1997). By contrast, the α7 subunit expressed in *Xenopus* oocytes was inhibited by ethanol with an EC_{50} of 33 mM (Yu et al., 1996). However, Covernton and Connolly (1997) observed slight potentiation or inhibition of the α7 subunit by 100 or 300 mM ethanol. These differences may be due to the different expression system and/or the different sources of the subunits used (human vs. rat).

3.3. General Anesthetics Modulate the ACh Receptor

In contrast to ethanol, which exerted little or no effect on ACh-induced currents in the human α3β2 subunit combination, halothane at 430 μM (~1 MAC) potently inhibited the current induced by either a high concentration (1 mM) or a low concentration (30 μM) of ACh in tsA201 cell line. However, propofol was much less potent even at 3 μM on the α3β2 subunit combination. Preliminary experiments showed that propofol was more potent on NGF-differentiated PC12 cells than undifferentiated PC12 cells, suggesting that it has a higher affinity for the α3β2 receptor than the α3β4 receptor.

Subunit-dependent modulation of the neuronal nicotinic ACh receptor was indeed demonstrated for isoflurane (Violet et al., 1997). The β2 subunit expressed in *Xenopus* oocytes in combination with any of the α2, α3 and α4 subunits was more sensitive to isoflurane than the β4 subunit combination with any of these α subunits. Isoflurane and propofol were also shown to act potently on the α4β2 subunit combination, but not on the α7 subunit (Flood et al., 1997).

These results clearly indicate that the neuronal nicotinic ACh receptor is an important target site of general anesthetics, and that different anesthetic sensitivities of different areas of brain may be explained by subunit dependence.

ACKNOWLEDGMENTS

The works described in this chapter were supported by grants from the NIH R01 NS14144 (T. N.), R01 AA07836 (T. N.), F32 AA05447 (G. A.), and R01 NS11323 (J. M. L.), the Muscular Dystrophy Association (J. M. L.) and the Smokeless Tobacco Research Council (J. M. L.). We thank Nayla Hasan for technical assistance and Julia Irizarry for secretarial assistance.

REFERENCES

Aguayo, LG (1990) Ethanol potentiates the $GABA_A$-activated Cl^- current in mouse hippocampal and cortical neurons. Eur J Pharmacol 187:127–130.

Aguayo LG, Pancetti FC (1994) Ethanol modulation of the γ-aminobutyric $acid_A$- and glycine-activated Cl^- current in cultured mouse neurons. J Pharmacol Exp Ther 270:61–68.

Armstrong CM, Binstock L (1964) The effects of several alcohols on the properties of the squid giant axon. J Gen Physiol 48:265–277.

Barker JL, Harrison NL, Lange GD, Owen DG (1987) Potentiation of γ-aminobutyric-acid-activated chloride conductance by a steroid anaesthetic in cultured rat spinal neurones. J Physiol (Lond) 386:485–501.

Burt DR, Kamatchi GL (1991) $GABA_A$ receptor subtypes: from pharmacology to molecular biology. FASEB J 5:2916–2923.

Colquhoun D, Ogden DC (1988) Activation of ion channels in the frog end-plate by high concentrations of acetylcholine. J Physiol (Lond) 395:131–159.

Covernton PJO, Connolly JG (1997) Differential modulation of rat neuronal nicotinic receptor subtypes by acute application of ethanol. Br J Pharmacol 122:1661–1668.

Crews FT, Morrow AL, Criswell H, Breese G (1996) Effects of ethanol on ion channels. Internat Rev Neurol 39:283–367.

Diamond I , Gordon AS (1997) Cellular and molecular neuroscience of alcoholism. Physiol Rev 77:1–20.

Flood P, Ramirez-Latorre J, Role L (1997) $\alpha 4\beta 2$ neuronal nicotinic acetylcholine receptors in the central nervous system are inhibited by isoflurane and propofol, but $\alpha 7$-type nicotinic acetylcholine receptors are unaffected. Anesthesiology 86:859–865.

Forman SA, Righi DL, Miller KW (1989) Ethanol increases agonist affinity for nicotinic receptors from *Torpedo.* Biochim Biophys Acta 987:95–103.

Franks NP, Lieb WR (1994) Molecular and cellular mechanisms of general anaesthesia. Nature 367:607–614.

Franks NP, Lieb WR (1996) An anesthetic-sensitive superfamily of neurotransmitter-gated ion channels. J Clin Anesthesia 8:3S-7S.

Hamill OP, Marty A, Neher E, Sakmann B, Sigworth FJ (1981) Improved patch-clamp techniques for high-resolution current recording from cells and cell-free membrane patches. Pflüegers Arch 391:85–100.

Harris RA, Mihic J, Dildy-Mayfield JE, Machu TK (1995a) Actions of anesthetics on ligand-gated ion channels: role of receptor subunit composition. FASEB J 9:1454–1462.

Harris RA, Proctor WR, McQuilkin SJ, Klein RL, Mascia MP, Whatley V, Whiting PJ, Dunwiddie TV (1995b) Ethanol increases $GABA_A$ responses in cells stably transfected with receptor subunits. Alcohol Clin Exp Res 19:226–232.

Harris RA, Mihic SJ, Brozowski S, Hadingham K , Whiting PJ (1997) Ethanol, flunitrazepam, and pentobarbital modulation of $GABA_A$ receptors expressed in mammalian cells and *Xenopus* oocytes. Alcohol Clin Exp Res 21:444–451.

Henderson LP, Gdovin MJ, Liu C, Gardner PD , Maue RA (1994) Nerve growth factor increases nicotinic ACh receptor gene expression and current density in wild-type and protein kinase A-deficient PC12 cells. J Neurosci 14:1153–1163.

Ma JY, Reuveny E, Narahashi T (1994) Terbium modulation of single γ-aminobutyric acid-activated chloride channels in rat dorsal root ganglion neurons. J Neurosci 14:3835–3841.

Macdonald RL, Olsen RW (1994) $GABA_A$ receptor channels. Ann Rev Neurosci 17:569–602.

Macdonald RL, Rogers CJ, Twyman RE (1989) Barbiturate regulation of kinetic properties of the $GABA_A$ receptor channel of mouse spinal neurones in culture. J Physiol (Lond) 417:483–500.

Marszalec W, Kurata Y, Hamilton BJ, Carter DB, Narahashi T (1994) Selective effects of alcohols on γ-aminobutyric $acid_A$ receptor subunits expressed in human embryonic kidney cells. J Pharmacol Exp Ther 269:157–163.

McGehee DS, Role LW (1995) Physiological diversity of nicotinic acetylcholine receptors expressed by vertebrate neurons. Ann Rev Physiol 57:521–546.

Mihic SJ, Harris RA (1996) Inhibition of $\rho 1$ receptor GABAergic currents by alcohols and volatile anesthetics. J Pharmacol Exp Ther 277:411–416.

Mihic SJ, Ye Q, Wick MJ, Koltchine VV, Krasowski MD, Finn SE, Mascia MP, Valenzuela CF, Hanson KK, Greenblatt EP, Harris RA, Harrison NL (1997) Sites of alcohol and volatile anaesthetic action on $GABA_A$ and glycine receptors. Nature 389:385–389.

Moore JW, Ulbricht W, Takata M (1964) Effect of ethanol on the sodium and potassium conductances of the squid axon membrane. J Gen Physiol 48:279–295.

Nagata K, Aistrup GL, Huang C-S, Marszalec W, Song J-H, Yeh JZ , Narahashi T (1996) Potent modulation of neuronal nicotinic acetylcholine receptor-channel by ethanol. Neurosci Lett 217:189–193.

Nakahiro M, Arakawa O, Narahashi T (1991) Modulation of γ-aminobutyric acid receptor-channel complex by alcohols. J Pharmacol Exp Ther 259:235–240.

Nakahiro M, Arakawa O, Nishimura T, Narahashi T (1996) Potentiation of GABA-induced Cl^- current by a series of n-alcohols disappears at a cutoff point of a longer-chain n-alcohol in rat dorsal root ganglion neurons. Neurosci Lett 205:127–130.

Nakahiro M, Yeh JZ, Brunner E, Narahashi T (1989) General anesthetics modulate GABA receptor channel complex in rat dorsal root ganglion neurons. FASEB J 3:1850–1854.

Nayeem N, Green TP, Martin IL, Barnard EA (1994) Quaternary structure of the native $GABA_A$ receptor determined by electron microscopic image analysis. J Neurochem 62:815–818.

Nicoll RA (1972) The effects of anaesthetics on synaptic activation and inhibition in olfactory bulb. J Physiol (Lond) 223:803–814.

Nishio M, Narahashi T (1990) Ethanol enhancement of GABA-activated chloride current in rat dorsal root ganglion neurons. Brain Res 518:283–286.

Olsen RW, Tobin AJ (1990) Molecular biology of $GABA_A$ receptors. FASEB J 4:1469–1480.

Osmanovic SS, Shefner SA (1990) Enhancement of current induced by superfusion of GABA in locus coeruleus neurons by pentobarbital, but not ethanol. Brain Res 517:324–329.

Papke RL, Craig AG, Heinemann SF (1994) Inhibition of nicotinic acetylcholine receptors by bis (2,2,6,6-tetramethyl-4-piperidinyl) sebacate (Tinuvin 770), an additive to medical plastics. J Pharmacol Exp Ther 268:718–726.

Peoples RW, Li C; Weight FF (1996) Lipid vs protein theories of alcohol action in the nervous system. Ann Rev Pharmacol Toxicol 36:185–201

Reynolds JN, Prasad A (1991) Ethanol enhances $GABA_A$ receptor-activated chloride currents in chick cerebral cortical neurons. Brain Res 564:138–142.

Rogers SW, Mandelzys A, Deneris ES, Cooper E, Heinemann S (1992) The expression of nicotinic acetylcholine receptors by PC12 cells treated with NGF. J Neurosci 12:4611–4623.

Sanna E, Mascia MP, Klein RL, Whiting PJ, Biggio G, Harris RA (1995) Action of the general anesthetic propofol on recombinant human $GABA_A$ receptors: influence of receptor subunits. J Pharmacol Exp Ther 274: 353–360.

Sieghart W (1992) $GABA_A$ receptors: Ligand-gated Cl^- ion channels modulated by multiple drug-binding sites. Trends Pharmacol Sci 13:446–450.

Study RE, Barker JL (1981) Diazepam and (-)-pentobarbital: fluctuation analysis reveals different mechanisms for potentiation of gamma-aminobutyric acid responses in cultured central neurons. Proc Natl Acad Sci (USA) 78(11):7180–4

Tanguy J, Yeh JZ, Hamilton BJ, Carter DB, Brunner EA (1995) GABA-activated response of recombinant rat $GABA_A$ receptors and its modulation by volatile anesthetics. Progress in Anesthetic Mechanism 3 (Special Issue), 82–91.

Tatebayashi H, Motomura H, Narahashi T (1998) Alcohol modulation of single $GABA_A$ receptor-channel kinetics. Neuro Report 9:1769–1775

Twyman RE, Macdonald RL (1992) Neurosteroid regulation of $GABA_A$ receptor single-channel kinetic properties of mouse spinal cord neurons in culture. J Physiol (Lond) 456:215–245.

Twyman RE, Rogers CJ, Macdonald RL (1989) Differential regulation of γ-aminobutyric acid receptor channels by diazepam and phenobarbital. Ann Neurol 25:213–220.

Vicini S, Mienville J-M, Costa E (1987) Actions of benzodiazepine and β-carboline derivatives on γ-aminobutyric acid-activated Cl^- channels recorded from membrane patches of neonatal rat cortical neurons in culture. J Pharmacol Exp Ther 243:1195–1201.

Violet JM, Downie DL, Nakisa RC, Lieb WR, Franks NP (1997) Differential sensitivities of mammalian neuronal and muscle nicotinic acetylcholine receptors to general anesthetics. Anesthesiology 86:866–874.

White G, Lovinger DM, Weight FF (1990) Ethanol inhibits NMDA-activated current but does not alter GABA-activated current in an isolated adult mammalian neuron. Brain Res 507:332–336.

Whitten RJ, Maitra R, Reynolds JN (1996) Modulation of $GABA_A$ Receptor Function by alcohols: effects of subunits composition and differential effects of ethanol. Alcohol Clin Exp Res 20:1313–1319.

Yeh HH, Kolb JE (1997) Ethanol modulation of GABA-activated current responses in acutely dissociated retinal bipolar cells and ganglion cells. Alcohol Clin Exp Res 21:647–655.

Yu D, Zhang L, Eisele J-L, Bertrand D, Changeux J-P, Weight FF (1996) Ethanol inhibition of nicotinic acetylcholine type α7 receptors involves the amino-terminal domain of the receptor. Mol Pharmacol 50:1010–1016.

5

ALCOHOL AND THE 5-HT_3 RECEPTOR

David M. Lovinger and Qing Zhou

Department of Molecular Physiology and Biophysics
Department of Pharmacology
Vanderbilt University Medical School
Nashville, Tennessee 37232-0615

1. INTRODUCTION

Studies performed over the last ten years indicate that the 5-hydroxytryptamine$_3$ (5-HT_3) receptor for the neurotransmitter serotonin is involved in the acute and chronic actions of alcohol (see Grant, 1995 for review). Furthermore, experiments at the cellular and molecular level have indicated that ethanol and other alcohols may have direct actions on this ligand-gated ion channel (Lovinger, 1991a; Lovinger and Zhou, 1994; Machu and Harris, 1994; Barann et al., 1995). In this chapter, I will review this body of information. In addition, I will provide a rationale for examining the 5-HT_3 receptor not only to understand alcohol actions on this protein, but to use the receptor as a model protein for understanding alcohol effects on a larger group of related ligand-gated ion channels that are important targets for alcohol actions.

I will also discuss studies in which alcohol effects on receptor-channel kinetics are examined using whole-cell patch-clamp recording combined with rapid agonist application. Examining ligand-gated ion channel function using these techniques allows the investigator to examine receptor function on a time scale similar to that of synaptic transmission. By examining these receptors in isolated cells, the investigator can eliminate any presynaptic effects of applied drugs, and examine receptor-channel kinetics in a model "postsynaptic" structure. This approach allows the experimenter to examine changes in kinetic behavior of ligand-gated ion channels within a meaningful time domain. This can lead to insights into the way in which allosteric drugs, including alcohols, alter the function of these molecules. Kinetic simulation and modeling techniques can be used to help provide a more complete picture of the changes in receptor-channel function in the presence of a drug such as ethanol. We have used combined whole-cell recording and rapid drug application, along with kinetic simulations and modeling of 5-HT_3 receptor-channel function, to characterize alcohol effects on this important neuronal protein.

The "Drunken" Synapse, edited by Liu and Hunt.
Kluwer Academic / Plenum Publishers, New York, 1999.

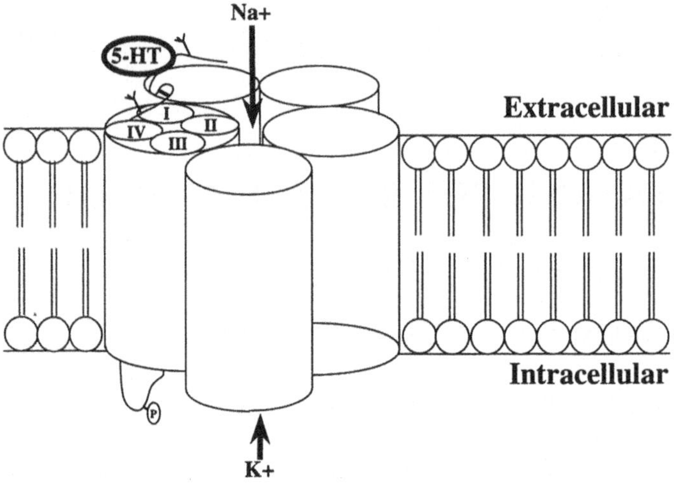

Figure 1. Schematic illustration of the 5-HT_3 receptor-channel complex. The receptor is believed to exist as a pentamer and is fully functional as a homopentamer. Serotonin binding to the n-terminal extracellular domain leads to gating of the nonspecific cation channel. Amino acid residues in the second transmembrane domain are thought to line the channel pore and control ion permeability.

2. THE 5-HT_3 RECEPTOR

The 5-HT_3 receptor is the only serotonin receptor that is a ligand-gated ion channel (Figure 1). The structure and function of the receptor have been reviewed in detail elsewhere (Jackson and Yakel, 1995). Thus, I will not deal with this subject except where it is necessary to review aspects of receptor structure, function, or pharmacology that bear on the subject at hand, namely the actions of alcohol on the receptor.

To date, only one gene product that can form a functional 5-HT_3 receptor has been discovered and named, the 5-HT_3 receptor A subunit (Maricq et al., 1991). Splice variants of this subunit have also been identified and characterized (Hope et al., 1993). The predicted structure of the 5-HT_3 receptor A subunit clearly places the receptor within the nicotinic aceytylcholine (nACh) receptor-like subclass of ligand-gated ion channels. Indeed, the 5-HT_3 receptor is most closely related in structure, function and pharmacology to certain ACh receptor subunits, and functional chimeric receptors can be formed by combining elements of both receptors (Eisele et al., 1993). It is clear that the 5-HT_3 receptor functions quite well as a homomultimeric receptor/channel (Maricq et al., 1991; Hope et al., 1993; Lovinger and Zhou, 1994; Machu and Harris, 1994). Evidence from electron microscopic examination indicates that the purified receptor has a homopentameric structure (Green et al. 1995).

The pharmacology of the 5-HT_3 receptor has been reasonably well characterized. Potent and highly selective antagonists have been developed (see Jackson and Yakel, 1995; Grant 1995 for review), and these compounds appear to be competitive antagonists. Some agonists have also been synthesized, but these compounds have lesser potency than the antagonists do, and all have relatively poor specificity for the receptor (Grant 1995). Tubocurarine, an ACh receptor blocker, acts as an antagonist at the mouse receptor with nanomolar potency (Downie et al., 1994). Agents have also been identified that alter 5-HT_3 receptor/channel kinetics and channel function in a voltage-dependent manner (Kooyman et al., 1994; Lovinger 1991b). However, no unequivocal open-channel blockers have yet been identified for this receptor. Other allosteric effectors of the 5-HT$_3$ receptor have also been reported, but none of these interactions have been characterized in great detail.

The 5-HT$_3$ receptor-gated channel is a nonspecific ligand-gated ion channel that shows preference for monovalent cation permeability under physiological conditions. However, the receptor channel is Ca^{++} permeable and thus activation of the receptor appears to contribute to rises in intracellular calcium (Reiser et al. 1992).

3. 5-HT$_3$ RECEPTORS ARE IMPLICATED IN THE NEURAL EFFECTS OF ALCOHOL

A variety of lines of evidence indicate that 5-HT$_3$ receptors have a role in the neural effects of alcohol and other drugs of abuse (see Grant 1995 for review). Different 5-HT$_3$ receptor antagonists reduce alcohol intake in animal models (Fadda et al., 1991; LeMarquand et al., 1994b; Tomkins et al., 1995). These studies have been performed with animals given free access to alcohol and total intake has been measured. Thus, it is not clear if alcohol drinking is reduced under all experimental conditions. Studies of human alcoholics also indicate decreased alcohol consumption after treatment with 5-HT$_3$ receptor antagonists (Johnson et al., 1993; LeMarquand et al., 1994a; Sellers et al., 1994).

Antagonists at the 5-HT$_3$ receptor also prevent the subjective effects of acute alcohol in some drug-discrimination paradigms. This has been demonstrated in pigeons (Grant and Barrett, 1991) and for some antagonists in rats (Grant and Colombo, 1993). In these experiments, most of the effects on drug discrimination were not associated with changes in alcohol metabolism. However, effects of MDL 72222 in the rat appear to be secondary to altered ethanol metabolism (Grant and Colombo, 1993). Thus, there is some evidence for involvement of the 5-HT$_3$ receptor in subjective cues associated with acute intoxication in animal models, but these effects will have to be carefully separated from effects on alcohol metabolism if and when human studies are performed.

The 5-HT$_3$ receptor may also interact with alcohol and other drugs of abuse via its role in stimulating release of the neurotransmitter dopamine (DA) in the nucleus accumbens and other targets of the mesolimbic and mesocortical dopaminergic pathways (reviewed in Grant 1995). Activation of receptors leads to increased extracellular dopamine in these brain regions, which might be due to the actions of presynaptic 5-HT$_3$ receptors on dopaminergic terminals. Dopaminergic transmission in the nucleus accumbens appears to play a central role in the reinforcing and addictive effects of a number of drugs of abuse, including ethanol. Thus, it is possible that potentiation of 5-HT$_3$ receptor function during alcohol intoxication might lead to enhanced dopaminergic transmission and this would constitute a mechanism by which the receptor could participate in the reinforcing effects of ethanol.

4. ALCOHOLS AND ANESTHETICS HAVE POTENT ACTIONS ON THE 5-HT$_3$ RECEPTOR

The effects of alcohols and several anesthetic agents on 5-HT$_3$ receptor function have been examined in isolated cells using single cell electrophysiological approaches as well as measurements of flux of radiolabelled ions (Lovinger, 1991a; Zhou and Lovinger,

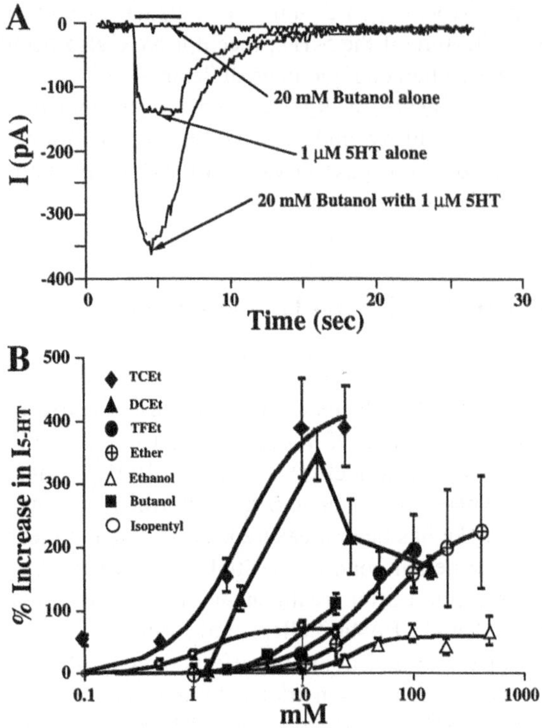

Figure 2. Potentiation of $5\text{-}HT_3$ receptor-mediated current by alcohols and diethyl ether. A) Current traces showing responses to butanol, to 5-HT and to 5-HT and butanol. Note that application of butanol alone does not elicit ion current even at a concentration that strongly potentiates current activated by 5-HT. The line above the current traces indicates the duration of drug application. B) Concentration-response curves for all of the alcohols tested as well as diethyl ether. Note the different maximal efficacies of the compounds. Note also the biphasic DCEt concentration-response curve. Values are mean ± SEM. (Reprinted with permission from Zhou and Lovinger 1996).

1996; Machu and Harris, 1994; Barann et al., 1995; Jenkins et al., 1996). The short-chain alcohols and several volatile anesthetic agents have been shown to potentiate $5\text{-}HT_3$ receptor function at concentrations in the range at which their actions are thought to occur *in vivo* (Figure 2, Lovinger, 1991a; Zhou and Lovinger, 1996; Jenkins et al., 1996). More will be said about the interactions of these compounds with the receptor, and their mechanisms of action, later in this chapter.

Alcohols with carbon chain lengths longer than five have been reported to have either no effect (Fan and Weight, 1994) or an inhibitory effect (Jenkins et al., 1996) on $5\text{-}HT_3$ receptors. The inhibitory action may be similar to the channel blocking effects of these anesthetics and alcohols on the nACh receptor (Forman et al., 1995). Effects of a range of injectable anesthetic agents on $5\text{-}HT_3$ receptor function have also been examined. Barbiturates inhibit receptor-mediated ion current at moderate to high micromolar concentrations (Lovinger and Peoples, 1993; Jenkins et al., 1996). The injectable general anesthetic propofol also has actions on the receptor (Barann et al., 1993), but it is doubtful that the clinically relevant actions of this agent have anything to do with effects on the $5\text{-}HT_3$ receptor. Likewise, certain benzodiazepines inhibit receptor function, but only at relatively high concentrations (Lovinger and Peoples, 1993). Thus, the simple alcohols and volatile anesthetics are the major classes of sedative/hypnotic/anesthetic drugs that have effects on the $5\text{-}HT_3$ receptor which are likely to play a role in their *in vivo* pharmacology. These findings suggest that alterations in $5\text{-}HT_3$ receptor function cannot explain all sedative/hypnotic or anesthetic actions. Based on the findings discussed in the last section, it is probable that the $5\text{-}HT_3$ receptor is involved in intoxication and alcohol reinforcement, and may participate, along with other molecules such as the $GABA_A$ receptor, in anesthetic effects of alcohols and volatile agents.

The actions of ethanol, other short chain alkanols, and volatile anesthetics on 5-HT$_3$ receptors have been characterized in considerable detail. Our own work has focused mainly on the effects of three alcohols, ethanol, butanol and trichloroethanol (TCEt) (Lovinger, 1991a; Lovinger and Zhou, 1994; Zhou et al., 1998), and thus I will focus most of my discussion on the effects of these three agents while noting similar findings with other compounds. The rationale for examining ethanol effects is straightforward, especially in light of the evidence for a role of the receptor in ethanol's neural actions reviewed above. We have also examined butanol since this short-chain alkanol has actions similar to ethanol, but with greater potency and efficacy. This has allowed us to use butanol as a closely related compound for comparison with the actions of ethanol. The rationale for examining TCEt actions is threefold. First, TCEt is a halogen modified ethanol analog, and thus the use of this compound allows us to probe alcohol structure-activity relationships in more detail. Second, TCEt is the active metabolite of the general anesthetic chloral hydrate, and is responsible for most of the anesthetic actions after chloral hydrate administration (Lovinger and Zhou, 1993). Thus, the compound is a useful tool for examining general anesthetic actions on the receptor. Finally, TCEt is the most efficacious of the alcohols we have examined to date with respect to potentiation of 5-HT$_3$ receptor function (Figure 2, Zhou and Lovinger, 1996), and thus we can use TCEt to analyze large changes in receptor-channel kinetics in detail.

All three of these alcohols potentiate current mediated by 5-HT$_3$ receptors at low 5-HT concentrations (EC50 and below). Potentiation is not observed in the presence of higher agonist concentrations. Thus, the alcohols produce a parallel leftward shift in the agonist concentration-response curve, which is indicative of increased agonist potency (Zhou and Lovinger, 1996; Zhou et al., 1998). The effective concentrations of ethanol and TCEt are within the range of concentrations at which these alcohols produce intoxication and general anesthesia, respectively (Lovinger, 1991a; Lovinger and Zhou, 1993).

In one study, we examined the effects of a variety of alkanols and halogen substituted alcohols (Zhou and Lovinger, 1996). All of the alcohols potentiated receptor function at a 5-HT concentration that is ~EC_{10} (Figure 2). The potency with which alcohols potentiated 5-HT$_3$ receptor function was related to hydrophobicity in a manner consistent with the Meyer-Overton relation (*i.e.*, potency increased with increasing hydrophobicity). Potentiation by several of the alcohols appeared to saturate at relatively high concentrations. The maximal efficacy of potentiation by alcohols varied considerably among the alcohols, with TCEt exhibiting greater efficacy than butanol, which in turn was greater than ethanol. However, maximal efficacy did not correlate with hydrophobicity of the alcohols, suggesting that the actions of alcohols on the receptor are not determined simply by the concentration of the alcohol that is present in the cell membrane.

Using a combination of whole-cell patch-clamp recording and rapid agonist application, we have recently examined the alterations in 5-HT$_3$ receptor-channel kinetics produced by TCEt, butanol and ethanol to gain a better understanding of alcohol actions on different aspects of the function of this protein (Zhou et al. 1998). The 5-HT$_3$ receptor is quite amenable to this method of analysis, since receptor-channel kinetics are reasonably slow compared to other ligand-gated ion channels (Zhou et al. 1998). Furthermore, the NCB-20 neuroblastoma cells used for these studies were relatively small, allowing for reasonably rapid solution exchange around the cell.

All three alcohols examined produced increases in the rate of receptor-channel activation, decreases in the rate of desensitization and decreases in the deactivation rate (also known as the closing or agonist unbinding rate). These effects tended to stabilize and favor the receptor-channel remaining in the open, ion conducting state. To determine which

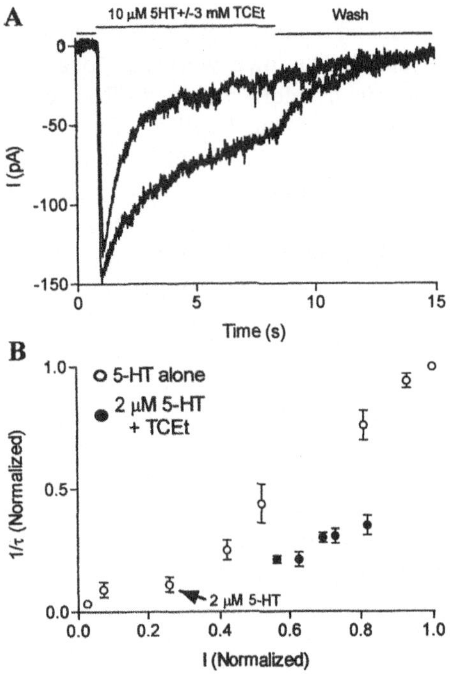

Figure 3. TCEt slows the desensitization of 5-HT$_3$ receptors. A) Two representative current records in the presence of 10 μM 5-HT with or without 3 mM TCEt. TCEt clearly slowed the decay rate in the continuous presence of this high agonist concentration. B) Current amplitude (normalized) and 1/τ (normalized) were calculated by dividing the amplitude of the peak current or the decay rate of the current activated by the indicated concentration of 5-HT by the same measures made for responses to 40 μM 5-HT. Open circles represent current generated by 5-HT alone; closed circles represent current induced by 2 μM 5-HT in the presence of 1, 2, 4, 6, and 10 mM TCEt. (Reprinted with permission from Zhou et al. 1998).

of these effects are likely to contribute to potentiation of peak ion current, we used both current simulation and kinetic modeling approaches. It is possible that potentiation of peak current results simply from an increase in receptor channel activation that favors the open channel state. However, simulations of current produced by activation of a receptor with behavior described by a simple three state kinetic scheme indicated that some of the changes that we observed in the presence of alcohols could not be accounted for by a simple increase in activation rate (Zhou et al., 1998). For example, while the rate of receptor desensitization was increased in the presence of alcohols, this increase was not as large as would have been predicted based on the observed increase in peak current amplitude (Figure 3).

Furthermore, we used non-linear global analysis techniques to fit data generated by receptor activation in the presence and absence of alcohols and generate estimates of changes in receptor/channel rate constants produced by alcohols (Figures 4 and 5). This analysis indicated that potentiation by alcohols involved changes in receptor-channel activation/deactivation rate constants as well as changes in desensitization/resensitization rate constants (Table 1). All of these changes appeared to be necessary to generate proper fits to the experimental data, suggesting that these changes are crucial for alcohol-induced potentiation of receptor function.

$$R+A \underset{k_{-1}}{\overset{2^*k_1}{\rightleftharpoons}} AR+A \underset{2^*k_{-1}}{\overset{k_1}{\rightleftharpoons}} A_2R \underset{\alpha}{\overset{\beta}{\rightleftharpoons}} A_2O \underset{k_{-d}}{\overset{k_d}{\rightleftharpoons}} A_2D$$

Figure 4. In this model, R, AR, and A$_2$R represent ligand free, monoliganded closed, and doubly liganded closed receptors. A represents agonist. A$_2$O and A$_2$D represent the open and desensitized channel states respectively. The constants k_1 and k_{-1} are the receptor association (or binding) and dissociation (or unbinding) rate constants while β and α are the opening and closing rate constants for the liganded receptor, and k_d and k_{-d} are the desensitization rate constants.

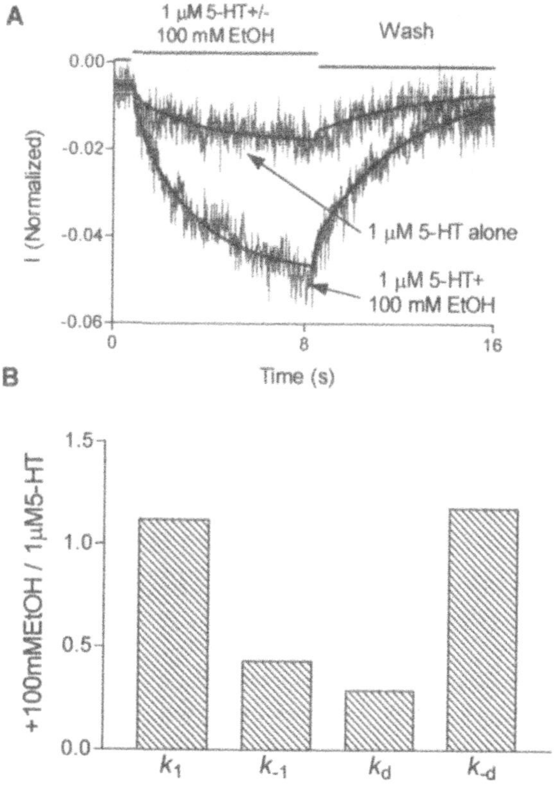

Figure 5. Fits of currents generated by 5-HT+/-EtOH and EtOH-induced changes in estimated channel kinetic parameters. A) The smaller amplitude current (shown in grey) was induced by 1 μM 5-HT alone and the larger amplitude current (also shown in grey) was generated by 1 μM 5-HT + 100 mM EtOH. Both currents are averages of responses generated in four different cells and normalized to the current generated by a saturating concentration of 5-HT (10 μM), as described in Zhou et al. (1998). The continuous lines represent the corresponding fits. B) Ethanol increased the estimated receptor activation rate constant and estimated channel resensitization rate constant, and decreased the estimated receptor deactivation rate constant and estimated channel desensitization rate constant. (Reprinted with permission from Zhou et al., 1998).

It has proven difficult to examine single 5-HT_3 receptor-channels in any detail due to the small conductance of these channels, especially in neuroblastoma cell lines (<1 pS, Yang 1990, Shao et al. 1991, Hussy et al. 1994). However, several findings indicate that alcohol potentiation of 5-HT_3 receptor-mediated currents does not result from an increase in single channel conductance. First, alcohols do not potentiate current activated by a saturating concentration of 5-HT, a result that is inconsistent with increased channel conductance. Second, estimates of single channel conductance from non-stationary noise analysis indicated no change in conductance in the presence of ethanol. Estimated conductance in three NCB-20 cells averaged 0.27 ± 0.03 pS for current activated by 2 μM 5-HT alone and 0.3±0.03 pS for current activated by 2 μM 5-HT in the presence of 100 mM EtOH. These observations suggest that alcohols do not potentiate current by enhancing single channel conductance.

Table 1. EtOH effects on estimated transition rate constants

Parameters	1μM 5-HT Alone	1μM 5-HT+100 mM EtOH
K_1 (1/M*s)	840750	941600
K_{-1} (1/s)	5.63	2.42
β (1/s)	937.5	937.5
α (1/s)	732.1	732.1
K_d (1/s)	49.7	14.2
K_{-d} (1/s)	1.6	2.83

Data were generated from fits to current records from four cells. Adapted with permission from Zhou et al., 1998.

Neurotransmitter release at synapses may lead to rapid increases in extracellular neurotransmitter concentrations that reach quite high levels. In this case, it might appear that alcohols would have no effect on synaptic responses, given the lack of alcohol potentiation of maximal current amplitudes evoked at high agonist concentrations. However, alcohols reduce receptor desensitization and deactivation rate constants, and thus could produce prolongation of synaptic responses in the absence of any change in peak synaptic current amplitude. Indeed, both ethanol and trichloroethanol have been shown to prolong $GABA_A$ receptor-mediated IPSCs at synapses in the hippocampus (Lovinger et al., 1993; Wan et al., 1996). Similar effects might be observed at serotonergic synapses containing 5-HT_3 receptors.

The observation that alcohols increase the potency with which 5-HT activates the 5-HT_3 receptor might indicate that alcohols increase agonist affinity for the receptor. However, a similar increase in potency would occur if the probability of channel opening was increased, an effect that would not be apparent at high agonist concentrations where probability of opening is already high. It is difficult to distinguish between these two mechanisms of action since separating agonist affinity from probability of channel opening is not straightforward (Culquhoun and Farrant, 1993). This is a question that needs to be addressed in future experiments.

5. THE MOLECULAR SITE OF ALCOHOL ACTION ON THE 5-HT_3 RECEPTOR

Several lines of evidence indicate that there may be specific sites of alcohol action on the 5-HT_3 receptor and related ligand-gated ion channels. Evidence indicates that alcohols can act specifically on proteins in the absence of lipids, indicating that the lipid membrane is not necessary for alcohol effects (Franks and Lieb, 1994). Alcohol effects on the 5-HT_3 receptor are observed in a variety of cell lines including non-neuronal heterologous expression systems such as *Xenopus Laevis* oocytes (Lovinger and Zhou, 1994; Machu and Harris, 1994). Thus, alcohol actions are not dependent on a particular cellular lipid environment, but seem to be conferred by the protein itself. The observation of "cutoff" of effects on particular receptors when the alcohol carbon chain exceeds a certain length also supports the idea of a protein site of alcohol action, since this sort of cutoff effect would not be expected with a purely lipid site of alcohol action (Li et al., 1994; Peoples and Weight, 1995; Dildy-Mayfield et al., 1996; Franks and Lieb, 1994). Finally, recent evidence indicates that alcohol effects on the 5-HT_3, glycine and $GABA_A$ receptor can be altered in chimeric receptors or by mutagenesis of single amino acids in the membrane spanning domains of the glycine and $GABA_A$ receptors (Yu et al., 1996; Mihic et al., 1997; Wick et al. 1998). This is perhaps the strongest evidence to date of a potential site of action of alcohols on a protein. The fact that all of these ligand-gated ion channels are similar in structure, suggests the possibility that alcohol sensitivity of the 5-HT_3 receptor may involve amino acids in the region of the 5-HT_3 receptor that corresponds to the amino acids identified in the glycine and $GABA_A$ receptors. Indeed, this research may lead to identification of a generalized alcohol sensitive site on many members of the nACh receptor-related ligand-gated ion channel subfamily.

It is possible, however, that the amino acids identified as conferring alcohol sensitivity in the glycine and $GABA_A$ receptors may be involved in transduction of the effect of alcohol, rather than being sites of direct alcohol interaction with the protein. It is difficult to provide direct physical evidence for alcohol-protein interactions. Alcohols have low affinity for their sites of action, precluding the use of traditional radioligand binding studies. One challenge in future studies will be to use biophysical techniques to examine interactions of alcohols with alcohol-sensitive ligand-gated ion channels (Lovinger, 1997).

Pharmacological approaches have been used to address the question of whether alcohols interact with ligand-gated ion channels in a manner consistent with the existence of alcohol-binding sites on the receptor protein itself. As mentioned above, the cutoff effect observed with alkanols indicates that the alcohol binding site has limited physical dimensions that do not seem to fit with actions on the lipid region of the membrane. This constitutes good evidence that alcohol's actions involve some interaction with the receptor protein. However, it is not clear that the different alcohols used in cutoff effect studies act on the same site associated with the receptor-channel complex. Indeed, attempts to demonstrate the expected competition for an alcohol binding site on the 5-HT$_3$ receptor have not provided evidence consistent with a single site of alcohol action (Zhou and Lovinger, 1996). This evidence suggests that alcohols may act at several different sites on this receptor-channel complex. Thus, if a single mutation is found that blocks the effects of several alcohols at this receptor, it may reflect prevention of transduction of the alcohol-protein interaction into an allosteric effect on the receptor-channel, rather than elimination of an alcohol binding site. On the other hand, recent studies of the glycine and GABA$_A$ receptors indicate that mutations of key residues in the TM2 region that change the volume of particular amino acids in this region of these ligand-gated ion channels can alter the carbon chain length at which cutoff is observed (Wick et al., 1998). The most compelling explanation of these observations is that cutoff is determined by interactions with amino acids in this region of the protein which constitute an alcohol binding site. It will be interesting to determine if this idea will be supported by more direct physical evidence, and if similar findings will be forthcoming from studies of other ligand-gated ion channels.

6. SUMMARY

For the last several years, we and others have examined alcohol effects on the 5-HT$_3$ ligand-gated ion channel. This receptor-channel has been implicated in acute alcohol intoxicating actions and alcohol consummatory behavior, and thus it is important to understand the direct alcohol effects on the receptor. Furthermore, this receptor has several desirable characteristics of a model protein for study of the nACh receptor-like subfamily of ligand-gated ion channels, many of which are very sensitive to alcohols. We have observed that ethanol and other alcohols potentiate 5-HT$_3$ receptor function in a number of neuronal cell types and in heterologous expression systems. We have used whole-cell recording coupled with rapid agonist application to examine alcohol effects on receptor-channel kinetics over a meaningful time scale. In these studies we observed that alcohols stabilize and favor the open channel state. The results of this analysis are likely to be useful in understanding alcohol's effects on other members of the receptor subfamily. Several lines of evidence indicate that these effects on ligand-gated ion channels stem from direct interactions with the receptor proteins themselves. Future research will be directed at understanding if an alcohol binding site(s) can be identified on the 5-HT$_3$ receptors.

ACKNOWLEDGMENTS

This work was supported by grants from NIAAA (AA08986) and the Alcoholic Beverage Medical Research Foundation. The authors would like to thank Yingchun Yu for her technical assistance with the projects described in this paper.

EDITOR'S NOTE

A recent publication (Davies et al. 1999) indicates that another subunit termed the 5-HT3B subunit can coassemble with human 5-HT3RA subunits and alters receptor-channel properties. This new subunit is expressed in brain, and thus, some brain 5-HT3 receptors may exist as heteromultimers.

REFERENCES

Barann M, Gothert M, Fink K, Bonisch, H (1993) Inhibition by anaesthetics of ^{14}C-guanidinium flux through the voltage-gated sodium channel and the cation channel of the 5-HT_3 receptor of N1E-115 neuroblastoma cells. Naunyn-Schmiedeberg's Arch Pharmacol 347(2):125–132.

Barann M, Ruppert K, Göthert M, Bonisch H (1995) Increasing effect of ethanol on 5-HT_3 receptor-mediated ^{14}C-guanidinium influx in N1E-115 neuroblastoma cells. Naunyn-Schmiedebergs Arch Pharm 352(2):149–156.

Colquhoun D, Farrant M (1993) The binding issue. Nature 366:510–511.

Davies PA, Pistis M, Hanna MC, Peters JA, Lambert JJ, Hales TG, Kirkness EF (1999) The 5-HT3B subunit is a major determinant of serotonin-receptor function. Nature 397(6717):359–363.

Dildy-Mayfield JE, Mihic SJ, Liu Y, Deitrich RA, Harris RA (1996) Actions of long chain alcohols on $GABA_A$ and glutamate receptors: relation to *in vivo* effects. Br J Pharmacol 118(2):378–384.

Downie DL, Hope AG, Lambert JJ, Peters JA, Blackburn TP, Jones BJ (1994) Pharmacological characterization of the apparent splice variants of the murine 5-HT_3 R-A subunit expressed in Xenopus laevis oocytes. Neuropharmacol 33(3–4):473–482.

Eisele J-L, Bertrand S, Galzi J-L, Devillers-Thiery A, Changeux J-P, Bertrand D (1993) Chimaeric nicotinic-serotonergic receptor combines distinct ligand binding and channel specificities. Nature 366:479–483 .

Fadda F, Garau B, Marchei F, Colombo G, Gessa GL (1991) MDL 72222, a selective 5-HT_3 receptor antagonist, suppresses voluntary ethanol consumption in alcohol-preferring rats. Alcohol Alcohol 26:107–110.

Fan P, Weight FF (1994) Alcohols exhibit a cutoff effect for the potentiation of 5-HT_3 receptor-activated current. Soc Neurosci Abst 20:1127.

Forman SA, Miller KW, Yellen G (1995) A discrete site for general anesthetics on a postsynaptic receptor. Mol Pharmacol 48(4):574–581.

Franks NP, Lieb WR (1994) Molecular and cellular mechanisms of general anaesthesia. Nature 367(6464):607–614.

Grant KA (1995) The role of 5-HT_3 receptors in drug dependence. Drug Alcohol Depend 38:155–171.

Grant KA, Barrett JE (1991) Blockade of the discriminative stimulus effects of ethanol with 5-HT_3 receptor antagonists. Psychopharm 104:451–456.

Grant KA, Colombo G (1993) The ability of 5-HT_3 antagonists to block an ethanol discrimination: Effect of route of administration. Alcohol Clin Exp Res 17:497.

Green T, Stauffer KA, Lummis SCR (1995) Expression of recombinant homo-oligomeric 5-hydroxytryptamine3 receptors provides new insights into their maturation and structure. J Biol Chem 270(11):6056–6061.

Hope AG, Downie DL, Sutherland L, Lambert JJ, Peters JA, Burchell B (1993) Cloning and functional expression of an apparent splice variant of the murine 5-HT_3 receptor A subunit. Eur J Pharmacol 245(2):187–192.

Hussy N, Lukas W, Jones KA (1994) Functional properties of the cloned 5-HT_3A receptor. J Physiol 481:311–323.

Jackson MB, Yakel JL (1995) The 5-HT_3 receptor channel. Ann Rev Physiol 57:447–468.

Jenkins A, Franks NP, Lieb WR (1996) Actions of general anaesthetics on 5-HT_3 receptors in N1E-115 neuroblastoma cells. Br J Pharm 117(7):1507–1515.

Johnson BA, Campling GM, Griffiths P, Cowen PJ (1993) Attenuation of some alcohol-induced mood changes and the desire to drink by 5-HT_3 receptor blockade: A preliminary study in healthy male volunteers. Psychopharmacol 112:142–144.

Kooyman AR, van Hooft JA, Vanderheijden PM, Vijverberg HP (1994) Competitive and non-competitive effects of 5-hydroxyindole on 5-HT_3 receptors in N1E-115 neuroblastoma cells. Br J Pharmacol 112(2):541–546.

LeMarquand D, Pihl RO, Benkelfat, C (1994a) Serotonin and alcohol intake, abuse, and dependence: clinical evidence. Biol Psychiat 36(5):326–337.

LeMarquand D, Pihl RO, Benkelfat C (1994b) Serotonin and alcohol intake, abuse, and dependence: findings of animal studies. Biol Psychiat 36(6):395–421.

Li C, Peoples RW, Weight FF (1994) Alcohol action on a neuronal membrane receptor: Evidence for a direct interaction with the receptor protein. Proc Natl Acad Sci (USA) 91:8200–8204.

Lovinger DM (1991a) Ethanol potentiates 5-HT$_3$ receptor-mediated ion current in NCB-20 neuroblastoma cells. Neurosci Lett 122:54–56.

Lovinger DM (1991b) Inhibition of 5-HT$_3$ receptor-mediated ion current by divalent metal cations in NCB-20 neuroblastoma cells. J Neurophysiol 66(4):1329–1337.

Lovinger DM (1997) Alcohols and neurotransmitter gated ion channels: past, present and future. Naunyn-Schmiedeberg's Arch Pharmacol 356:267–282.

Lovinger DM, Peoples RW (1993) Actions of alcohols and other sedative/hypnotic compounds on cation channels associated with glutamate and 5-HT$_3$ receptors, In: Alling C, Diamond I, Leslie SW, Sun GY, Wood WG (eds) Alcohol, Cell Membranes and Signal Transduction in Brain, Plenum Press, New York pp. 157–168.

Lovinger DM, Zimmerman SA, Levitin M, Jones ML, Harrison, NL (1993) Trichloroethanol potentiates synaptic transmission mediated by GABA$_A$ receptors in hippocampal neurons. J Pharmacol Exp Ther 264:1097–1103.

Lovinger DM, Zhou Q (1993) Trichloroethanol potentiation of 5-hydroxytryptamine$_3$ receptor-mediated ion current in nodose ganglion neurons from adult rat. J Pharmacol Exp Ther 265:771–777.

Lovinger DM, Zhou Q (1994) Alcohols potentiate ion current mediated by recombinant 5-HT$_3$RA receptors expressed in a mammalian cell line. Neuropharmacol 33:1567–1572.

Machu TK, Harris RA (1994) Alcohols and anesthetics enhance the function of 5-hydroxytryptamine3 receptors expressed in *Xenopus laevis* oocytes. J Pharmacol Exp Ther 271(2):898–905.

Maricq AV, Peterson AS, Brake AJ, Myers RM, Julius D (1991) Primary structure and functional expression of the 5HT3 receptor, a serotonin-gated ion channel. Science 254 (5030):432–437.

Mihic SJ, Ye Q, Wick MJ, Koltchine VV, Krasowski MD, Finn SE, Mascia MP, Valenzuela CF, Hanson KK, Greenblatt EP, Harris RA, Harrison NL (1997) Sites of alcohol and volatile anaesthetic action on GABA$_A$ and glycine receptors. Nature 389:385–389.

Peoples RW, Weight FF (1995) Cutoff in potency implicates alcohol inhibition of N-methyl-D-aspartate receptors in alcohol intoxication. Proc Natl Acad Sci (USA) 92(7):2825–2829.

Reiser G, Donie F, Ginmoller F-J (1992) Serotonin regulates cytosolic Ca^{2+} activity and membrane potential in a neuronal and in a glia cell line via 5-HT$_3$ and 5-HT2 receptors by different mechanisms. J Cell Sci 93:545–555.

Sellers EM, Toneatto T, Romach MK, Some G.R, Sobell LC, Sobell MB (1994) Clinical efficacy of the 5-HT$_3$ antagonist ondansetron in alcohol abuse and dependence. Alcohol Clin Exp Res 18:879–885.

Shao XM, Yakel JL, Jackson MB (1991) Differentiation of NG 108–15 cells alters channel conductance and desensitization kinetics of the 5-HT$_3$ receptor. J Neurophysiol 65:630–638.

Tomkins DM, Le AD, Sellers EM (1995) Effect of the 5-HT$_3$ antagonist ondansetron on voluntary ethanol intake in rats and mice maintained on a limited access procedure. Psychopharmacol 117(4):479–485

Wan FJ, Berton F, Madamba SG, Francesconi W, Siggins GR (1996) Low ethanol concentrations enhance GABAergic inhibitory postsynaptic potentials in hippocampal pyramidal neurons only after block of GABA$_B$ receptors. Proc Natl Acad Sci (USA) 93(10):5049–5054.

Wick MJ, Mihic SJ, Ueno S, Mascia MP, Trudell JR, Brozowski SJ, Ye Q, Harrison NL, Harris RA (1998) Mutations of γ-aminobutyric acid and glycine receptors change alcohol cutoff: Evidence for an alcohol receptor? Proc Natl Acad Sci (USA) 95(11):6504–6509.

Yang J (1990) Ion permeation through 5-hydroxytryptamine-gated channels in neuroblastoma N18 cells. J Gen Physiol 96:117–119.

Yu D, Zhang L, Eisele JL, Bertrand D, Changeux JP, Weight FF (1996) Ethanol inhibition of nicotinic acetylcholine type alpha 7 receptors involves the amino-terminal domain of the receptor. Mol Pharmacol 50(4):1010–1016.

Zhou Q, Lovinger DM (1996) Pharmacological characteristics of potentiation of 5-HT$_3$ receptors by alcohols and diethyl ether in NCB-20 neuroblastoma cells. J Pharmacol Exp Ther 278:732–740.

Zhou Q, Verdoorn TA, Lovinger DM (1998) Alcohols potentiate the function of 5-HT$_3$ receptor-channels on NCB-20 neuroblastoma cells by favouring and stabilizing the open channel state. J Physiol 507:335–352.

6

QUESTIONS AND ANSWERS OF SESSION I

Synaptic Transmission

1. Q&As BETWEEN AUDIENCE AND INDIVIDUAL SPEAKERS

1.1. Q&As between Audience and Dr. Tsien

1.1.1. Presynaptic Release Dynamics and Postsynaptic Receptor Saturation

AUDIENCE MEMBER: You showed us data from two sets of experiments. One is related to the dynamics of transmitter release from the presynaptic terminal. Another one is related to the saturation of a transmitter, glutamate, on its postsynaptic receptors. But, in your conclusion, you didn't mention these two systems have any relationship. If you change the presynaptic release, would that influence the saturation of the postsynaptic receptor?

DR. TSIEN: I obviously left that as a possibility. One could very well imagine that if the process of vesicle fusion was somehow modifiable—in other words, if the size of the fusion pore were under control by something like phosphorylation—the profile of glutamate that came out into the cleft could actually be seen very differently by the glutamate receptors. If the AMPA receptors are not at saturation, then there is the possibility for changing the actual response that you get as a result of a presynaptic mechanism like modification of the fusion pore. I showed you that the duration of the fusion events before the endocytosis ends is variable, but what I didn't show you was how alterations in fusion pore dynamics might lead to changes in glutamate concentration that would matter to AMPA receptors as opposed to NMDA receptors. That remains to be done, and we're actively working on this issue.

1.1.2. Calcium Dependence of Transmitter Release

AUDIENCE MEMBER: I have a question about the presynaptic portion of your talk. You showed that in 8 mM calcium there was less release than in the lower amount of calcium. Do you think there might be a surface charge effect of calcium to change vesicle fusion?

In other words, the charge along the membrane could change, which in turn will influence release.

DR. TSIEN: I don't think so. The way that we could really convince you of that is to measure transmitter release under the same conditions that we did in that experiment. In fact, when we do that, we find that 8 mM calcium causes more transmitter release than 1 mM calcium, typical of what you would expect at a central synapse. So we are convinced that it is not due to some fancy effect of surface charge shifting the activation of calcium channels. Our work indicates that the effect of calcium permeating more rapidly through the channels actually overwhelms the effect of the surface charge shifting the voltage dependence. Therefore, I believe that is not the explanation of the result. Rather, I think it has to do with how long the fusion pore stays open. The fusion pore is staying open for a shorter length of time in the high calcium. We can see it with FM dyes, even though we would not be able to see it with a fast transmitter, which escapes from the vesicle on a millisecond time scale, long before the fusion pore has time to close.

1.2. Q&As between Audience and Dr. Treistman

1.2.1. Measurement of Hormone Release

AUDIENCE MEMBER: My question regards your last slide. How was the sample collected during the measurement of hormone release?

DR. TREISTMAN: What that sample number represents is the time at which each sample is collected. So the first samples are collected as a baseline measure of release, then the samples are collected in the presence of high potassium.

AUDIENCE MEMBER: Where do you measure the release?

DR. TREISTMAN: We've done it both with isolated terminals and with the intact posterior pituitary, and both gave the same result.

AUDIENCE MEMBER: Did you say that following chronic ethanol exposure, the release function decreases?

DR. TREISTMAN: No. In fact, after chronic exposure, the release is increased. And that would make sense if it was compensating for the presence of the drug that is inhibiting release acutely.

AUDIENCE MEMBER: So how long is the total time that you collect the sample? One hour? Two hours?

DR. TREISTMAN: The measurements are on a seconds time scale.

AUDIENCE MEMBER: Have you measured for a longer time period, longer than seconds?

DR. TREISTMAN: I guess we could, but we haven't.

1.2.2. Voltage and Calcium Dependence of Calcium-Activated Potassium Channels

DR. TSIEN: These are well coordinated and logical experiments. Dr. Treistman, I just wanted to ask you about the change in voltage dependence and the change in calcium dependence. Where in one case, there seemed to be an upward scaling; in the other case, there seemed to be a convergence of the curves. Lots of people are working on kinetic models of that very same type of channel. Rick Aldrich's model (Cox et al., 1997a & b; Cui et al., 1997) would lead to the suggestion of the following experiment. Try to see what happens to the ethanol effect on the calcium dependence—on the voltage dependence, both at low calcium and also at high calcium.

DR. TREISTMAN: We did not test the voltage dependency at both low and high calcium.

DR. TSIEN: Do you mean that the opening probability is increased by ethanol regardless of how strongly you depolarized the membrane?

DR. TREISTMAN: That is correct.

DR. TSIEN: That's really quite remarkable. The finding might be interesting for theoretically inclined people, because it puts constraints on their models. And it's also very interesting for what ethanol might be doing mechanistically as well. Thank you.

1.2.3. Properties of Presynaptic Calcium and Calcium-Activated Potassium Channels

DR. APELL: (Sarah Apell from University of Illinois). This is a question for Dr. Treistman. Can you make comment on the differences between M-slow and B-slow? Not knowing that system well, I was very intrigued by the increase and decrease in current of those two different channels.

DR. TREISTMAN: Yes. Most of the differences occur in the tail region, although there's some point differences throughout. But there's a sequence that's very different, and there's also a difference in calcium sensitivity of the two that appears to be related to the region.

DR. APELL: So they differ in calcium sensitivity?

DR. TREISTMAN: Maybe.

DR. APELL: Also I have a question regarding the EC_{50}s or the effect on the L-type calcium channel versus the BK channel. That went by really fast, but it looked like 9 mM was a very low EC_{50} for the calcium channel?

DR. TREISTMAN: That was the IC_{50} for the L-type calcium channel.

DR. APELL: And that's a G-protein-linked effect?

DR. TREISTMAN: We don't know that for the terminal channel. We haven't really worked it out well enough to know if there's a G-protein component to that. There may well be.

DR. TSIEN: But aren't most L-type channels immune to G-protein modulation? I'm not saying that would be true in your system, but it would be quite surprising and exceptional if it turned out that there was a heavy G-protein modulation.

DR. TREISTMAN: Except in PC12 cells, where we actually did find modulation of L-type channels by G_i. But you're right. In general, people think of other channel types being G-protein sensitive. What we do know is that even for the $K_{(Ca)}$ channel, the onset of the potentiation is very fast, but the offset during wash takes many minutes, suggesting something else is going on. But we don't know what that is right now.

DR. APELL: But the EC_{50} for potentiation of BK is quite a bit higher than what you found for the L-channel inhibition.

DR. TREISTMAN: Yes. It was about 20 mM.

DR. APELL: Oh, so that's not too much higher. Thank you.

1.3. Q&As between Audience and Dr. Lovinger

1.3.1. Dopamine and 5-HT$_3$ Receptor

DR. TSIEN: I didn't quite understand the argument that you made using dopamine as a partial agonist. It would seem that if the TCEt was increasing the affinity for the ligand, and it was doing so with 5-HT, then it could do so for dopamine as well and shift its dose-response curve to the left.

DR. LOVINGER: It could. But it is at a concentration of dopamine that completely occupies the agonist-binding site. In other words, it's a maximum concentration that can completely compete with 5-HT, the concentration at the top of the dopamine dose-response curve. So even though that might shift the dose-response curve leftward, you wouldn't expect it to increase current at that maximum effective concentration.

DR. TSIEN: So, to use an old-fashioned pharmacological term, it doesn't appear like a mere change in affinity, but actually an increase in efficacy.

DR. LOVINGER: Yes.

DR. TSIEN: And whether it has to do with gating or not is another question, but it seems to be efficacy rather than affinity.

DR. LOVINGER: Yes. I certainly agree with the idea of increased efficacy.

1.4. Q&As between Audience and Dr. Narahashi

1.4.1. Ethanol and the α_7 ACh Receptor

AUDIENCE MEMBER: Have you looked at any ethanol-α_7 interactions?

DR. NARAHASHI: No. I haven't done that yet. We are working on it.

1.4.2. Ethanol Effect on ACh Receptors and Alcohol-Related Behavior

AUDIENCE MEMBER: Is there a behavioral correlation?

DR. NARAHASHI: Good question. We do not have enough data even to infer correlation.

1.4.3. Supersensitivity of Nicotinic ACh Receptors and Alcohol

DR. BILL LANDS: (Bill Lands from NIAAA). Dr. Narahashi, you emphasized the super sensitivity of acetylcholine receptors to ethanol. Nice work. I have two questions related to the information you gave us. The first question: What sort of physiologic neural synaptic system would have those supersensitive receptors? What sort of physiologic systems do we know of that have that kind of acetylcholine signaling? The second question: Do we know of any biological phenomenon that is influenced by sub-millimolar of ethanol? So, first, what about the physiologic neuronal role? And, secondly, what about the phenomenology?

DR. NARAHASHI: I think these two are very reasonable questions that I rather anticipated.

The first question: What kind of physiological role? As I indicated, the $\alpha_3\beta_4$ neuronal acetylcholine receptors are highly sensitive and the $\alpha_3\beta_2$ receptors are not as sensitive. So any neurons containing $\alpha_3\beta_4$ subunits should be highly sensitive to ethanol. On the other hand, any neurons containing $\alpha_3\beta_2$ subunits are not expected to be sensitive. The question is: Which neuron contains which subunit combination? Although some studies have been done, the data in the literature are not enough to draw conclusions. Although the brain contains the α_4 subunit, the α_3 subunit is also known to be present, and α_7 is one of the predominant subunits of acetylcholine receptors in the brain.

The second question was the physiological role or systemic effect. We're also asking that question ourselves. But for the moment, I don't have any clear-cut answer to that. Micromolar concentration of ethanol probably means that if you drink half a glass of beer, you may get the micromolar concentration, and that is probably the effect you can see. Beyond that, I have no idea.

1.4.4. Alcohol and Nicotine

DR. NARAHASHI: Here is Bill Marszalec, my collaborator.

DR. MARSZALEC: There seems to be a correlation between smoking and ingestion of alcohol. So a low concentration of alcohol may not necessarily lead to inebriation *per se*, but could perhaps facilitate the desire to smoke a cigarette, by some mechanism.

AUDIENCE MEMBER: What is the mechanism?

DR. MARSZALEC: It is probably something that involved in the dopaminergic system. There are nicotinic receptors on the dopaminergic system that seem to promote the release of dopamine. The nicotinic acetylcholine receptor may reinforce the desire to smoke. If alcohol could interact with these receptors, it may facilitate this reinforcement.

1.4.5. Special Receptors for Alcohol

AUDIENCE MEMBER: Are there any special receptors for alcohol?

DR. NARAHASHI: No. It's the nicotinic acetylcholine receptor, not alcohol receptor*.

AUDIENCE MEMBER: Since you said no, I have a related question: Is the action of alcohol in the brain and the nerves system universal or specific?

DR. NARAHASHI: I don't think it is specific. Here I only emphasized the subunit specificity of the acetylcholine receptor to alcohol, which means any nicotinic acetylcholine receptors containing $\alpha_3\beta_4$, possibly some other subunit combinations, will be sensitive to alcohol.

1.4.6. Direct or Indirect Effect of Alcohol on ACh Receptors

AUDIENCE MEMBER: Is there any possibility that this alcohol effect is secondary rather than direct action? In other words, is the effect direct or indirect?

DR. NARAHASHI: You mean, effect of ethanol is either direct or indirect? Or is there any secondary effect as a result of ethanol-nicotinic receptor interaction? Was that your question?

AUDIENCE MEMBER: Yes. Your recording showed that the nicotinic receptors have some change after ethanol treatment. Is this kind of treatment related to direct effects from ethanol or through some indirect pathways?

DR. NARAHASHI: We studied direct effects. However, the direct effect does not necessarily mean direct ethanol-receptor interaction. It could be that ethanol interacts with a receptor via intracellular component, such as a PKC. This issue is still under debate.

1.4.7. "Amplification" of Ethanol Sensitivity

DR. SIGGINS: A related question: Your slide showed the rather high EC_{50}s or IC_{50}s for ethanol. A lot of those seem derived from isolated systems. I'm wondering if those IC_{50}s or EC_{50}s could be driven down quite a bit by conditioning, by post-translational modification, such as phosphorylation, of those receptors that would occur in a more intact system. Would that make those receptors much more sensitive to ethanol? Is that possible?

DR. NARAHASHI: I have an answer to something related to your question, but not directly. This question is always asked: What is the significance of this very high concentration and very low potency of ethanol on receptors other than the nicotinic acetylcholine

---

* Editors' note: There is no evidence for a specific "alcohol receptor" (see Hunt & Liu in this book).

receptors? I don't have much time to explain it in detail. We developed a concept of the amplification of the effect, which was based on our previous study on the effect of pyrethroid insecticides on voltage-gated sodium channels in prolonging their open time. In order for the pyrethroids to cause hyper-excitation of the entire nervous system, we need to modify only 1% of sodium channels. When one-hundredth of sodium channels are modified, hyper-activity is produced, the potency of which is translated into approximately EC_1 instead of EC_{50}. The same concept is applicable to the opposite direction, such as anti-epileptic drugs. Although that concept for anti-epileptic drugs has yet to be experimentally demonstrated, the concept is certainly applicable. The same thing could happen to synaptic transmission. If one synapse is affected by low concentrations of ethanol, whether it is NMDA or GABA receptors, that effect could be amplified through a cascade of events of synaptic transmission, including excitatory and inhibitory synapses, so the end result could be enormous. You do not need EC_{50} concentrations to cause systemic effects.

DR. LOVINGER: I have one comment in relation to what was just said: The concept that there is amplification of responses beyond the current that goes through the channel itself. I think we have to be a little bit cautious, when we use indirect measures that aren't measuring current through the channels themselves, that we don't get fooled about changes in apparent potency of alcohol or changes in conditions that change the apparent potency, because in some of these systems, you are amplifying responses. If you're measuring release of transmitter downstream from activation of the receptor, you could really fool yourself into thinking that you have found a way to change ethanol sensitivity. I think it's most important that you measure the most direct effects on the receptor that you want to study.

2. DISCUSSION BETWEEN AUDIENCE AND SPEAKERS OF SESSION I

2.1. Direct or Indirect Effect of Ethanol on Membrane Proteins

DR. LIU: Dr. Treistman, from your experiments, do you have any comment on the earlier question about whether the effect of ethanol is direct or indirect?

DR. TREISTMAN: Yes. I think from our work, we have two different classes of responses. We feel that the calcium channel effect is primarily mediated, or at least a portion of it is mediated, by G-proteins. It is very interesting that we also saw, as Toshio Narahashi did, a difference in calcium channel sensitivity to ethanol between undifferentiated and differentiated PC12 cells that we published a while ago. The G-protein component of the response appears to be lost after differentiation. But at this point, we do not know the reason for that.

For the calcium-activated potassium channels, I think that the whole point to the increasing reductionism was coming to a situation where we feel that there is a direct interaction with the protein itself. Although it is becoming more and more apparent that even what we thought to be an isolated protein is in fact a protein complex, so it remains a little more complicated. But in that case, it certainly does not involve G-proteins. Because there are no nucleotides in the mix, and it is about 20 minutes after the excision when we are doing the experiments. So I think there are some cases where direct interaction is much more likely than in others.

DR. LIU: I just want to add one more comment on that. It is a common phenomenon in the alcohol research field—if you use oocytes or some other expressing systems, you need higher concentrations of ethanol to activate the channel. So it seems that there are probably several different mechanisms. Some of them might be direct; some of them might be indirect. There is a lot of evidence now indicating some signal transduction processes might be involved. I think maybe Dr. Yeh wants to give a comment or ask a question on that.

2.2. Direct Measurement of Synaptic Release

DR. YEH: Actually I was. Although I probably should leave this particular comment to my own presentation. But at this point, I'd like to direct the discussion back to the synapse itself. We do have some questions here I believe we need to address. Part of the value of having somebody like Dr. Tsien here is to provide us with some insights regarding cutting-edge work with synapses. Then we need to think about how to apply that to the work that Steve Treistman does, for example.

Dr. Treistman, I have a question with regards to your isolated synaptic terminal experiments, which I think are really neat experiments. But to carry it one step beyond, besides using FM1–43 and related compounds to be able to correlate the actions that you see with ethanol on the channels, where can you take that particular preparation to gain some insight into how the effect that you see might be correlated with neurotransmitter release? Is there a direct way of being able to measure release, peptide release, from your isolated terminal preparations?

DR. TREISTMAN: Yes. The last series of slides I showed was, indeed, populations of isolated terminals, so we can measure release from isolated terminals as a group. Individually? Yes, we are actually developing fluorescence imaging now to be able to go that route. Although it is not there yet, it will be there soon enough. And we are also developing capacitance techniques to look at release from individual terminals, if that's what you mean.

DR. YEH: Yes. I think that the challenge is to try to take the preparations that we have now, that we think are valuable, one step further—being able to measure the peptide release from the synaptic terminals and tightly correlate that with your mechanistic study.

DR. TREISTMAN: Right. That's the idea. Another direction that we have thought about going but have not been very successful in pursuing is to reconstitute release in an expression system, where we can couple a smaller number of elements and see what is happening. That is not so easy to do, but it certainly is a direction that might be worth pursuing.

2.3. Concentrations of Transmitters at the Synapse

DR. YEH: One last quick question for Dave Lovinger and Toshio Narahashi: With regard to the modulatory effect of ethanol on the prolonged decay phase, as regards to both the nicotinic acetylcholine receptors and the 5-HT_3 receptor. Dave, you mentioned that this might actually be physiological, if we were to know what the concentration of transmitter would be at the synaptic site. Is that a pretty fair paraphrasing?

DR. LOVINGER: Yes. I think there are a couple of things you can imagine. One is that the concentration of, say, GABA, which is easier to deal with, than 5-HT...

DR. YEH: I think so, too.

DR. LOVINGER: ...is going to be very high at the synapse, then I don't think you're going to see much potentiation of the peak current.

DR. YEH: What concentration are you expecting at the synapse?

DR. LOVINGER: I'm talking about maybe 1 mM. Now, that really is derived from the glutamatergic literature. I have not followed the GABAergic literature as closely. Maybe you can comment a little more on that.

But there are two things that are going on. With time, the concentration of transmitters in the synapse is going to fall, of course. If potentiation is greater at a lower agonist concentration, you might see more potentiation there. But the other thing you'll have is probably slowing of unbinding or closing of channels that could contribute to a prolonged synaptic response. So I think that is more likely to be the case for the effects of ethanol. Or let us put it this way. The total charge that is moving through the synaptic channel is probably going to be affected more by that prolongation than by an increase in the peak current. Now, if you look at the effects of steroids or pentobarbital, you do see amazingly long prolongation of these synaptic responses, so I think it is an important component of the effects at the synapse.

DR. NARAHASHI: Just to add a comment to David's point. Additional importance is the possible effect of repetitive activity. Generally, neurons do not produce just one action potential, and repeated activities are going on all over. So the effect of desensitization or a very brief transmitter release, which causes slowing of the decay phase could be even more significant.

DR. ALGER: Dr. Tsien, does your work on factors underlying quantal variability have implications for understanding of the concentration of neurotransmitters in the synaptic cleft? Using your model and the data you reported this morning, have you calculated the concentrations of glutamate in the cleft that would correspond to small and large quantal events, for example?

DR. TSIEN: We don't actually measure the concentration directly. I don't think anybody has. But there's some beautiful work done by Clements, Jahr, and Westbrook (Clements et al., 1992), which suggests that glutamate at excitatory synapses can reach millimolar concentrations for about a millisecond. This meant to be a kind of rough approximation. It is quite conceivable that saturation of the AMPA type of glutamate receptors simply does not occur, not because the concentration does not get high enough, but because the high concentration does not last long enough to allow transmitter receptors to equilibrate. So this illustrates what David Lovinger mentioned earlier, the importance of knowing the exact concentration profile and its timing. So there is really no discrepancy between the conclusions that we are reaching and those of Clements, Jahr and Westbrook.

DR. ALGER: Is it possible to create a model that would give something like the effective transmitter concentration?

DR. TSIEN: If you make a realistic model of the synaptic cleft—and Bill Holmes (Holmes 1995) is an example of a scientist who has done so—and put a reasonable number of molecules in the vesicle and really describe the geometry of the cleft, it's very easy to imagine that you will not see receptor saturation. While the concentration in the vesicle is high enough to allow cleft concentration to reach millimolar levels, diffusion out of the cleft dissipates the transmitter concentration so quickly that ligand receptor agonist reaction need not reach equilibrium, so you can get pretty much any result, ranging from saturation to non-saturation.

Getting back to alcohol research, I think all the work that we are doing, whether it be at the level of a ligand and a receptor channel or at the level of a vesicle releasing transmitter, reemphasizes the importance of understanding the kinetics of reactions. Several of the speakers focused on all aspects that involved gating and time dependence. It matters how long the voltage pulse lasts; it matters how long the ligand pulse lasts. And I think alcohol research is not immune to those considerations.

2.4. Concentration of Ethanol at the Synapse

DR. TSIEN: This morning we had a session where one person (RWT) who obviously knows almost nothing about alcohol talked about synaptic transmission; and three people who know a lot about alcohol, hardly spoke about synaptic transmission. Let's consider ligand-gated receptors, which is what this session was really about. It was obvious that you can see alcohol effects at reasonably low concentrations on a whole variety of ligand-gated channels. The effects mostly involve gating and not selectivity, and the effects are extremely subunit-dependent, and they raise all sorts of interesting effects on kinetics. Let me start off with a specific question for Dr. Narahashi: To me the most striking effect you showed was that 90 μM ethanol drastically modifies the deactivation of nicotinic acetylcholine receptors. Both you and your colleague emphasized the point that before you get legally drunk and can't walk the line for your Breathalyzer test, you can feel the effects of the beer. Can you calculate for me, at an effective concentration that strikingly affects deactivation, what the concentration of ethanol molecules will be in the membrane itself? Have people really measured the concentration of the alcohol in the membrane? And how often would there be collisions between ethanol molecules and a molecule of the size of the nicotinic acetylcholine receptor?

DR. NARAHASHI: You mean in the membrane or in the serum?

DR. TSIEN: I'd like to know in the membrane itself. Because, I'm looking to see whether there might be an effect within the membrane bilayer. So it was brought up that the composition of the bilayer might be important, PE versus PE/PS. Let's just start off at 90 μM, not 10 mM but 90 μM. Does the ethanol hit the channel often enough for it to be of interest?

DR. NARAHASHI: Well, then I would ask a question whether ethanol gets access to those receptors from outside, not through membrane, or gets access through the receptor via the lipid phase of the membrane. Actually both ways are possible. I don't know the ethanol concentration within the membrane—and you're probably talking about the lipid phase of the membrane.

DR. TSIEN: Right.

DR. NARAHASHI: I don't have any straightforward answer.

2.5. Access Site of Ethanol to Its Protein Targets: Lipid vs. Aqueous Phase (I)

DR. TSIEN: Supposing we can't make the calculation right away. Do people in the field discount the idea that ethanol might work through the lipid phase? Do they automatically assume that it is acting in free solution?

DR. NARAHASHI: Are you talking about the so-called "lipid vs. protein" theory?

DR. TSIEN: No. I believe that you folks are studying the gating of proteins, and that proteins are the ultimate readout. But there are two ways to access the protein, one in the aqueous phase and one partitioned into the bilayer. Maybe Steve can comment on that?

DR. TREISTMAN: Maybe I can. One of the problems is that the lipid matrix is not a homogeneous mix. It's certainly not inconceivable that different channel proteins segregate differently amongst different lipid compartments. It is an extraordinarily hard thing to say, if you no longer think of it as a homogeneous compartment and the alcohol may well partition differentially into one compartment versus another, which may contain one channel population versus another.

DR. NARAHASHI: One possible answer is: We demonstrated, like some other people did, that the effect of alcohol in potentiating the GABA receptors is linearly related to their carbon chain length, which is linearly related to lipid solubility. This tells the importance of lipid solubility in exerting a potent effect, and means that access via lipid membrane is certainly one of the ways, although it might not be the only way.

DR. LOVINGER: But I think all of the evidence in the literature is really consistent with the idea that the site is hydrophobic. But that does not necessarily mean the hydrophobic site of the action has to be the lipids. It could be hydrophobic portions of the protein.

DR. NARAHASHI: Yes. I'm not talking about the type of action. I'm talking about the access to the site.

DR. LOVINGER: Right. Access. And access, of course, is still going to be limited by the same problems when you are talking about a hydrophobic site in a protein.

DR. TSIEN: So to recapitulate: It is possible that the ethanol actually interacts with lipids and that the lipids influence the proteins. It is possible that the alcohol partitions into the lipid membrane and affects a hydrophobic region of the protein, including the transmembrane segment. It is even possible—and you have not mentioned this—that the increase in potency is simply due to the fact that alcohol gets anchored in the membrane, and therefore its local concentration near an extra-lipid, extra-hydrophobic area of the protein is higher.

All of those are possible, and it would be interesting to devise good experiments to test this. Steve Treistman is in a great position to do this, because he has got M-slow and B-slow channels and they have opposite responses to alcohol. So you could presumably make chimeras to find out what critical region it is for the site, and I will bet you that the critical region is one in a transmembrane-spanning domain.

DR. NARAHASHI: The same concept has been developed by Franks & Lieb (Franks & Lieb, 1996) in England, who studied general anesthetics. Their study calls for the binding of those anesthetic molecules to the pocket of the protein via lipid phase.

DR. LOVINGER: Yes. I think there are a couple of points. One is, you could even imagine the case where it's lipid-independent, and Franks & Lieb have examples of that. Now, they're somewhat farfetched. You can have the Meyer-Overton relation for actions of alcohols and anesthetics on luciferase in the absence of any lipid, and their idea is that there's a hydrophobic pocket in the protein.

But the other point is related to the chimeric idea. There is a paper that has just come out in *Nature* (Mihic et al., 1997) about the chimeric receptors of $GABA_\rho$ and glycine receptors. It showed that there is some site in the membrane-spanning domain that seems to be crucial for the actions of alcohol, whether they are potentiation or inhibition on glycine versus $GABA_\rho$. Now, it's hard to say that that is the binding site of the alcohol, but it certainly suggests it, and it's an important site.

DR. TSIEN: I think that's Neil Harrison's work.

DR. LOVINGER: Yes.

2.6. Multiple Action Sites of Ethanol

DR. PEOPLES: (Bob Peoples from NIAAA). I just wanted to make a comment: I think we could expect there to be very different sites of action on different receptors like the GABA receptor. It's possibly with the site that's been shown in the recent *Nature* paper. There's some work in our lab using chimeric receptors between the $5\text{-}HT_3$ and neuronal nicotinic receptor. The site on the neuronal nicotinic receptor is on the n-terminal extracellular region. So it's clearly not a membrane effect. But I don't think there's any reason to expect that it would be exactly the same site among different proteins.

DR. TREISTMAN: There may even be multiple sites on a given protein. I think one of the differences between B-slow and M-slow is that they both share one site, an inhibitory site, but M-slow actually has the excitatory site as well. So my guess is that there are probably multiple sites.

DR. LOVINGER: Yes. I actually believe that for both the nicotinic and the $5\text{-}HT_3$ receptors, there are sites involved in potentiation that are separate from sites only for longer chain alcohols and channel-blocking effects. That has been worked out much better for the nicotinic receptor where you have an actual identified site, probably in the pore where you have channel blocking.

DR. TSIEN: Let me follow up by asking Dr. Peoples which might simply have a "Yes" or "No" answer. Since you have a molecularly well-defined site of action that's not a transmembrane-spanning domain, did you examine the chain-length dependence to demonstrate that the same rules apply for alcohol actions, even when the target site is not in the plane of the bilayer?

DR. PEOPLES: No. We only did that on the native NMDA receptors but not on the chimeric receptors yet.

DR. TSIEN: But it would seem possible that there could be actions on targets, which are not actually within the bilayer, which, nevertheless, depend upon the local concentration of alcohol there, and that could depend upon the structure of the molecule.

2.7. Access Site of Ethanol to Its Protein Targets: Lipid vs. Aqueous Phase (II)

DR. SIGGINS: A quick question for Steve Treistman. The L-channel effect, is it membrane-delimited? I mean, you have the ability to pull off the patch?

DR. TREISTMAN: No. We don't have too much ability to do it, because they last such a short time. That's why we actually spent so much time with $K_{(Ca)}$ channels. It's not so easy to study the calcium channels in that way.

DR. TSIEN: I think George's idea is that you use the cell-attached configuration with a L-type channel in the patch, then apply the ethanol to the rest of the cell. That's a classical way of studying modulation of L-type channel.

DR. TREISTMAN: Yes. In that situation, we get an effect, but I don't think that says too much, because ethanol is going to move through that membrane very, very quickly.

DR. LIU: I think Dr. Tsien asked a very good question about how ethanol accesses its target. I would like to make a comment on that.

Whether ethanol is working directly on lipid or on protein has been debated since the last century. The original hypotheses was that ethanol disturbs the lipid membrane. Then, later, people developed the hypothesis that ethanol is working on a hydrophobic pocket on the protein. Accumulating evidence is now showing that the target probably is the protein. As physiologists or biophysicists, we often think about a ligand and its target and the interaction between them. In that model, the lipid layer of the membrane is not permeable to the ligand, and the interaction only happens between the ligand and the protein. If we apply this picture to ethanol and its target, it's not correct, because ethanol can penetrate the membrane, can anchor in the membrane, and also can go inside the membrane. So it's not a clear-cut picture, like a particular ligand with a particular ethanol receptor. I would like to encourage the audience, especially the people who are not very familiar with alcohol research to keep this point in mind. This is one of the biggest challenges in alcohol-related neuroscience research compared to neuroscience research on other ligands or abused substances.

2.8. Future Direction of Research on Synaptic Transmission

DR. LIU: My last question is for Dr. Tsien. As we all saw this morning, Dr. Tsien has made a significant contribution to our meeting by giving a wonderful presentation and asking so many challenging questions. Dr. Tsien, I would like to ask you to comment as a pioneer in the synaptic research field: What aspects of synaptic research should we apply into the alcohol field in the future?

DR. TSIEN: Thanks for the compliment. Clearly, it would make sense for a few investigators to reexamine the cycling of vesicles and to see if ethanol or TCEt had an effect. This is just one example of a general class of experiments. However, it seems to me that the most interesting problems in the field go beyond issues of molecular mechanisms as such, but involve their relationship to behavioral consequences.

To highlight just one last thing—I found it fascinating to learn about Steve Treistman's experiment with tolerance, where feeding the animals alcohol leads to a change in the IC_{50} for the alcohol effect. This seems like a wonderful starting point for research to pinpoint specific parts of the brain involved in the alcohol response. If you're seeing such dramatic effects on the ability of the ethanol to act, there are almost certainly novel and intriguing basic mechanisms to be uncovered in the future.

REFERENCES

Clements JD, Letser RA, Tong G, Jahr CE, Westbrook GL (1992) The time course of glutamate in the synaptic cleft. Science 258:1498–1501

Cox DH, Cui J, Aldrich RW (1997) Allosteric gating of a large conductance Ca-activated K^+ channel. *J Gen Physiol* 110(3):257–281

Cox DH, Cui J, Aldrich RW (1997) Separation of gating properties from permeation and block in mslo large conductance Ca-activated K^+ channels. *J Gen Physiol* 109(5):633–646

Cui J, Cox DH, Aldrich RW (1997) Intrinsic voltage dependence and $Ca2^+$ regulation of mslo large conductance Ca-activated K^+ channels. *J Gen Physiol* 109(5):647–673

Franks NP, Lieb WR (1996) An anesthetic-sensitive superfamily of neurotransmitter-gated ion channels. J Clin Anesth 8(3 Suppl):3S-7S

Holmes WR (1995) Modeling the effect of glutamate diffusion and uptake on NMDA and non-NMDA receptor saturation. Biophys J 69(5):1734–47

Mihic SJ, Ye Q, Wick MJ, Koltchine VV, Krasowski MD, Finn SE, Mascia MP, Valenzuela CF, Hanson KK, Greenblatt EP, Harris RA, Harrison NL (1997) Sites of alcohol and volatile anaesthetic action on GABA(A) and glycine receptors. Nature 389:385–9

Section II

SYNAPTIC MODULATION

7

DEPOLARIZATION-INDUCED SUPPRESSION OF INHIBITION (DSI) INVOLVES A RETROGRADE SIGNALING PROCESS THAT REGULATES $GABA_A$-MEDIATED SYNAPTIC RESPONSES IN MAMMALIAN CNS

Bradley E. Alger

Department of Physiology
School of Medicine and Program in Neuroscience
University of Maryland, Baltimore
655 West Baltimore Street
Baltimore, Maryland 21201

1. INTRODUCTION

Regulation of gamma-aminobutyric acid A ($GABA_A$)-receptor-mediated inhibition is important for the control of neuronal excitability in the brain. Decreases in inhibition facilitate the induction of long-term potentiation (LTP) (Wigstrom and Gustafsson, 1983) and long-term depression (LTD) (Wagner and Alger, 1995; Bear and Abraham, 1996). However, if pronounced in magnitude or duration, decreases in $GABA_A$ergic inhibition can cause various pathophysiological conditions such as epileptic seizures (Meldrum, 1975) and excitotoxicity (Thompson et al., 1996). A physiologically useful type of inhibitory modulation might be one in which regulation of specific $GABA_A$ergic influences would be limited in time and space.

This chapter will describe a regulatory process recently discovered in hippocampal pyramidal cells and cerebellar Purkinje cells that has features that would enable it to control GABA inhibition in precise spatially and temporally delimited ways. The process is called "depolarization-induced suppression of inhibition" and is abbreviated DSI. DSI is expressed as a suppression of $GABA_A$-receptor-mediated IPSCs that follows a brief depolarization of a pyramidal cell (see Figure 1). This review will present data concerning DSI induction and expression, focusing on evidence that it is mediated by a retrograde signaling mechanism and recent work on the nature of the retrograde signal. Because several aspects of the DSI process are known to be altered by alcohol, it may be important to consider DSI when seeking to understand the effects of alcohol on the brain.

The "Drunken" Synapse, edited by Liu and Hunt.
Kluwer Academic / Plenum Publishers, New York, 1999.

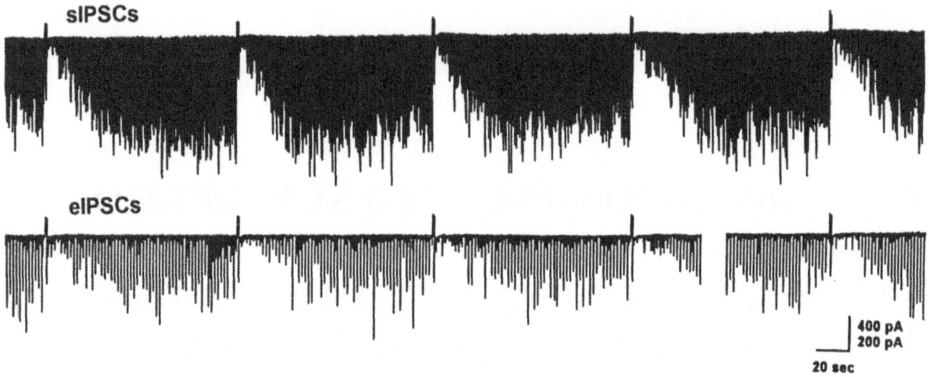

Figure 1. Depolarization-induced suppression of inhibition (DSI) of spontaneous IPSCs (sIPSCs) and monosynaptic evoked IPSCs (eIPSCs). The inward currents, downward deflections, are $GABA_A$-mediated IPSCs recorded using high-[Cl^-]-containing, whole-cell patch electrodes in two representative hippocampal CA1 pyramidal cells in the presence of the ionotropic glutamate receptor antagonists APV and CNQX. The cholinergic receptor agonist, carbachol (3 μM) was also present in the bathing solution for the cell illustrated in the upper trace to induce the high degree of sIPSC activity. One-second depolarizing voltage steps from the holding potential of -70 mV to 0 mV were delivered as indicated by the upward deflections occurring regularly at two-minute intervals. The decreases in both sIPSC and eIPSC amplitudes following the voltage steps represent DSI periods. (From L. A. Martin, Ph.D. thesis, unpublished.)

There are many kinds of intercellular communication (Jessell and Kandel, 1993). Among neurons the predominant form occurs at a traditional synapse in which a presynaptic element releases a chemical neurotransmitter at a morphologically distinct active zone. This mode of communication from pre- to postsynaptic cell is referred to as "orthograde" (sometimes "anterograde") synaptic transmission. There are also several kinds of nonsynaptic interactions in which cells can influence one another in less specific ways, for example by altering the concentration of ions in the extracellular space, or through extracellular electrical fields. In these instances the identification of pre- and postsynaptic elements is less clear.

In another mode of communication a cell, defined as postsynaptic by virtue of its receiving a chemical synaptic input, sends a signal in the reverse direction to a presynaptic cell, i.e., one that provides synaptic input to the postsynaptic cell. This is called "retrograde" transmission. The signaling mechanisms involved in retrograde transmission are much less well understood than those in orthograde transmission. Indeed the existence of a retrograde signaling process is generally difficult to detect, as many other possibilities must be eliminated before the identification can be made. In the mammalian CNS there are only a few instances in which retrograde signaling has clearly been shown to be involved in rapid communication between cells, although its involvement in slower trophic processes is well established (Jessell and Kandel, 1993).

GABA is considered the most prevalent inhibitory neurotransmitter in the mammalian brain (Macdonald and Olsen, 1994). It acts mainly on two receptor subtypes: $GABA_A$ and $GABA_B$. The $GABA_A$ receptor subtype is chiefly a mediator of postsynaptic inhibition. Its activation opens a Cl^--permeable pore, which is an integral part of the macromolecular receptor complex, and the resultant decrease in membrane resistance and influx of negatively charged ions tends to decrease cell excitability. (Evidence for a bicarbonate permeability of the $GABA_A$ ionic channel and the consequent depolarizing action of GABA will not be dealt with here. The reader is referred to Taira et al. (1997) and Staley et al. (1995) for information on this topic.) Activation of the $GABA_A$ receptor-channel is inhibited by bicuculline and picrotoxin. The $GABA_B$ receptor subtype affects ion channels indirectly through the activation of a pertussis-toxin-sensitive G protein (Bowery, 1993). Thus far DSI has only been shown to affect $GABA_A$- receptor-mediated neurotransmission. In the experiments described below the $GABA_B$-receptor-mediated events have been blocked pharmacologically and will not be further considered.

2. EXPERIMENTAL CONDITIONS FOR INVESTIGATING HIPPOCAMPAL DSI

The experiments on DSI in the hippocampus have all been performed on pyramidal cells in the CA1 region of the adult rat hippocampal slice preparation. The slices were prepared according to conventional techniques, using a vibratome to section the tissue. Afterwards, individual slices were studied while they were submerged in a physiological saline maintained at 30°C in a constant-perfusion chamber (Nicoll and Alger, 1981). We mainly used the blind, whole-cell recording technique (Blanton et al., 1989), although in some experiments intracellular recordings with high-resistance microelectrodes were used. The pipette solution usually had a high Cl^- concentration (~150 mM) so that E_{Cl^-} was near 0 mV, and activation of the $GABA_A$ conductance near resting membrane potentials produced an efflux of Cl^- and an inward current under voltage clamp (or a depolarization under current clamp). GABA-releasing interneurons were either allowed to fire spontaneously, sometimes under the influence of muscarinic receptor agonists (see Figure 1, top), or were stimulated by placing an extracellular stimulating electrode in their vicinity and evoking single "monosynaptic" IPSCs (Davies et al., 1990) (see Figure 1, bottom).

The IPSCs were considered to be monosynaptic because they were elicited in the presence of the ionotropic glutamate receptor antagonists CNQX (20 μM) and APV (50 μM), and thus polysynaptic pathways involving fast glutamate synapses were blocked. Ordinarily the extracellular stimulation was sufficiently large to evoke IPSCs several hundred picoamps in amplitude. On the basis of measurements of quantal GABA responses (e.g., Edwards et al., 1990) and anatomical investigations of the numbers of synaptic terminations of the interneurons onto pyramidal cells (Miles et al., 1996; Buhl et al., 1994), it can be concluded that these large responses typically result from the activation of numerous individual interneurons and many synapses. In cases where electrode position was optimized and weak stimulating currents were given, all-or-none "minimal" IPSCs, approximately 20 pA in amplitude, each thought to represent the stimulation of a single GABA synapse, could be activated. A minimal evoked IPSC appears to represent the response of one quantum of GABA. As noted, to exclude a role for ionotropic glutamate receptors, and many neuronal circuit effects, we did all of our experiments in the presence of the ionotropic glutamate receptor blockers. Because kainic acid acts on receptors other than the AMPA class of glutamate receptors and can block IPSCs presynaptically (Clarke et al., 1997; Rodriguez-Moreno et al., 1997; Fisher and Alger, 1984), it was important to test for their involvement in DSI. In some experiments we increased CNQX to 100 μM, or used NBQX at 50 μM, to block the kainate receptors, which otherwise would not be entirely suppressed by 20 μM CNQX. The more potent antagonist treatments failed to affect DSI, indicating that kainate receptors also do not play a role in DSI. The interval of time following a DSI-inducing voltage step during which IPSCs are reduced in control conditions is referred to as the "DSI period." To determine if experimental treatments affect DSI, we compared IPSCs occurring during the DSI period in control to those occurring

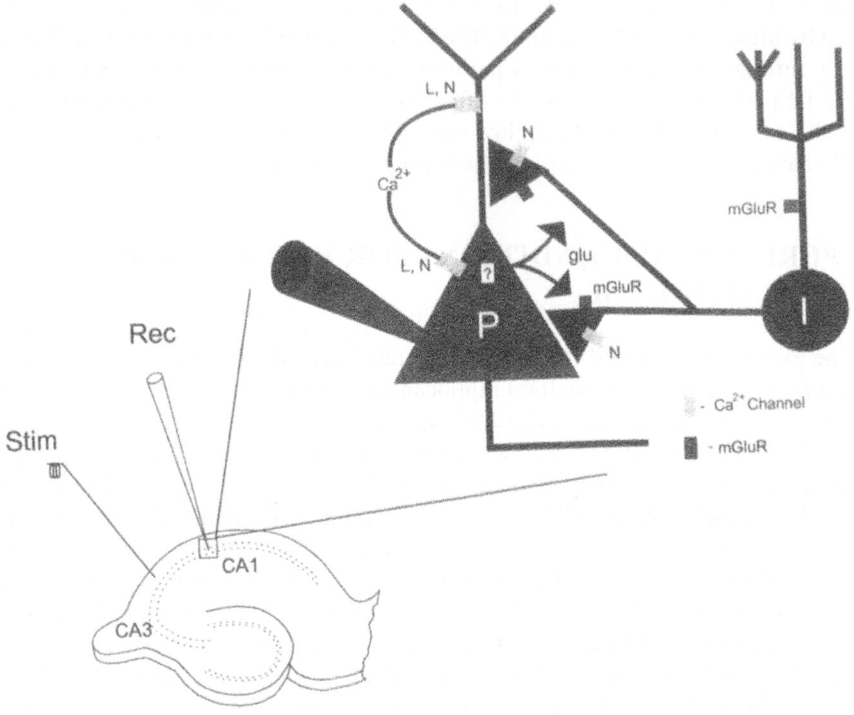

Figure 2. Experimental setup for studying, and hypothetical mechanism of, hippocampal DSI. Left diagram shows a hippocampal slice with typical placement of stimulating and recording electrodes. The physiological saline was comprised of (in mM): NaCl 120, KCl 3.5, NaH_2PO_4 1.25, $NaHCO_3$ 25, $CaCl_2$ 2, $MgSO_4$ 2 and glucose 10; pH = 7.4, temperature 29–31°C. In most experiments except as noted, CNQX 20 μM, and APV 50 μM, were in the extracellular saline to block ionotropic glutamate responses. Patch electrodes (2–4 MΩ) were filled with (in mM): $CsCH_3CO_4$ 100, CsCl 55, HEPES 10, BAPTA 2, $CaCl_2$ 0.2, $MgCl_2$ 1, MgATP 1, Tris-GTP 0.3 and QX-314 5 (pH adjusted to 7.25 with KOH). In some experiments high resistance electrodes (50–90 MΩ) containing 3 M KCl were used. Right portion illustrates a pyramidal cell (P) and a GABAergic interneuron (I). Recordings of inhibitory postsynaptic currents (IPSCs) were made from pyramidal cells only. Interneurons were either allowed to fire spontaneously or were stimulated with extracellular electrodes. Features of the cells on the right illustrate the hypothetical mechanism for DSI induction and expression for which evidence is presented in this chapter. Depolarization of the pyramidal cell opens N-, or in some cases, L-type, voltage dependent Ca^{++} channels in the somatic/dendritic regions of this cell. Increase in intracellular $[Ca^{++}]$ leads to the release of glutamate or a glutamate-like analog. The glutamate acts on metabotropic glutamate receptors (mGluR), a group I mGluR in CA1, on the interneuron and, through the action of a G-protein linked step (not shown) causes the release of GABA from the interneuron to be reduced for a period of many seconds.

during the same period of time in the experimental condition, and report the result as "percent DSI." Figure 2 shows schematically the recording conditions in the hippocampal slice and illustrates the main conclusions concerning the DSI mechanism summarized in this review.

3. INDUCTION AND BASIC PROPERTIES OF DSI

DSI can be induced in several ways. The DSI-inducing depolarization may be a direct-current pulse, delivered through the intracellular electrode, that initiates a series of action potentials (Pitler and Alger, 1992a), or a train of brief voltage steps (Llano et al., 1991), or it may be an epileptiform burst potential that occurs in a pyramidal cell (Le Beau and Alger, 1998) (see Figure 13).

For experimental convenience, however, in the hippocampus a 1-s (range 0.5–4 s) voltage-clamp step from the holding potential, usually -70 mV, to near 0 mV is typically used. With an elevated Cl^- concentration in the recording electrode prominent, $GABA_A$-receptor-mediated IPSCs that occur spontaneously (sIPSCs) can be readily recorded in pyramidal cells under control conditions (see Figure 1, top trace). Large, spontaneous IPSCs can be induced to occur by bath application of muscarinic receptor agonists, which activate certain interneurons in the slice, causing them to fire frequently and release GABA (Martin et al., 1995; Martin and Alger, 1996). DSI seems to have a particularly strong effect on the sIPSCs whose occurrence is induced by muscarinic receptor activation (Martin et al., 1995; Pitler and Alger, 1992a, 1994) (see Figure 14). Nevertheless, for experimental convenience and control, we typically study monosynaptic IPSCs evoked by electrical stimulation (eIPSCs) in either the stratum radiatum or stratum oriens regions of CA1 (e.g., Alger et al., 1996). In studies of evoked monosynaptic IPSCs muscarinic agonists are **not** used, and, indeed, DSI of eIPSCs is independent of activation of muscarinic receptors, as atropine has no effect on it.

Regardless of the type of IPSCs studied, for a period lasting from 5 to 60 s after the depolarizing voltage step, they are suppressed in the great majority of cells tested. The percent of pyramidal cells in which DSI is observed ranges from 60 to 90, depending on various factors, not all of which are known, but which may depend in part on the experience of the experimenter, because cells that are damaged, e.g., have low resting potentials or high holding currents, often cannot be induced to undergo DSI. At its peak, the suppression usually amounts to a reduction of ~40% in IPSC amplitudes. DSI is often not maximal immediately at the end of the depolarization, but can take from 0.5 to 3 s to develop to its greatest extent (Pitler and Alger, 1994). Peak suppression lasts from 2 to 10 s, and the IPSCs recover gradually to control values over the DSI period, which can last up to one minute under our usual experimental conditions. DSI is a very robust phenomenon and can be induced at intervals of 90–120 s continuously for an hour or more in a stable cell, with little change in its magnitude or duration.

Although DSI can be initiated by action-potential firing in a pyramidal cell, action potentials are not needed for DSI induction, as it is easily evoked in cells recorded under whole-cell voltage clamp with patch electrodes containing the lidocaine derivative QX-314. QX-314 blocks voltage-dependent Na^+ channels (Hille, 1992), and from this we infer that action potential conduction down pyramidal cell axons is not involved in DSI. It is also unlikely that collateral axonal synapses electrotonically close to the soma are involved, because setting up antidromic action potentials in pyramidal cell axons does not cause DSI if the cell soma is voltage clamped and prevented from firing (Pitler and Alger, 1994). Inasmuch as the antidromic spike will invade pyramidal cell axon terminals, as well as those of axon collaterals, the failure to observe DSI under these conditions argues strongly against a role for release of transmitters or modulators from pyramidal cell synaptic terminals in DSI. In Purkinje cells, DSI is unaffected when axons are removed by lesioning (Vincent and Marty, 1993), thus demonstrating convincingly the independence of DSI from activation of axonal conduction. To simplify interpretation of the experiments in hippocampus, we typically have QX-314 in our recording electrodes. The ease of producing DSI under these conditions leads us to conclude that DSI induction occurs in the somatic/dendritic regions of pyramidal cells. Changes in extracellular ionic concentrations, which might also be thought to play a role in DSI, have largely been excluded as candidate mechanisms, as reviewed earlier (Alger and Pitler, 1995). Newer data, discussed below, provide further evidence against a causal role for extracellular ionic concentration changes in DSI.

4. ROLE OF INCREASES IN POSTSYNAPTIC CALCIUM IN DSI INDUCTION

Rather than Na^+-dependent action potentials, what does seem to be critical for DSI is an increase in intracellular $[Ca^{++}]$. Prevention of such increases by the inclusion of high concentrations of the Ca^{++} chelator BAPTA in the recording electrode prevents DSI (Vincent and Marty, 1993; Alger and Pitler, 1995). Because the BAPTA rapidly diffused from the electrode into the cells, these experiments precluded observation of DSI prior to its being blocked. In order to test the Ca^{++} dependence of DSI directly in a given cell, we have recently taken several additional experimental approaches.

Raising extracellular $[Ca^{++}]_o$ to 5 mM while decreasing $[Mg^{++}]_o$ markedly, and reversibly, increases evoked monosynaptic IPSCs and DSI (Lenz et al., 1998) (see Figure 3). Although both IPSCs and DSI were increased in high $[Ca^{++}]_o$, the increase in DSI was not dependent on the increased IPSC and remained enhanced even when the IPSCs were restored to their original amplitudes by decreasing stimulus intensity. Ohno-Shosaku et al. (1998) report that percent DSI in tissue culture increases with increasing duration of the voltage step, up to 5 s, which suggests the degree of DSI may be proportional to $[Ca^{++}]_i$ up to a point. Using rapid application of Ca^{++}-free solution, they also show that Ca^{++} must be present during the voltage step to produce DSI.

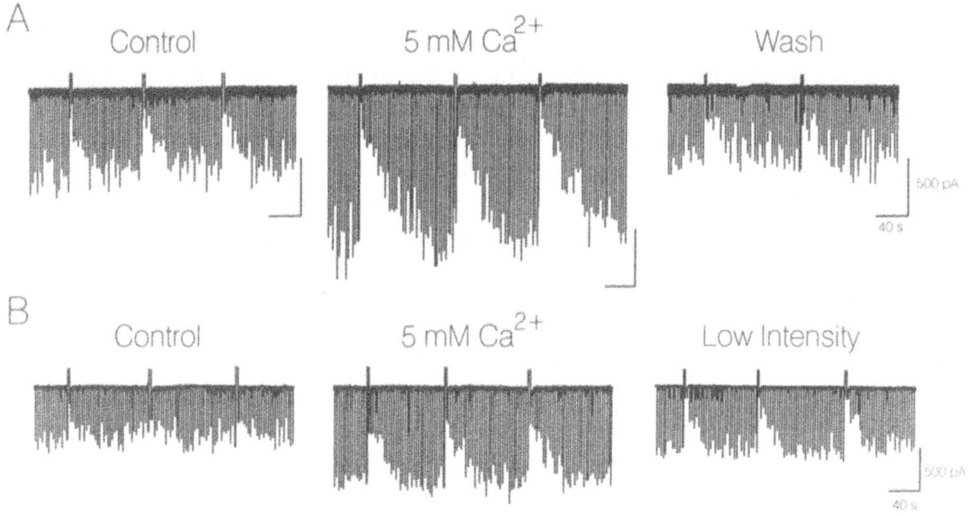

Figure 3. DSI is sensitive to changes in $[Ca^{++}]_o$. A) A 1-sec depolarizing step from the holding potential of -70 mV to -10 mV (upward deflection) resulted in DSI (41%) with 2.5 mM Ca^{++} and 2.0 mM Mg^{++} in the extracellular solution (left-hand trace). Perfusion with a saline containing high Ca^{++} (5 mM) and nominally 0 mM Mg^{++} greatly increased DSI (to 70%) and IPSC amplitudes in the same cell (middle trace). The effect of high Ca^{++} on DSI and IPSC amplitudes was reversible upon wash (right-hand trace; 43% DSI). B) Another cell in which DSI in control was small, but was markedly enhanced by raising extracellular $[Ca^{++}]$. DSI remained enhanced after lowering the stimulus intensity in high $[Ca^{++}]$ to elicit IPSCs of similar amplitude to those in control conditions (right-hand trace). (From Lenz et al., 1998, with permission.)

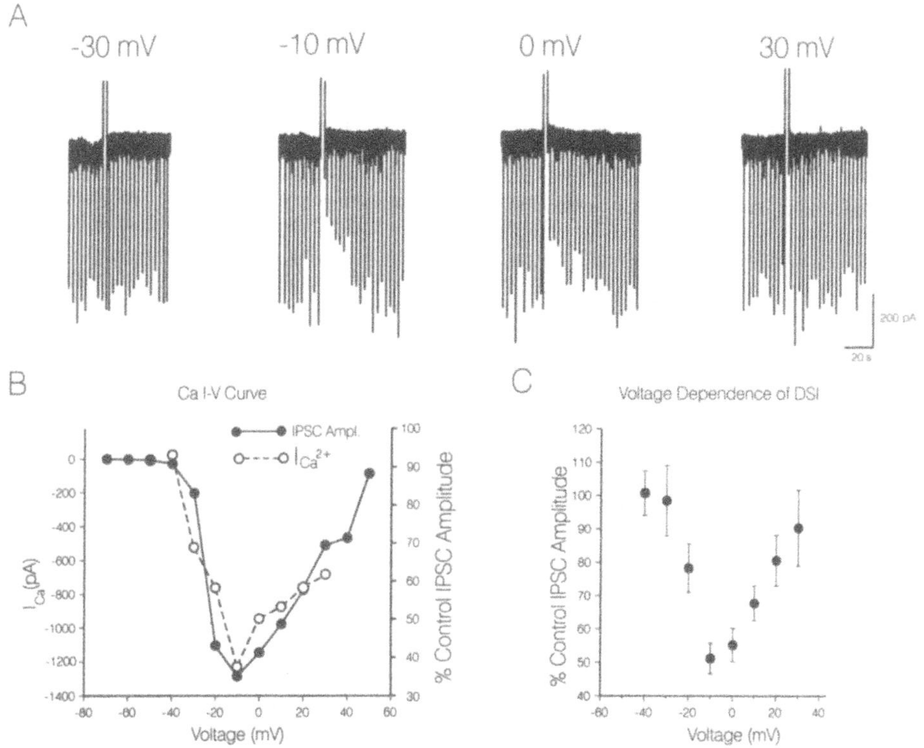

Figure 4. The voltage dependence of DSI parallels the voltage dependence of calcium channel activation. A) CA1 pyramidal cells were voltage clamped at -70 mV and stepped for 1 sec to the voltages indicated. Voltage steps to -30 mV did not result in DSI, whereas steps to -10 and 0 mV resulted in marked DSI. Interestingly, large voltage steps (+30 mV) approaching the Ca^{++} equilibrium potential did not cause any DSI. Traces across the top are all from the same cell. B) Plots of peak Ca^{++} current (I_{Ca}) (open circles) versus voltage and of percent of control IPSC amplitude versus the induction voltage step (filled circles) from the same cell are superimposed. The voltage dependences of I_{Ca} and of DSI are very similar, suggesting that the amount of DSI is related to the postsynaptic Ca^{++} influx during the voltage step. Because there is little or no DSI following very large voltage steps (when Ca^{++} influx is reduced), DSI cannot be due to the voltage step per se. C) The voltage dependence of DSI averaged from 6 cells is shown (mean ± S.E.M.).

Because of the evident Ca^{++} dependence of DSI and its induction by voltage-step protocols, we inferred that Ca^{++} might enter the cell through voltage-dependent Ca^{++} channels (VDCCs). This was tested by determining the percent DSI induced by voltage steps of different amplitudes (see Figure 4). Rather than using the conventional step from -70 mV to 0 mV, we stepped to potentials over the range of -30 to +30 mV. We also measured the peak amplitudes of the Ca^{++} current itself over the same range of activating voltages. There was a clear voltage dependence of DSI induction that paralleled the voltage dependence of high-voltage-activated Ca^{++} current. The percent DSI gradually increased with increasing steps until the step magnitude reached ~-10 mV; with greater steps percent DSI declined again.

Using selective antagonists of various voltage-dependent Ca^{++} channels (Dunlap et al., 1994), we addressed the question of which VDCCs could mediate DSI. Low-voltage-activated VDCCs (R- and T-type) are typically blocked by Ni^{2+} at concentrations of

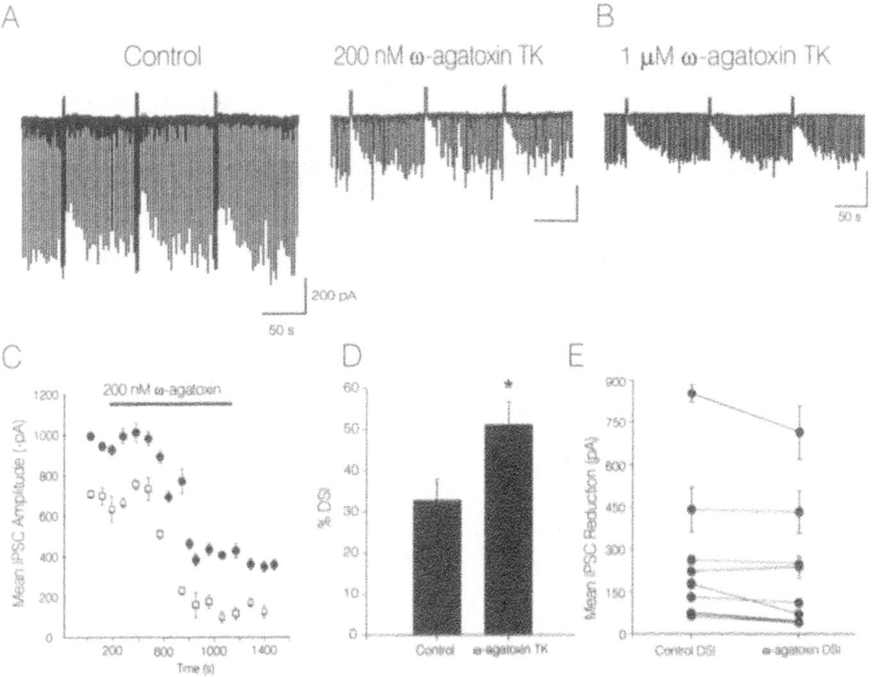

Figure 5. Activation of neither P- nor Q-type Ca^{++} channels is necessary for DSI. A) The selective P-type Ca^{++} channel blocker, ω-agatoxin TK (200 nM), greatly reduced the IPSC amplitudes, but did not block DSI. B) Preincubation of slices in 1 μM ω-agatoxin TK to block completely both P- and Q-type Ca^{++} channels did not block DSI. The cells from the preincubated slices were recorded with $CsCH_3SO_3$-filled electrodes, and thus the IPSCs were outward going in this experiment. For consistency of display, this trace was inverted, so the currents appear downward. C) Time course of a representative experiment where the solid circles represent the mean amplitude of the 8 IPSCs immediately preceding the DSI step and the open squares are mean amplitudes of the 5 IPSCs immediately following the DSI step. D) Group data (n = 8) showing that percent DSI in control is significantly less than in the presence of ω-agatoxin TK. E) The absolute reduction in IPSC amplitude during DSI is similar in control and ω-agatoxin TK, $p > 0.05$. (From Lenz et al., 1998, with permission.)

50–100 μM, and the high-voltage-activated channels, P- and Q-type, are blocked by ω-agatoxin at concentrations of 200 nM and 1 μM, respectively. Neither Ni nor ω-agatoxin had any effect on DSI (see Figure 5), although they did reduce inhibitory synaptic transmission by 30–50%.

Under our standard DSI-inducing voltage protocol, i.e., a 1-s voltage step to 0 mV with a QX-314-containing, Cs^+-based electrode solution, DSI of eIPSCs appears to be mediated entirely by ω-conotoxin-sensitive, N-type, VDCCs. At 250 nM this toxin almost completely blocked DSI (see Figure 6). Conotoxin also has profound effects on inhibitory transmission (Poncer et al., 1997; Potier et al., 1993). IPSCs are reduced by about 85%. The L-type channel blocker, nifedipine, 10 μM, had no effect on DSI induced under the conventional protocol (data not shown).

It might seem that, by blocking eIPSCs so markedly, ω-conotoxin could simply prevent the observation of DSI; i.e., if the IPSCs that are susceptible to DSI do not occur, then DSI cannot be seen. Nevertheless, much of the DSI-blocking effect of ω-conotoxin appears to be the result of decreasing Ca^{++} influx into the pyramidal cell through postsynaptic N-type VDCCs. We have two reasons for drawing this conclusion: 1) If the ω-conotoxin effect were exclusively presynaptic, then IPSCs during the DSI period could never be larger in the presence of ω-conotoxin than they were during DSI in control conditions; i.e., if ω-conotoxin only acted presynaptically, the postsynaptic induction of DSI would be unaltered, and the DSI process in ω-conotoxin should be able to reduce the IPSCs to the same extent that it did in the absence of ω-conotoxin. Nevertheless, in several instances we have observed the IPSCs during the DSI period in ω-conotoxin to be larger than they were during the DSI period in control conditions in the same cells (see, e.g., Figure 6A). For example, in 3 of 7 cells the mean IPSC amplitude during the DSI period in ω-conotoxin was larger (256 pA) than it was in the same cells during the DSI period in control (186 pA). This IPSC increase during the DSI period is best explained as a reduction in postsynaptic N-channel-mediated Ca^{++} influx. 2) An exclusively presynaptic site of action of ω-conotoxin is also incompatible with observations that it can reduce DSI significantly

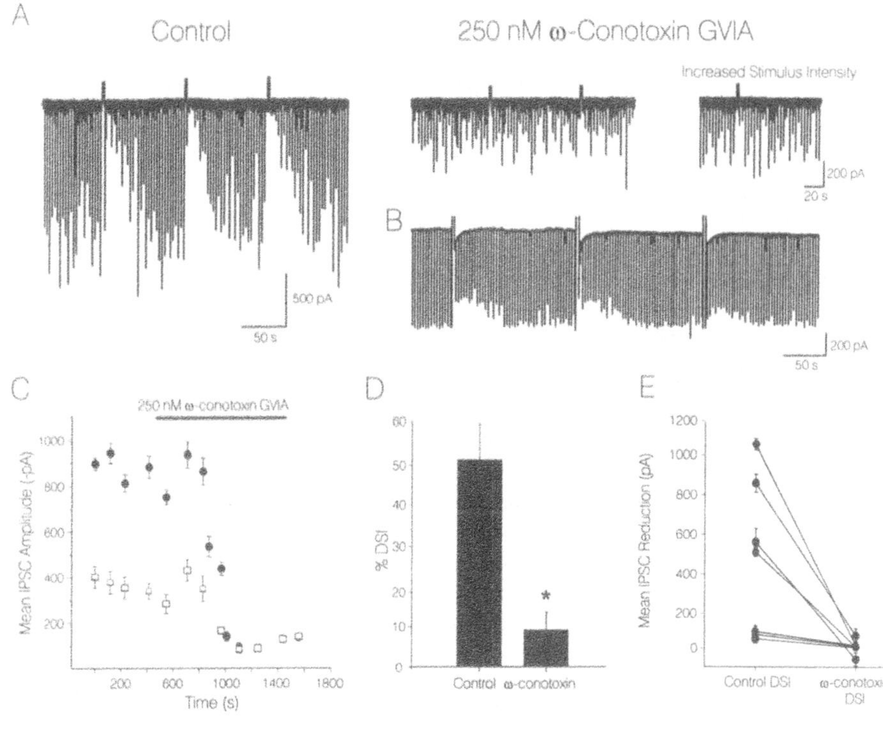

Figure 6. N-type Ca^{++} channel activation is necessary for DSI. A) At concentrations that selectively block N-type Ca^{++} channels, ω-conotoxin GVIA (250 nM) greatly reduced the IPSC amplitude and virtually abolished DSI. Increasing the stimulus amplitude to recover the IPSC amplitude partially did not recover DSI. Note that, although the scales were changed, IPSCs, 100–200 pA in amplitude, followed the depolarizing step in ω-conotoxin (middle trace in A), but no IPSCs greater than baseline noise are visible immediately after the step in control (left trace in A). B) ω-Conotoxin can cause a progressive block of DSI prior to reducing IPSC amplitudes. ω-Conotoxin was applied at a slow rate that allowed a progressive reduction of DSI from 38% after the first depolarizing step to 26% after the third. IPSC amplitudes were not altered over this period. Continued application of ω-conotoxin to this cell resulted in a complete block of DSI and a greater reduction in IPSC amplitude (not shown). C) The time course of a typical experiment with ω-conotoxin demonstrates that the toxin rapidly reduced IPSC amplitude as well as DSI. D) Control DSI was significantly larger than the DSI in ω-conotoxin ($p = 0.005$, $n = 7$). E) The absolute reduction in IPSC amplitude was profoundly greater during DSI in control than in ω-conotoxin. (From Lenz et al., 1998, with permission.)

before it reduces eIPSC amplitudes (see, e.g., Figure 6B). If the effect of ω-conotoxin were solely presynaptic, then DSI reduction should be tightly linked to IPSC reduction.

Thus, N-type VDCCs clearly play some role in DSI other than blocking transmitter release presynaptically, and the most likely interpretation is that ω-conotoxin also blocks postsynaptic N-channel-dependent Ca^{++} influx necessary to induce DSI. In view of evidence that DSI is dependent on VDCC-mediated Ca^{++} influx, and the lack of effect of L-, R-, P-, Q- and T-type channel antagonists (under our standard DSI-induction protocol), the efficacy of ω-conotoxin in blocking DSI seems most easily accounted for by a block of postsynaptic N-type channels.

Nevertheless, it cannot be concluded that only N-channel-mediated Ca^{++} influx can mediate DSI. CA1 pyramidal cells are heavily invested with L-type VDCCs, which mainly exist on the soma and proximal dendrites of these cells (Westenbroek et al., 1990). Even though L-type channels appeared to play no role in DSI induced by the standard protocol, we wondered whether or not they might play a role under conditions in which Ca^{++} influx through L-type channels, relative to Ca^{++} influx through other channels, was maximized. When the DSI-inducing conditions are changed such that QX-314 is omitted from the pipette and small (20–30 mV) rather than large (70 mV) voltage steps are used to elicit DSI, an unclamped action potential, probably mediated by both Na^+ and Ca^{++} currents, is induced (Lenz et al., 1998). QX-314, in addition to blocking Na^+ channels, blocks a variety of ion channel types (Perkins and Wong, 1995; Nathan et al., 1990; Connors and Prince, 1982) and probably promotes the passive spread of voltage from the somatic into the dendritic region. The omission of QX-314 and the use of small voltage steps may preferentially enhance the contribution of somatic Ca^{++} influx via L-type VDCCs to DSI. Indeed, unlike the DSI caused by the standard protocol, the DSI induced by this unclamped, somatic Ca^{++} spike was significantly reduced by nifedipine (control DSI 22.5 ± 2.0%, nifedipine DSI 10.9 ± 1.0%, $p = 0.03$, $n = 4$, data not shown). When, in these same cells, standard 1-s, 70-mV depolarizing pulses to 0 mV were used, nifedipine no longer antagonized DSI (control DSI 31.8 ± 8.3%, nifedipine DSI 27.9 ± 2.6 %, $p > 0.4$, $n = 4$, data not shown). Thus, the effectiveness of nifedipine is dependent on the induction protocol, not the cells. This variability in the contributions of N- and L-type VDCCs to DSI following the different induction protocols is probably related to the differences in the regions of the pyramidal cells activated by the voltage steps. With QX-314 in the electrode, the large voltage steps usually employed will be very effective in depolarizing the dendritic regions in which the majority of N-type channels are found. Without QX-314, smaller voltage steps would primarily depolarize the soma and nearby proximal dendrites, where L-type VDCCs are mainly localized. It is not clear why large voltage steps produce DSI with only ω-conotoxin-sensitive DSI. It could be that the ω-conotoxin-sensitive Ca^{++} influx is sufficient to saturate the DSI process such that the L-type component is only seen when the N-type component is submaximal. Alternatively, it may be that only N-type current is capable of inducing DSI and that the role of L-type current, under a protocol that is suboptimal for activating N-type channels directly, is to boost the dendritic voltage such that it reaches and effectively activates the N-type VDCCs. Nifedipine would block DSI under these conditions by preventing this booster effect of L-channel current. Which of these mechanisms is involved, as well as whether N- or L-type channels are the more important under physiological conditions, remain to be determined.

It is clear from the experiments with VDCC antagonists that the degree of DSI induced by influx through the various VDCCs is not a simple function of total Ca^{++} influx (Lenz et al., 1998). Thus, nifedipine had no effect on DSI under our standard induction protocol, but nevertheless reduced the peak calcium current by ~60%. Conotoxin and agatoxin were essentially equipotent in reducing I_{Ca}; both reduced it by ~25%, but, whereas the former was very effective in reducing DSI, the latter was completely ineffective. Ni^{++} had no effect on DSI despite decreasing I_{Ca} by ~10%. VDCCs are differentially localized to distinct regions of the pyramidal cell. Our data suggest that, rather than peak I_{Ca}, what may be more important for DSI is Ca^{++} influx localized to particular cellular regions. The data indicating a dendritic localization of N-channels would be compatible with a dendritic locus for DSI induction.

We have also considered the possibility that Ca^{++} released from intracellular stores plays a role in DSI, but thus far have found no evidence to support this idea. Triggered by voltage-dependent Ca^{++} influx, calcium-induced calcium release (CICR) from intracellular stores is a significant contributor to increased $[Ca^{++}]_i$ in many neuron types (Llano et al., 1994; Lipscombe et al., 1988; McBurney and Neering, 1987). The Ca^{++}-ATPase inhibitor, cyclopiazonic acid (CPA), depletes internal Ca^{++} stores in hippocampal cells (Garaschuk et al., 1997). It does this because the store contents are determined by a dynamic equilibrium between constant Ca^{++} leakage and store refilling via a Ca^{++}-ATPase. In blocking the refilling process, CPA allows Ca^{++} leakage to become dominant, and the stores empty within a few minutes. We found that CPA, bath-applied at concentrations that deplete measured Ca^{++} stores in hippocampal cells in culture, had no effect on DSI (Lenz et al., 1998). Thus, at this point it appears that Ca^{++} influx through VDCCs is sufficient to induce DSI.

Other divalent cations, notably Sr^{++} (Miledi, 1966), can substitute for Ca^{++} in causing transmitter release, albeit with a lower efficiency than Ca^{++}. It was interesting to note that substitution of 4 mM Sr^{++} for 4 mM external Ca^{++} allowed us to observe DSI (Morishita and Alger, 1997) (see Figure 7). Provided the duration of the DSI-inducing voltage step was doubled, DSI comparable to that induced in Ca^{++} was produced. Extracellular EGTA, 1 mM, could be added to the perfusate to chelate any residual Ca^{++} (EGTA binds Sr^{++} with a much lower affinity than it does Ca^{++}), so it was clear that an influx of Sr^{++} was sufficient to induce DSI. However, whether internal Sr^{++} actually substituted for Ca^{++} in the DSI mechanism, or whether it simply raised internal Ca^{++}, perhaps by displacing it from other binding sites, was not clear.

5. EXPRESSION OF DSI AS A DECREASE IN GABA RELEASE

Although DSI is induced by an increase in internal $[Ca^{++}]$, it does not appear to be mediated by a decrease in postsynaptic $GABA_A$ receptor responsiveness. This was at first surprising, because a great deal of other evidence had shown that $GABA_A$ receptor sensitivity can be down-regulated by increases in $[Ca^{++}]_i$ (Stelzer, 1992), probably through a Ca^{++}-dependent dephosphorylation reaction (Stelzer et al., 1988). On the other hand, this work had, in general, found that Ca^{++} influx through NMDA receptor channels was most important in decreasing $GABA_A$ receptor function and that Ca^{++} influx through VDCCs did not have this effect (but cf. Inoue et al., 1986, for an exception). Actually, in the cerebellum, increases in internal $[Ca^{++}]$ increase the sensitivity of $GABA_A$ receptors to applied GABA, even as Ca^{++} induces DSI (Llano et al., 1991). Increased $GABA_A$ receptor sensitivity occurred with synaptically released, as well as iontophoretically applied, GABA and was seen as an increase in IPSCs that outlasted the DSI period for many minutes (Vincent et al., 1992). The cause of the increased receptor responsiveness in cerebellum is not known. In any case, these results rule out the possibility that cerebellar DSI is related to decreased $GABA_A$ receptor sensitivity.

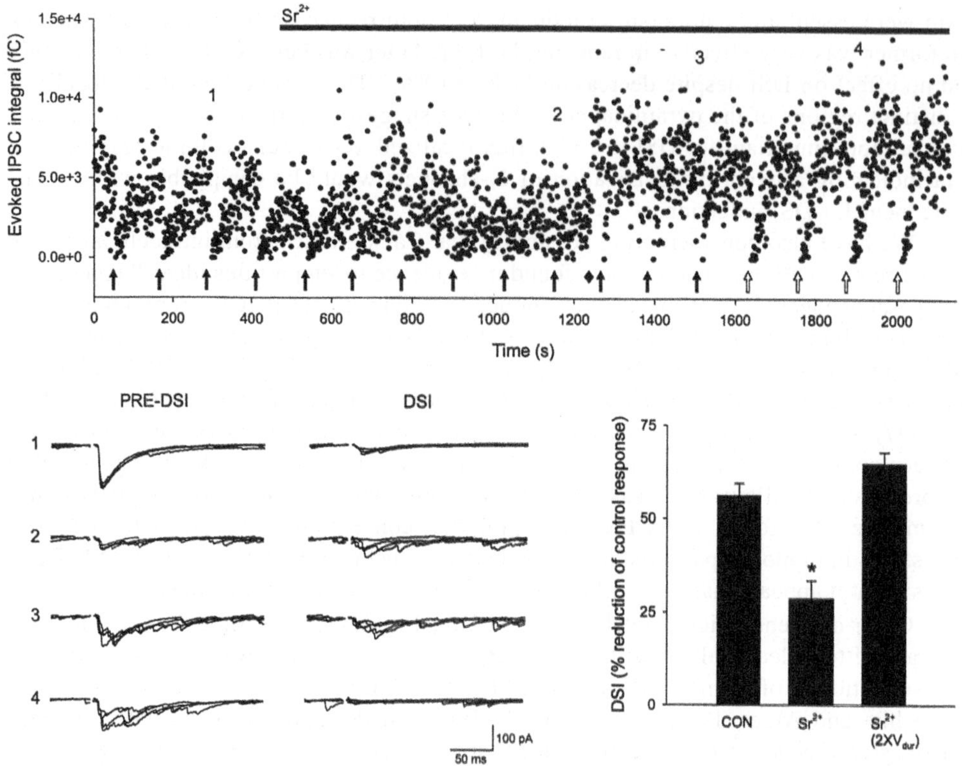

Figure 7. Extracellular Sr^{++} can substitute for Ca^{++} in the DSI induction process. DSI of evoked IPSCs is reduced in the presence of Sr^{++}, but can be recovered by increasing the duration of the depolarizing voltage step. The top graph represents a typical experiment where DSI was evoked at regular intervals by depolarizing the postsynaptic membrane potential to 0 mV for 2 s from a holding potential of -75 mV (arrows). Each point represents the integral of the IPSC calculated over 200 ms. Notice that DSI (depression in IPSC integrals immediately after the arrows) is reduced in the presence of Sr^{++} (duration of application indicated by solid bar). Increasing the extracellular stimulation intensity (diamond) produced larger IPSCs and a partial recovery of DSI. Only when the duration of the voltage step was doubled (open arrows) did DSI recover to pre- Sr^{++} levels. Below the graph are traces (comprised of IPSCs elicited by 4 consecutive stimuli) illustrating the IPSCs before (PRE-DSI) and 4 s following the depolarizing voltage step (DSI) recorded at the indicated points on the graph above. Stimulation artifacts are blanked for clarity. The bar graph to the right of the traces shows the amount of DSI obtained in control (CON), in Sr^{++} and in Sr^{++} when the duration of the depolarizing voltage step was doubled ($2xV_{dur}$, n = 10). Asterisk indicates significant difference from the control value as calculated by a Student's paired *t*-test ($p < 0.05$).

We tested $GABA_A$ receptor sensitivity during DSI in CA1 with iontophoretic GABA application (Pitler and Alger, 1992a) and amplitude analysis of spontaneous, TTX-insensitive miniature IPSCs (mIPSCs) (Pitler and Alger, 1994). The former technique assesses $GABA_A$-receptor sensitivity throughout the cells; both synaptic and nonsynaptic receptors are activated by iontophoretic application. Miniature IPSCs by definition represent activation of synaptic $GABA_A$ receptors, and therefore their amplitudes reflect the ability of truly synaptic $GABA_A$ receptors to be activated. Neither iontophoretic $GABA_A$ responses nor mIPSC amplitudes was altered during the DSI period, suggesting that $GABA_A$ receptor sensitivity was unaffected by DSI. Nevertheless both of these measurement techniques have limitations. In the hippocampal CA1 region, the mIPSC frequency is also not altered during DSI, whereas, in cerebellar Purkinje cells, the frequency of TTX-insensitive mIPSCs is reduced during DSI (Llano et al., 1991). We believe this represents a fairly minor difference between the expression mechanisms of hippocampal and cerebellar DSI, rather than fundamental differences in the DSI process itself. As discussed below, a difference in the proposed retrograde signaling systems between CA1 and cerebellum can, in principle, reconcile these disparate effects.

Since their introduction by del Castillo and Katz (1954), the methods of quantal analysis have been the standards for investigating the mechanism of changes in synaptic efficacy. Regulation of the numbers of quanta (each "quantum" taken to represent the transmitter contents of a single synaptic vesicle) released by a nerve terminal, the quantal content of the response, is a function of the presynaptic terminal. The quantal size, however, the magnitude of the response a quantum of neurotransmitter produces on the postsynaptic cell, is a function of the postsynaptic membrane generally, either of the state of the neurotransmitter receptors, or of the driving force on the ions involved. Although occasionally debate arises about the applicability of the assumptions underlying traditional quantal analysis to cases other than the neuromuscular junction (e.g., Faber and Korn, 1991), by and large, the reasoning has been supported when directly tested, and the assumptions are accepted as a reasonable first approximation.

Investigation of the locus of expression of DSI using several types of quantal analysis to determine whether it is pre- or postsynaptic has revealed that DSI is caused by a decrease in GABA release from presynaptic nerve terminals. This conclusion is supported by failure analysis of evoked minimal, presumably quantal, IPSCs, coefficient of variation analysis of large eIPSCs, and by direct counting of evoked asynchronous quantal release in Sr^{++}.

In no case has evidence for an alteration in postsynaptic responsiveness been found, and, on the contrary, in every case quantal content was found to be reduced. The proportion of failures of evoked transmission dramatically increased during the DSI period, while the amplitudes of the evoked quantal events themselves did not change (Alger et al., 1996; Vincent et al., 1992). A failure of transmission occurs when the presynaptic action potential arrives at the nerve terminal, but does not induce transmitter release. Occurrence of success or failure of transmission is determined presynaptically, and an increase in failures is evidence for presynaptic action. The coefficient of variation (CV) analysis of large IPSCs showed that the changes in these IPSCs likewise could be accounted for by a pre- but not a postsynaptic mechanism (Alger et al., 1996). Sr^{++} can substitute for Ca^{++} in the extracellular saline in the quantal release process (Miledi, 1966). However, in Sr^{++} the release of multiple quanta becomes desynchronized, spread out in time (Goda and Stevens, 1994; Miledi, 1966) such that the actual numbers of quanta released by a presynaptic action potential can be counted directly. In the presence of Sr^{++} it is clear that numbers of quanta released by GABAergic nerve terminals are markedly decreased during DSI; however, there is no change in the single quantal size (Morishita and Alger, 1997) (see Figure 8). The conclusion of these analyses is that the DSI process reduces IPSCs by decreasing GABA release from interneurons.

Additional novel support for the role of a retrograde signal process in cerebellar DSI was provided by Vincent and Marty (1993), who recorded from pairs of Purkinje cells simultaneously. They delivered a DSI-inducing voltage protocol to only one of the two cells and asked if DSI would thereby be produced simultaneously in the passive cell (i.e., the one not receiving the DSI-inducing protocol). The pairs were selected such that both Purkinje cells received GABA inputs from a common set of interneurons. Under these circumstances, the investigators found that delivering a DSI-induction protocol to one cell produced a suppression of IPSCs in both cells. DSI in this case obviously could not be solely a postsynaptic effect, because two distinct postsynaptic cells were involved. The DSI had to be

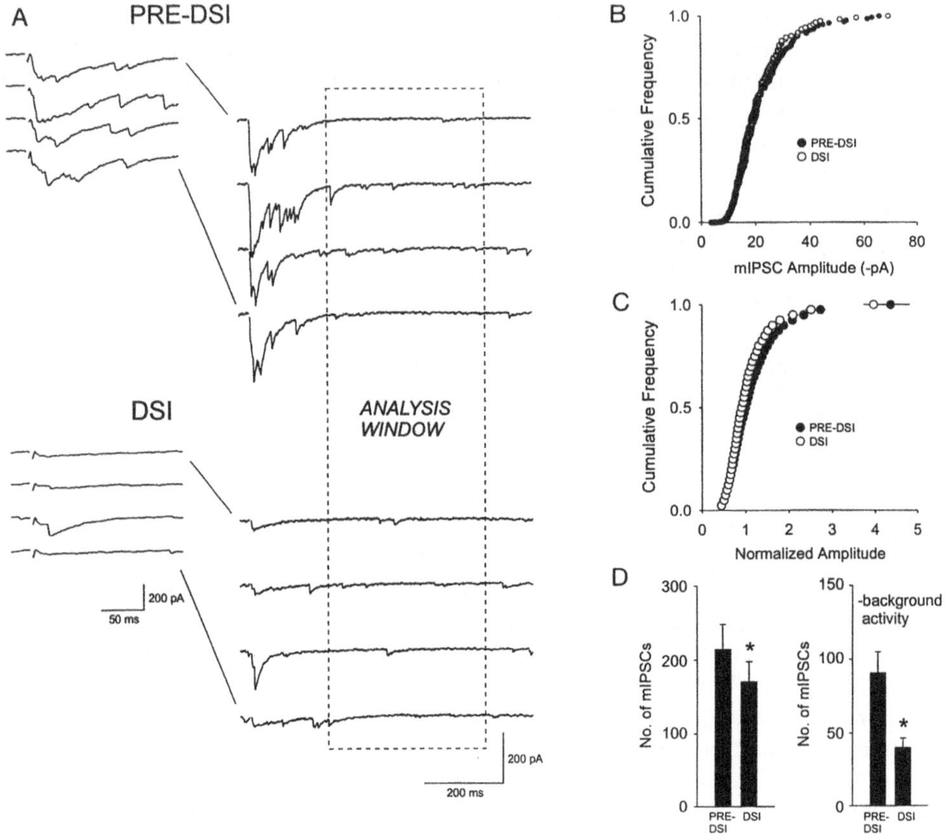

Figure 8. DSI decreases quantal content without reducing quantal size. A) illustrates a series of evoked IPSCs in the presence of Sr^{++} in control and during DSI (the IPSCs are shown on an expanded time scale at the far left). Stimulation artifacts are blanked for clarity. The amplitudes and numbers of mIPSCs were measured in a 400-ms-wide analysis window. B) compares the amplitude distribution obtained from a cell prior to DSI (278 events; mean, -17.2 ± 0.6 pA) and during DSI (241 events; mean, -17.1 pA ± 0.6, $p > 0.3$). C) shows the average amplitude distributions for pre-DSI and DSI periods ($n = 8$). The distributions of events before and during DSI in B and C are not statistically different from each other ($p > 0.3$ and $p > 0.01$, respectively) as determined by the Kolmogorov-Smirnov test. D) the first bar graph summarizes the average raw number of mIPSCs counted pre-DSI and during DSI ($n = 8$). To the right is a bar graph in which the background activity has been removed from the pre-DSI and the DSI mIPSC count. The difference in both the raw and corrected mIPSCs counts is significant (Student's paired *t*-test, $p < 0.02$ and $p < 0.001$, respectively). (From Morishita and Alger, 1997, with permission.)

mediated by a signal that originated in the first cell but affected other cells, probably by acting on the intervening interneurons. Vincent and Marty also found that TTX blocked only the DSI recorded in the passive Purkinje cell, but not the DSI of TTX-resistant mIPSCs in the Purkinje cell that received the DSI protocol. As discussed earlier, DSI of TTX-insensitive IPSCs is generally not seen in the CA1 region of the hippocampus, but is typical in cerebellum. That TTX blocked the propagation of DSI to the passive Purkinje cell implies the involvement of Na^+-dependent action potentials at some stage in the process. Inasmuch as DSI is blocked by TTX in CA1, it may be that this same TTX-sensitive step is the primary one there.

Taken together, the evidence indicates that DSI is induced by postsynaptic Ca^{++} influx into a pyramidal cell, and expressed as a decrease in GABA release from interneuron nerve terminals. Reconciling these findings appears to require the postulate of a retrograde signal process; that is, something from the pyramidal cell must influence the interneuron and induce it to release less GABA.

6. PROPERTIES OF THE DSI PROCESS

Testing this hypothesis thoroughly demands that the signal itself be identified. We began by studying the properties of the DSI process, reasoning that the putative signal process must have, or mimic, the same properties. An important physiological observation involved the kinetics of DSI. A lag of about 1.5 s after the end of the DSI-inducing pulse precedes the development of maximal DSI (Pitler and Alger, 1994). This time course, slow compared to the usual events of synaptic transmission, suggested the participation of a second messenger. Moreover, DSI appeared to be in part dependent on a G protein, because pertussis toxin (Pitler and Alger, 1994), and the sulfhydryl-alkylating agent N-ethylmaleimide (NEM) (Morishita et al., 1997) (which inhibits pertussis-sensitive G-protein-dependent events, e.g., Shapiro et al., 1994), both virtually abolish DSI. The relevant G protein is probably associated with the GABA-releasing interneuron, since disturbance of postsynaptic G-protein function, either by including various non-hydrolyzable GTP analogs in the recording pipette, or by omitting GTP from the recording electrode, had no effect on DSI. Moreover, postsynaptic injection of NEM sufficient to block the G-protein-dependent $GABA_B$ current has no effect on DSI (Lenz and Alger, unpublished observations).

Other evidence, in addition to quantal analysis, that DSI is expressed presynaptically as a decrease in GABA release came from pharmacological demonstrations that DSI can be reduced by agents that can only act presynaptically. Bath application of the K^+ channel blocker 4-aminopyridine (4-AP) at 50 μM or veratridine, which slows Na^+ channel inactivation, at 250 nM reduces DSI, reversibly in the case of 4-AP (Alger et al., 1996). Although bath-applied, both 4-AP and veratridine appeared to act presynaptically because postsynaptic K^+ and Na^+ channels had already been blocked by the Cs^+ and QX-314, respectively, in the recording solution. Thus, it seems that the DSI process may act via a presynaptic G-protein-dependent process that could affect GABA release by acting on K^+ channels, or Na^+ channels, or on both.

DSI expression, despite its presynaptic locus, is not simply the result of a conventional presynaptic inhibition process. One piece of evidence is that DSI does not alter the frequency of TTX-insensitive mIPSCs (Alger et al., 1996; Pitler and Alger, 1994), although some presynaptic inhibitory transmitters do (Thompson, 1994). Another unusual feature of DSI is that paired-pulse modifications of neurotransmitter release are not altered during DSI of eIPSCs (Morishita and Alger, 1997; Alger et al., 1996). Typically, delivering two identical stimuli to GABA-releasing nerve terminals at an interval of ~200 ms leads to elicitation of a smaller eIPSC by the second pulse than by the first (Lambert and Wilson, 1994; Davies et al., 1990). This is called paired-pulse depression (PPD). In general paired-pulse properties are taken to reflect determinants of the probability of presynaptic release (Martin, 1977). Influences that reduce transmitter release by, e.g., low $[Ca^{++}]$, or presynaptic inhibitory neurotransmitters such as the $GABA_B$ antagonist baclofen (Davies et al., 1990), reduce PPD, probably by altering the probability of transmitter release by the first stimulus and, as a consequence, the ability of the second stimulus to release transmitter. During the DSI period IPSCs are reduced to an extent comparable to that caused by baclofen. However, PPD is not affected by DSI; both IPSCs in the pair are reduced pro-

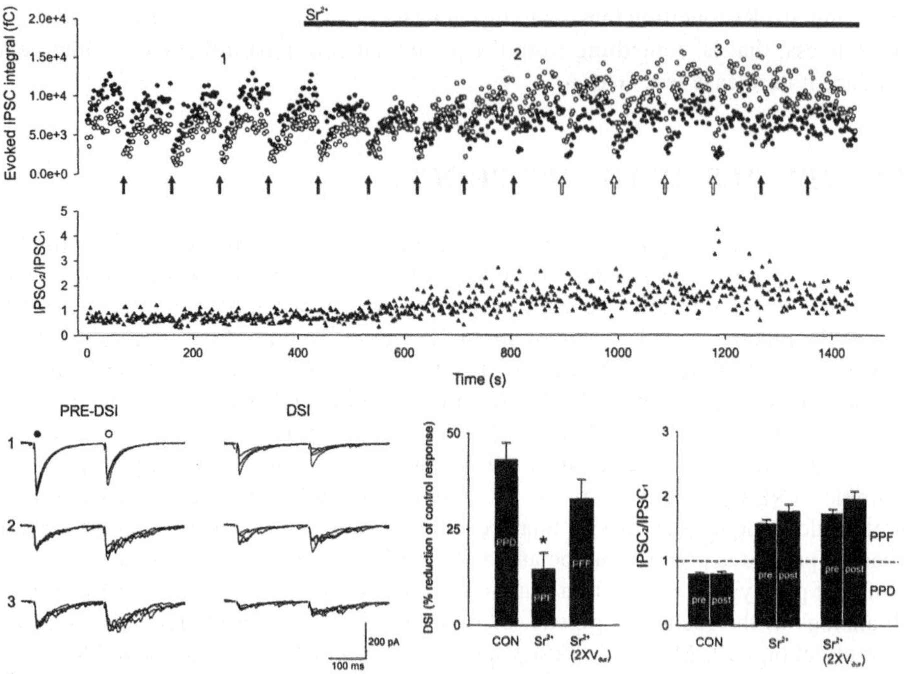

Figure 9. Neither PPD nor Sr^{++}-induced PPF of IPSCs is altered during DSI. The top graph represents the magnitude of the IPSCs evoked by the first stimulation pulse ($IPSC_1$, filled circles) and those elicited by an identical pulse 200 ms later ($IPSC_2$, open circles) before and during application of Sr^{++} (duration of application indicated by solid bar). Because quantal release is desynchronized in Sr^{++}, IPSCs were integrated over 200 ms to measure their magnitude. The graph representing the $IPSC_2/IPSC_1$ ratios (filled triangles) is shown below. Notice that both IPSCs can undergo a DSI following the voltage step (arrows) and that DSI is reduced in Sr^{++}, but can be significantly increased when the duration of the voltage step is doubled (open arrows). The traces below the graphs (4 consecutive traces in each series) were recorded at the indicated time points immediately before (PRE-DSI) and 6 s following the depolarizing voltage step (DSI). The bar graph to the right of the traces shows the amount of DSI of the first pulse in the pair in Ca^{++} (CON), in Sr^{++}, and in Sr^{++} when the duration of the voltage step is doubled ($2xV_{dur}$) (n = 9). The bar graph at the far right compares the $IPSC_2/IPSC_1$ ratio before (pre) and following (post) the depolarizing voltage step in Ca^{++} and in the presence of Sr^{++} and again in Sr^{++} when the duration of the voltage step is doubled (n = 9). Notice that neither PPD in Ca^{++} nor PPF in Sr^{++} is significantly altered during the DSI period (Student's paired *t*-test, $p > 0.07$). (From Morishita and Alger, 1997, with permission.)

portionately so that their ratio is unaltered (see Figure 9). When Sr^{++} is substituted for Ca^{++} in the external saline, IPSCs are reduced and PPD is changed into a paired-pulse facilitation (PPF), as expected if Sr^{++} decreases the probability of GABA release. In the presence of Sr^{++}, PPF is also unaffected during DSI (Morishita and Alger, 1997). Hence DSI reduces GABA release by acting at some step that does not affect probability of release as usually assessed. DSI could, e.g., block action potential conduction in fine preterminal branches, completely suppress Ca^{++} influx into the nerve terminal, or prevent the presynaptic Ca^{++} sensor from responding to Ca^{++}. Provided that transmitter is not released by the first pulse, then there would be no change in the properties of the second response. Both responses would become smaller, but the paired-pulse ratio would not change. (Interestingly, the DSI process in tissue-cultured neurons (Ohno-Shosaku et al., 1998) is accompanied by an increase in the PPD ratio, suggesting a different expression mechanism for

DSI may operate in those cells.) In any event, while the reason for the lack of interaction between DSI and PPD or PPF is not yet known (and will be important to understanding the actual mechanism by which DSI suppresses release), the observation can be used as part of the diagnostic profile to identify the DSI signal.

To summarize, a candidate DSI signal must: 1) have the ability to mimic DSI (i.e., block the same IPSCs) and therefore occlude it, 2) not affect mIPSC amplitudes, 3) have IPSC-suppressing effects that are prevented by pertussis toxin, 4-AP and NEM, and 4) not affect PPD.

7. EVIDENCE FOR METABOTROPIC GLUTAMATE RECEPTORS IN DSI

With these criteria in mind we have recently re-examined a hypothesis put forward by Glitsch et al. (1996) to explain DSI in the cerebellum. These investigators propose that glutamate, or a glutamate-like substance, can be released from Purkinje cells and, by acting on presynaptic group II metabotropic glutamate receptors (subtypes mGluR2 and mGluR3) on interneurons, reduce the release of GABA. This conclusion is based on evidence that trans-ACPD, a broad-spectrum mGluR agonist, suppresses IPSCs (Llano and Marty, 1995). Glitsch et al., found that DCG-IV, a selective group II agonist (Hayashi et al., 1993), could mimic and occlude DSI by reducing IPSCs (see Figure 10). L-AP3, an mGluR antagonist that is effective on group II receptors, significantly inhibited DSI. Moreover, it was known that reduction in intracellular cAMP can reduce transmitter release, and that group II agonists typically reduce cAMP concentration (Pin and Duvoisin, 1995). The group II agonists could, therefore, suppress GABA release through this mechanism. Forskolin is a drug that activates adenylate cyclase and thus increases intracellular cAMP concentration. Glitsch et al. (1996) found that forskolin reduced DSI, and that this effect could not be explained simply by the increase in transmitter release that forskolin induced. It was concluded that the effect of forskolin could be consistent with the hypothesis that a group II mGluR mediates DSI.

We use CNQX and APV to prevent the occurrence of ionotropic glutamate responses, but glutamate can still be released and can act on presynaptic nerve terminals. Activation of mGluRs suppresses IPSCs in hippocampus (Gereau and Conn, 1995; Desai and Conn, 1991; Poncer et al., 1995). Involvement of a glutamate-like agonist on these receptors is a plausible mechanism for DSI. In hippocampal CA1 cells we found that the specific hypothesis of group II mGluR mediation of DSI was not supported; DCG-IV had no effect on IPSCs or DSI. This was different from what Poncer et al. (1995) had found in CA3, where DCG-IV reduces IPSCs. The lack of effect of DCG-IV on IPSCs in CA1 confirmed Gereau and Conn's (1995) observations. In view of the paucity of evidence that group II mGluRs exist in the CA1 region (Shigemoto et al., 1997), our results were not surprising, and we concluded that group II mGluRs are probably not involved in CA1 DSI. It remained possible that glutamate could play a role, because other mGluRs, most notably of the group I class (subtypes mGluR1 and mGluR5), are present in high concentrations in CA1.

We therefore tested the general hypothesis of mGluR involvement using the broad-spectrum agonist, 1S,3R-ACPD ("ACPD"), and found that, indeed, ACPD does mimic and occlude DSI by reducing eIPSCs (Morishita et al., 1998). ACPD's effects were similar in several ways to DSI. Both ACPD-induced reduction of IPSCs and DSI could be reversed by 4-AP and NEM. Neither ACPD nor DSI affects TTX-insensitive mIPSCs, and neither alters paired-pulse depression of IPSCs (cf. Barnes-Davies and Forsythe, 1995, who show

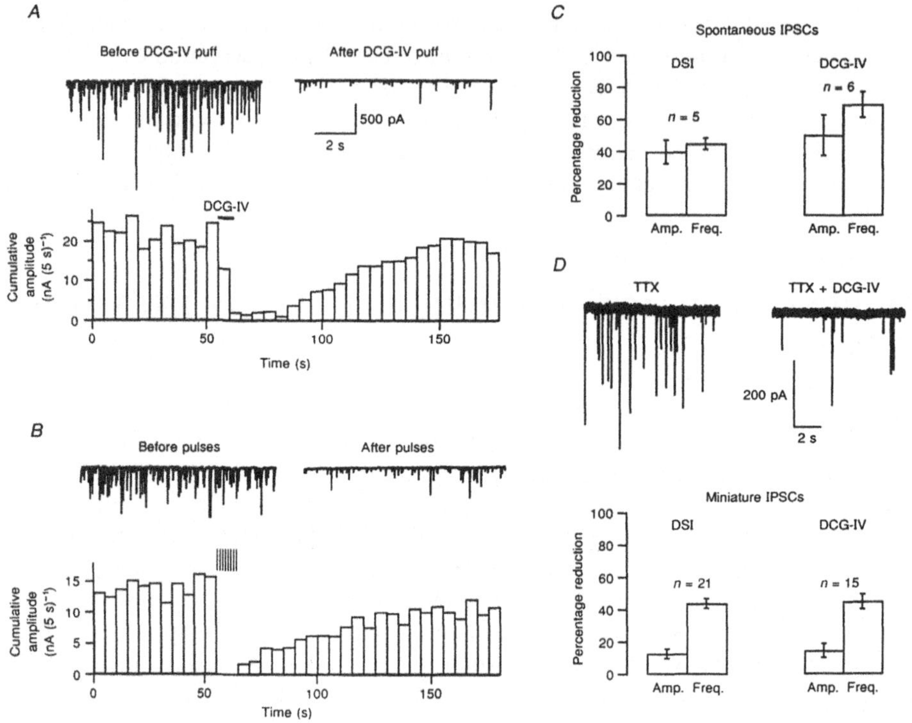

Figure 10. The group mGluR agonist DCG-IV mimics DSI in cerebellar Purkinje cells. A) a puff application of DCG-IV (concentration in application pipette, 5 μM; 6-s pressure application indicated by bar above histogram) reduces the mean frequency and amplitude of sIPSCs. Upper trace, records obtained about 5 s before and after the DCG-IV puff. Lower plot, cumulated amplitudes of sIPSCs over 5-s time bins. The effects of DCG-IV are maximal a few seconds after the end of the pulse, and they slowly recover over a 2-min period. B) DSI reduces sIPSCs similarly to a DCG-IV puff. Same cell as in A. Upper traces, records obtained about 5 s before and after applying a series of depolarizing pulses (8 × 100 ms steps to 0 mV at 1-s intervals) to the recorded cell. Lower plot, cumulated amplitudes of sIPSCs over 5-s time bins. The position of the train is indicated above the histogram. Same calibrations in A and in B. C) summarized results from 6 experiments similar to that illustrated in A and B. In 5 of these experiments both DSI and DCG-IV application results were available. The control results were taken over a period of 60 s before applying the DCG-IV or the voltage pulses; the test results were taken over a period of 10 s starting 5 s after the end of the stimulus. On average DCG-IV reduced the main amplitude and frequency of sIPSCs similarly to DSI, even though the mean event frequency was slightly more reduced by DCG-IV than by DSI. D) DCG-IV and DSI have similar effects on mIPSCs. Upper traces: mIPSCs recorded before and during bath application of 1 μM DCG-IV, in the presence of TTX. Lower plots, left, analysis of the effects of DSI on the mean amplitude and frequency of mIPSCs; right, results of bath application of 1–5 μM DCG-IV. (From Glitsch, et al., 1996, with permission).

that ACPD inhibits transmitter release, without affecting paired-pulse facilitation, at a giant excitatory synapse in rat brainstem). Thus, by all of the criteria developed so far to screen for putative DSI signals, glutamate, acting on an mGluR subtype, is a candidate.

If the glutamate hypothesis of DSI were true in CA1 then it should be possible to identify the mGluR subtype involved. A role for group II mGluRs had been excluded because the selective mGluR group II agonist, DCG-IV, has no effect on IPSCs in CA1, and group II mGluRs do not appear to be present in CA1. We tested the possibility of a role for group III mGluRs by using the selective group III agonist L-AP4 and antagonists M-AP4 and M-SOP.

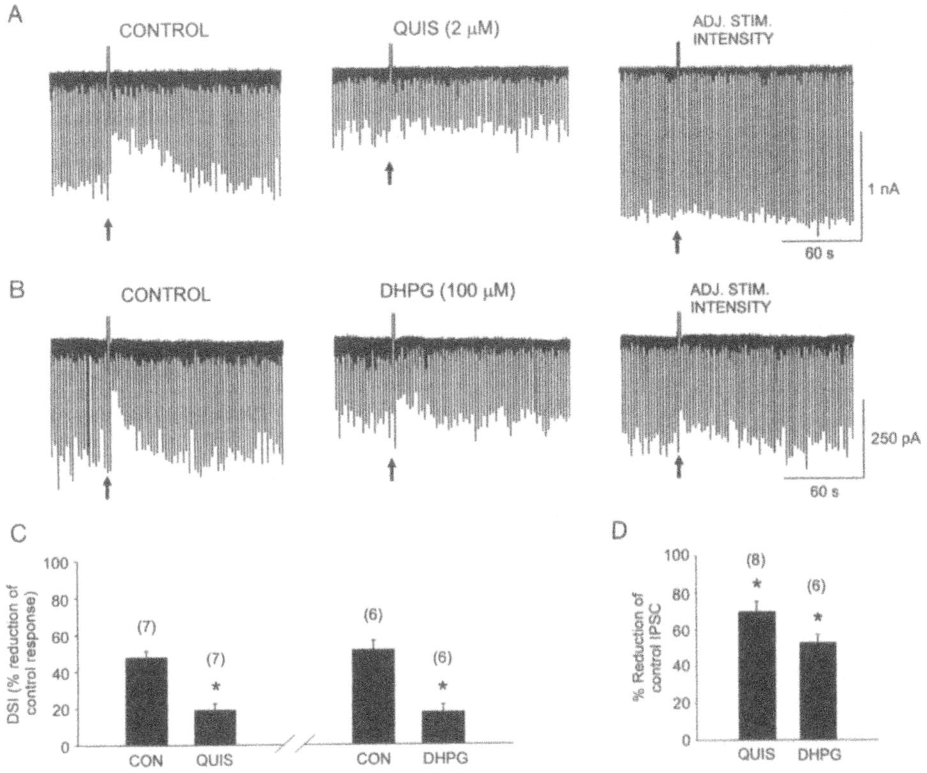

Figure 11. Group I mGluR agonists, L-quisqualate and DHPG, reduce the amplitude of evoked monosynaptic IPSCs and occlude DSI. A) the first trace shows DSI of IPSCs recorded in the control saline. The center trace shows IPSCs recorded during the 10th min of bath-application of L-quisqualate (QUIS). The right-hand trace shows a DSI trial still in L-quisqualate after the stimulation intensity had been increased (ADJ. STIM INTENSITY) to elicit IPSCs comparable to those in the control condition. B) illustrates the effects of DHPG on IPSCs and DSI; trace sequence as in A. Both L-quisqualate and DHPG suppress IPSCs and occlude DSI, and the effects persisted even after the stimulation intensity had been increased. C) a graph summarizing the effects of quisqualate and DHPG on DSI. D) a graph showing the effect of these agonists on IPSC amplitudes. Asterisks indicate significant differences from control values. (From Morishita et al., 1998).

Group III mGluRs include mGluR4, mGluR6, mGluR7 and mGluR8. Whereas mGluR6 is present only in the retina, mGluR4, mGluR7 and mGluR8 are present in CA1 (Shigemoto et al., 1997). We found that, although L-AP4 did cause a modest reduction in eIPSC amplitude, it did not occlude DSI; that is, the percent DSI was not altered by L-AP4, as it was by ACPD. Moreover, even though we confirmed that M-AP4 was an effective group III antagonist in our hands (it reversed the L-AP4-induced suppression of field potentials in control experiments), M-AP4 had no effect on DSI or on the ACPD-induced suppression of IPSCs. Similar results were obtained with M-SOP. Thus, while GABA release was reduced by the group III mGluR agonist, group III mGluRs seem not to be involved in DSI.

Both the selective group I agonists DHPG (100 μM) and quisqualate (2 μM) (at low concentrations quisqualate is selective for group I mGluRs) reduced eIPSCs and occluded DSI. Moreover, the broad-spectrum antagonist, MCPG, which blocks group I, but also to a lesser extent group II and III mGluRs, significantly reduced DSI and the ACPD-induced reduction in eIPSCs (see Figure 11).

Although high concentrations of MCPG, ≥ 1 mM, were necessary for substantial effects on DSI, this is not incompatible with a role for glutamate as the DSI signal. The potency of MCPG in blocking mGluR-mediated biochemical responses in cells is dependent on the agonist used to activate the receptor (Brabet et al., 1995; Littman and Robinson, 1994), with MCPG being more potent against ACPD-induced actions than against glutamate-induced actions on the same receptor. For example, Littman and Robinson (1994) report that, whereas 1 mM MCPG reduced PI hydrolysis in hippocampal tissue suspensions mediated by ACPD by ~80%, the same concentration of MCPG had negligible effects when glutamate was the agonist, and 3 mM MCPG was required to block the glutamate-induced effect by ~20%. Thus, the relatively weak effects of MCPG would be consistent with a role for glutamate in mediating DSI through an mGluR. In view of the negative effects of selective agonists and antagonists of groups II and III mGluRs, we interpret the effects of MCPG as being exerted mainly on group I mGluRs.

Our data would be consistent with the hypothesis that DSI is caused when glutamate, or a glutamate-like substance, is released in a Ca^{++}-dependent manner from pyramidal cells following voltage-dependent influx of Ca^{++} through N- and/or L-type VDCCs. Glutamate would diffuse to nearby GABA-releasing interneurons and, by activating group I mGluRs, prevent their release of GABA for a number of seconds. The pharmacological characteristics of DSI will vary with the properties of the mGluRs present on the target interneurons; hence, in cerebellum, DSI would be mediated by group II mGluRs, whereas, in hippocampal CA1, DSI would be mediated by group I mGluRs. This hypothesis is summarized in the schematic drawing of Figure 12.

Many details of this hypothesis remain to be worked out. The mechanism of glutamate release, its site and mechanism of action on interneurons, etc., are all at present unknown. An interesting puzzle regarding the mGluR hypothesis for DSI is that cerebellar Purkinje cells utilize GABA as their neurotransmitter, as do hippocampal interneurons in culture, which can also induce DSI (Ohno-Shosaku et al., 1998). Glutamate is the precursor of GABA (being converted to it by glutamic acid decarboxylase), so these GABAergic cells surely have copious quantities of intracellular glutamate available. Nevertheless, if glutamate is the retrograde signal, it cannot be from a neurotransmitter pool of glutamate in these cells. The mGluR hypothesis is testable, and testing it will provide important insights into a novel regulatory mechanism in the brain.

8. FUNCTIONAL IMPLICATIONS OF DSI

While the quantitative properties of DSI are undoubtedly different *in vivo* than they are *in vitro*, if we assume that the qualitative aspects are still present *in vivo*, then it is possible to speculate about the functional implications of DSI. During the DSI period, events that are normally suppressed by $GABA_A$ergic IPSCs will be enhanced. For example, when APV and CNQX are omitted from the bathing solution, the evoked IPSC overlaps, and truncates, an excitatory postsynaptic current (EPSC), and DSI then enhances the EPSC by reducing the IPSC (Wagner and Alger, 1996). The effect is blocked by BAPTA in the recording electrode and is clearly exerted on the IPSC itself, because no effect of the DSI-inducing voltage step is seen when bicuculline is added to the perfusate to block the IPSC. NMDA-receptor-mediated responses are often inhibited by $GABA_A$ergic IPSCs, and NMDA responses that occur during the DSI period should be enhanced relative to those responses that occur at other times. Because in CA1 LTP is induced through activation of NMDA receptors, glutamate-mediated responses during the DSI period should be more likely to induce LTP than the same responses occurring at other times.

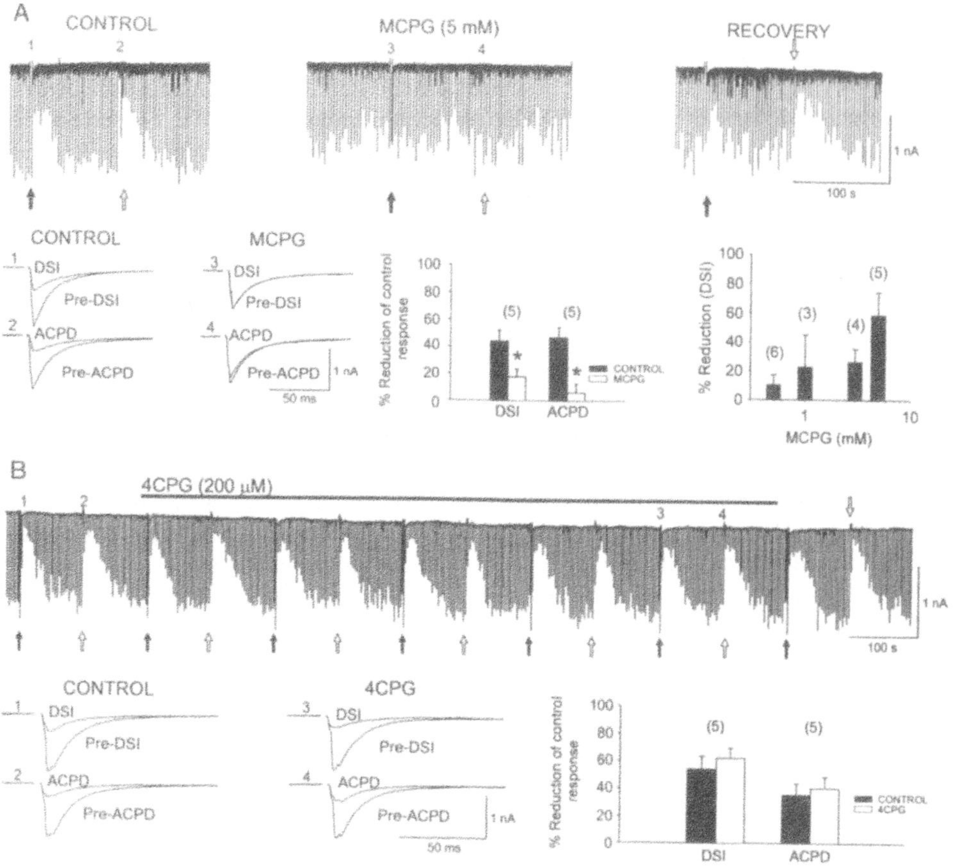

Figure 12. Selective block of DSI and the (1S,3R)-ACPD-induced suppression of IPSCs by (S)-MCPG, but not 4CPG. The traces in A illustrate the transient suppression of IPSCs during DSI (filled arrows) and following iontophoresis of (1S,3R)-ACPD (open arrows) and antagonism of both forms of IPSC suppression by (S)-MCPG. All traces at the top were recorded from the same cell. The trace in MCPG was recorded 18 min after starting MCPG application; the recovery trace was recorded 40 min after starting washout of MCPG. Below the traces are IPSCs recorded at the indicated time points before (pre-DSI) and during DSI (DSI) as well as before (pre-ACPD) and following (ACPD) iontophoretic application of (1S,3R)-ACPD. Recovery from (S)-MCPG took place during the gap (40 min) in the current trace. The bar graph in the center summarizes results from 6 cells. The bar graph to the extreme right shows the dose dependence of (S)-MCPG effects on DSI. B) shows a continuous record in which the evoked IPSCs were subjected to the same experimental protocol as in A. Below the record are IPSCs recorded at the indicated time points. Notice that 4CPG (duration of application indicated by the solid bar) does not antagonize DSI or the (1S,3R)-ACPD-induced suppression of IPSCs. The bar graph illustrates results from 5 cells. Individual IPSCs in A and B are averaged traces from 5 consecutive responses. (1S,3R)-ACPD was iontophoresed by a -155 nA, 2 s current. Asterisks indicate significant differences from the control values. (From Morishita et al., 1998, with permission).

The greater significance of DSI may be, however, not merely that it would enhance the probability of LTP induction, but that it could provide for temporal and spatial specificity of LTP induction. The temporal specificity would come from the brevity of the time window opened by DSI. Inputs occurring outside the window would be less likely to be potentiated, and, thus, timing of the input would be critical. Temporal coincidence detection is also implied, as the target pyramidal cell first would have to be sufficiently activated to open the DSI time window.

DSI can also serve to establish spatial specificity. In a typical *in vitro* extracellular stimulation paradigm $GABA_A$ergic inhibition is pharmacologically blocked, and large numbers of CA1 pyramidal cells are induced to undergo LTP simultaneously. However, under more physiological conditions, $GABA_A$ergic inhibition is operative, cells are not all synchronously activated, and DSI could provide a means for selecting only a certain subpopulation of pyramidal cells to undergo LTP; i.e., only those cells sufficiently activated recently to cause DSI undergo LTP in response to the potentially LTP-inducing stimulus. The same degree of excitatory stimulation applied to cells not experiencing DSI could be below threshold for LTP. DSI, in other words, could provide a mechanism for selectively addressing LTP spatially.

The idea of a spatially selective addressing function of DSI can be extended to apply to local regions of a single cell. Excitatory and inhibitory synapses are present in the dendritic regions of pyramidal cells, as are voltage-dependent Ca^{++} channels. It is conceivable that in a given cell DSI could be induced on one dendritic branch by excitatory activity on that dendritic branch and not on others. This could provide a mechanism for selective targeting of LTP to certain dendrites of a pyramidal cell, perhaps strengthening the influence of just one type of synaptic input on that cell. Local LTP addressing has recently been proposed to be mediated by dendritic 'A' type K^+ channels in pyramidal cells (Hoffman et al., 1997). Differences in the efficacy of the inhibitory influence of the A channels among different dendritic branches could allow LTP to be induced only in certain branches, and not on others. It is possible that A current regulation of voltage-dependent Ca^{++} influx could, similarly, control DSI in only certain branches, and thereby facilitate local NMDA responses. Modeling studies (Hoffman et al., 1997) confirm the possibility that LTP induction can be localized to single branches. The interaction between dendritic IPSPs and DSI could perform a similar local LTP-addressing function.

Much attention has recently been focused on the role of back-propagating dendritic action potentials, i.e., those originating in the soma and invading the dendrites (Markram and Tsodyks, 1996; Johnston et al., 1996) in neuronal physiology. EPSPs occurring concurrently with postsynaptic back-propagating action potentials are selectively induced to undergo LTP (Markram et al., 1997; Magee and Johnston, 1997). Wong and Prince (1979) and Miles et al. (1996) demonstrated a role for dendritic IPSCs in regulating active conductances in pyramidal cell dendrites. Using intradendritic recording it has been shown directly that dendritic IPSCs control the invasion of somatically induced action potentials into pyramidal cell (Tsubokawa and Ross, 1996) and mitral cell dendrites (Chen et al., 1997). DSI can regulate dendritic inhibition and should provide a window during which back-propagation of action potentials into dendrites is facilitated. Muscarinic receptor agonists enhance DSI, and it may be relevant that muscarinic agonists also enhance back-propagation of action potentials and dendritic Ca^{++} influx (Tsubokawa and Ross, 1997). LTP induction or other processes dependent on backpropagating action potentials should thereby be enhanced by DSI.

Ultimately, the issue of the functional implications of DSI will be determined by whether or not DSI actually occurs under physiological conditions. As a first step in addressing this issue, we asked if action potential discharges in CA1 cells can induce DSI (Le Beau and Alger, 1998). Following a single action potential, suppression of IPSPs is not obvious. Nevertheless, hippocampal pyramidal cells can fire in a "burst" mode under some physiological conditions (Kandel and Spencer, 1961), and burst firing is typical when inhibition is blocked by a $GABA_A$ receptor antagonist (Wong and Prince, 1979). A burst involves the activation of voltage-dependent Ca^{++} and Na^+ conductances, and we wondered therefore if a burst might induce DSI. Clearly, we could not use $GABA_A$ recep-

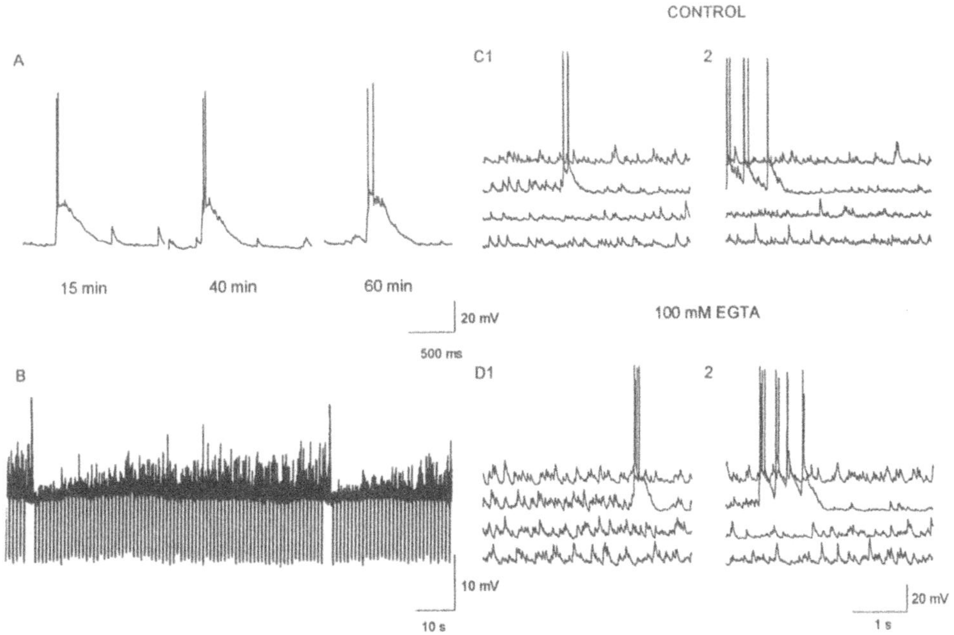

Figure 13. Epileptiform burst discharges in Mg^{++}-free saline cause a DSI-like phenomenon in hippocampal CA1 cells. A) Intracellular microelectrode recordings of spontaneous epileptiform burst discharges for one cell recorded 15, 40, and 60 minutes after commencing perfusion with Mg^{++}-free saline. Burst shape and magnitude remained constant over time. B) A second neuron fired spontaneous burst discharges represented by the tallest vertical lines (truncated for clarity) initially with an interburst interval of ~15–20 s after 22 minutes perfusion with Mg^{++}-free saline. A brief, but distinct, period of decreased synaptic activity is evident after each burst. (C,D) Ca^{++}-dependence of post-burst IPSP suppression. Four-second traces (consecutive in each group) for each of two cells, one recorded without EGTA in the electrode (C) and one with 100 mM EGTA in the electrode (D). For the cell recorded without EGTA a single burst produced a clear suppression of sIPSPs (C1), which was enhanced when three spontaneous burst discharges occurred in the same cell (C2). With 100 mM EGTA in the electrode, there is no suppression of sIPSPs following a single spontaneous burst (D1), but the same cell was capable of showing a reduction in sIPSPs following a series of four spontaneous epileptiform bursts (D2). Group data, (not shown) confirmed that no significant suppression occurred in EGTA-loaded cells (n = 6), whereas normal (~30% suppression) occurred in control cells (n = 7). Significant, though reduced suppression (~28% suppression) was present after clusters of bursts (3 or 4) in EGTA-loaded cells. (From Le Beau and Alger, 1998, with permission.)

tor antagonists when examining DSI, so we chose the low-Mg^{++} model of burst induction (Mody et al., 1987; Anderson et al., 1986) in which $GABA_A$ receptor-mediated inhibition remains active (Tancredi et al., 1990). In these experiments the extracellular saline contained nominally 0 mM Mg^{++}, and $CaCl_2$ was increased to 3.5 mM (neither CNQX nor APV was present). Carbachol, 1–5 μM, was used to induce spontaneous IPSPs and to suppress the Ca^{++} - K^+ conductances that normally follow a burst.

Within 15–20 min after switching to Mg^{++} -free saline, burst responses, consisting of a series of action potentials riding on a depolarizing wave, occurred spontaneously or could be evoked by single afferent stimulus pulses. Following the burst, spontaneous IPSPs were suppressed for ~10 s, and the mean maximal IPSP suppression in a given cell was ~30% from control values (see Figure 13). We performed several tests to determine if this suppression represented DSI. Our recording conditions prevented any change in resting input conductance from following the burst; hence the IPSP suppression could not be explained by shunting of the membrane resistance. NEM blocked the burst-induced suppression, as it did DSI, without affecting control IPSPs or burst magnitude. 4-AP increased the frequency of spontaneous burst firing so markedly that its effects on post-burst IPSP suppression could not be measured. We found that the suppression was Ca^{++} dependent, however, because inclusion of 100 mM EGTA in the recording pipette completely prevented IPSP suppression induced by a single burst (Figure 13C,D). If 2–4 bursts occurred in quick succession, a transient, albeit reduced, IPSP suppression still occurred. This showed that IPSP suppression could occur in EGTA-loaded cells; i.e., they were capable of producing it. The finding of IPSP suppression after multiple, but not single, bursts in EGTA-loaded cells suggests that the larger influx of Ca^{++} accompanying multiple bursts is able to overwhelm the buffer for a brief period.

Thus, post-burst IPSP suppression may be evidence that DSI can be induced by the burst-firing mode of pyramidal cell discharge. More work will have to be done to exclude the competing hypothesis that this suppression could be due to a postsynaptic $GABA_A$-receptor down-regulation (Chen and Wong, 1995). If post-burst IPSP suppression is in fact DSI, then it will be necessary to take its effects into consideration when trying to understand the firing patterns of cells that discharge in a burst mode under physiological or pathological conditions. It is expected that NMDA-dependent responses will be facilitated during the period of post-burst IPSP suppression.

9. IMPLICATIONS OF DSI FOR STUDIES OF ALCOHOL EFFECTS ON THE BRAIN

Alcohol inhibits L-type Ca^{++} channel activity acutely (Wang et al., 1994). In view of the lack of participation of Ca^{++} influx through L-type VDCCs, ethanol may not alter DSI as we normally induce it. However, (see section 4.; Lenz et al., 1998) under altered conditions of activation, Ca^{++} influx through L-type channels does contribute significantly to DSI; thus, ethanol should decrease DSI under those conditions. Decreased DSI means, effectively, an increased level of inhibition after pyramidal cell activity, and, hence, increased DSI could contribute to the acute suppressive effects of ethanol. Contrariwise, chronic ethanol treatment leads to an increase in VDCC function, perhaps related to an increased number of VDCCs (Messing et al., 1990), and chronic alcohol treatment could lead to enhanced DSI. Enhanced DSI, effectively greater reduction in $GABA_A$ergic inhibition following pyramidal cell activity, could play a role in the initiation of seizures induced by alcohol withdrawal.

Chronic ethanol treatment leads to disturbance of memory (see e.g., Cermak, 1993), perhaps mediated in part by the marked suppressive effects of long-term ethanol on the cholinergic projection originating in the medial septum/vertical limb of the nucleus of the diagonal band (Walker et al., 1993; Arendt et al., 1989). Induction of LTP, widely accepted to be the cellular substrate for learning, can be facilitated by activation of cholinergic pathways (Huerta and Lisman, 1995). Our conception of the physiological role of DSI is that it can play a role in the establishment of LTP. Disruption of DSI by impairment with cholinergic function could thus contribute to the alcohol-induced deficits in learning and memory. DSI is particularly prominent when muscarinic cholinergic agonists, either exogenously applied or released from cholinergic nerve terminals in the slice by electrical stimulation (Pitler and Alger, 1992a), activate spontaneous interneuronal activity (Martin et al., 1995; Pitler and Alger, 1994, 1992) (see Figure 14).

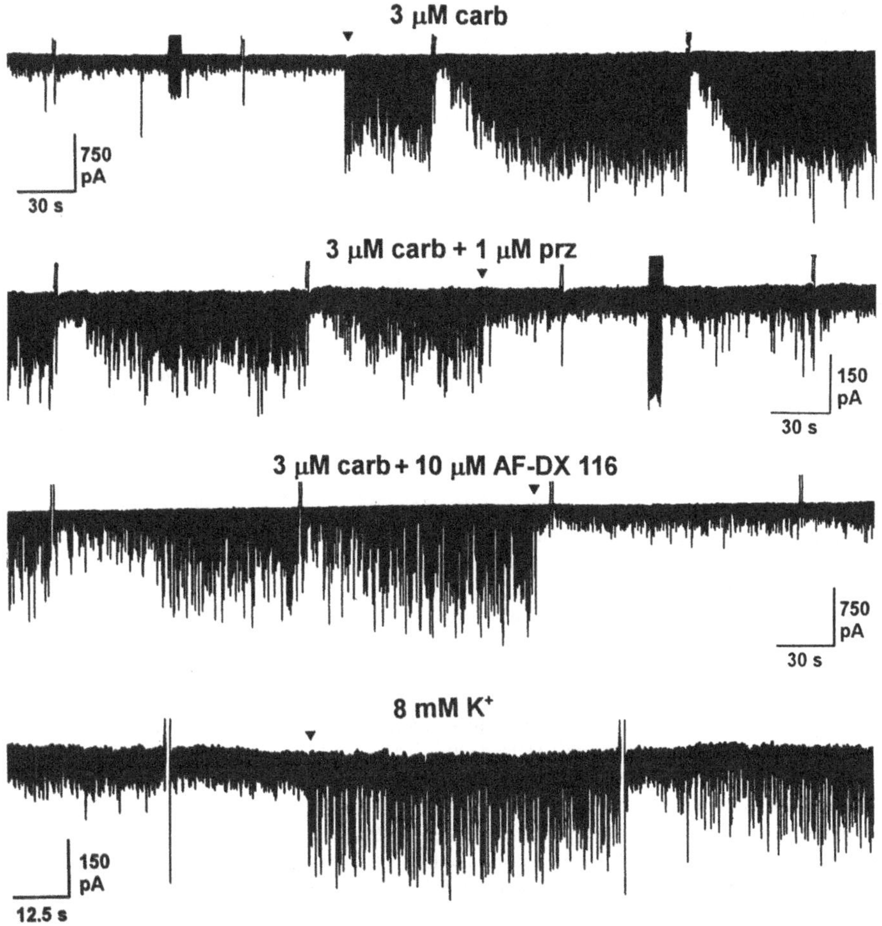

Figure 14. Large, spontaneous IPSCs induced by activation of muscarinic receptors are particularly sensitive to DSI. Voltage steps intended to induce DSI were given at the points indicated by the brief upward deflections along the current traces. The traces are from 4 different cells that displayed a sudden onset or offset of sIPSC activity. In the upper trace, note the abrupt onset of activity in carbachol (arrowhead). Small IPSCs on the baseline showed no sign of DSI, but large sIPSCs were dramatically reduced. This elevated activity could be equally abruptly eliminated during perfusion with muscarinic antagonists, pirenzepine (prz, second trace) and AF-DX 116 (third trace). This same sudden onset of relatively large amplitude IPSCs can also occasionally be observed when 8 mM extracellular K^+ is perfused (fourth trace, arrowhead). There is a strong resemblance between high-amplitude sIPSC activity and the appearance of DSI. Note that the level of sIPSC activity during maximal DSI is approximately the same amplitude as the baseline sIPSC activity (in the absence of DSI).

The reasons for the pronounced susceptibility of muscarinic-induced sIPSCs in DSI are not yet clear. Inasmuch as the disruption of cholinergic function results from chronic alcohol treatment, the hypothesis that DSI could be involved in some of these phenomena will have to be tested using chronic treatment procedures.

In conclusion, DSI represents a new and powerful means of regulating principal cell excitability in the brain. In view of the numerous interactions between alcohol and processes known to be involved in DSI initiation and alcohol, investigation of the influence of alcohol on DSI should prove rewarding.

ACKNOWLEDGMENTS

The work on hippocampal DSI in our laboratory was done by S.A. Kirov, F.E.N. Le Beau, R.A. Lenz, L.A. Martin, W. Morishita, T.A.Pitler, N. Varma and J.J. Wagner. We thank Evelyn Elizabeth for expert editorial and word-processing assistance. R.A. Lenz and L.A. Martin were supported in part by Training Grants NS07375 and T32-GM08181. The work is supported by PHS grants NS30219 and NS22010 to B.E.A.

REFERENCES

Alger BE, Pitler TA, Wagner JJ, Martin LA, Morishita W, Kirov SA, Lenz RA (1996) Retrograde signaling in depolarization-induced suppression of inhibition in rat hippocampal CA1 cells. J Physiol (Lond) 496:197–209.

Alger BE, Pitler TA (1995) Retrograde signaling at $GABA_A$-receptor synapses in the mammalian CNS. Trends Neurosci 18:333–340.

Anderson WW, Lewis DV, Swartzwelder HS, Wilson WA (1986) Magnesium-free medium activates seizure-like events in the rat hippocampal slice. Brain Res 398:215–219.

Arendt T, Allen Y, Marchbanks RM, Schugens MM, Sinden J, Lantos PL, Gray JA (1989) Cholinergic system and memory in the rat: effects of chronic ethanol, embryonic basal forebrain brain transplants and excitotoxic lesions of cholinergic basal forebrain projection system. Neuroscience 33:435–462.

Barnes-Davies M, Forsythe ID (1995) Pre- and postsynaptic glutamate receptors at a giant excitatory synapse in rat auditory brainstem slices. J Physiol (Lond) 488:387–406.

Bear MF, Abraham WC (1996) Long-term depression in hippocampus. Ann Rev Neurosci 19:437–462.

Blanton MG, Lo Turco JJ, Kriegstein AR (1989) Whole cell recording from neurons in slices of reptilian and mammalian cerebral cortex. J Neurosci Meth 30:203–210.

Bowery NG (1993) $GABA_B$ receptor pharmacology. Ann Rev Pharmacol Toxicol 33:109–147.

Brabet I, Mary S, Bockaert J, Pin J-P (1995) Phenylglycine derivatives discriminate between mGluR1- and mGluR5-mediated responses. Neuropharmacology 34:895–903.

Buhl EH, Halasy K, Somogyi P (1994) Diverse sources of hippocampal unitary inhibitory postsynaptic potentials and the number of synaptic release sites. Nature 368:823–828.

Cermak LS (1993) Memory deficits in alcoholic Korsakoff patients. In: Hunt WA, Nixon SJ, (eds), pp 157–171. U.S. Department of Health and Human Services.

Chen QX, Wong RKS (1995) Suppression of $GABA_A$ receptor responses by NMDA application in hippocampal neurones acutely isolated from the adult guinea-pig. J Physiol (Lond) 482:353–362.

Chen WR, Midtgaard J, Shepherd GM (1997) Forward and backward propagation of dendritic impulses and their synaptic control in mitral cells. Science 278:463–467.

Clarke VRJ, Ballyk BA, Hoo KH, Mandelzys A, Pellizzari A, Bath CP, Thomas J, Sharpe EF, Davies CH, Ornstein PL, Schoepp DD, Kamboj RK, Collingridge GL, Lodge D, Bleakman D (1997) A hippocampal GluR5 kainate receptor regulating inhibitory synaptic transmission. Nature 389:599–603.

Connors BW, Prince DA (1982) Effects of local anesthetic QX-314 on the membrane properties of hippocampal pyramidal neurons. J Pharmacol Exp Ther 220:476–481.

Davies CH, Davies SN, Collingridge GL (1990) Paired-pulse depression of monosynaptic GABA-mediated inhibitory postsynaptic responses in rat hippocampus. J Physiol (Lond) 424:513–531.

del Castillo J, Katz B (1954) Quantal components of the end-plate potential. J Physiol (Lond) 124:560–573.

Desai MA, Conn PJ (1991) Excitatory effects of ACPD receptor activation in the hippocampus are mediated by direct effects on pyramidal cells and blockade of synaptic inhibition. J Neurophysiol 66:40–52.

Dunlap K, Luebke JI, Turner TJ, Wheeler DB, Tsien RW, Randall A (1994) Identification of calcium channels that control neurosecretion. Science 266:828–831.

Edwards FA, Konnerth A, Sakmann B (1990) Quantal analysis of inhibitory synaptic transmission in the dentate gyrus of rat hippocampal slices: a patch-clamp study. J Physiol (Lond) 430:213–249.

Faber DS, Korn H (1991) Applicability of the coefficient of variation method for analyzing synaptic plasticity. Biophys J 60:1288–1294.

Fisher RS, Alger BE (1984) Electrophysiological mechanisms of kainic acid-induced epileptiform activity in the rat hippocampal slice. J Neurosci 4:1312–1323.

Garaschuk O, Yaari Y, Konnerth A (1997) Release and sequestration of calcium by ryanodine-sensitive stores in rat hippocampal neurones. J Physiol (Lond) 502:13–30.

Gereau RW, IV, Conn PJ (1995) Multiple presynaptic metabotropic glutamate receptors modulate excitatory and inhibitory synaptic transmission in hippocampal area CA1. J Neurosci 15:6879–6889.

Glitsch M, Llano I, Marty A (1996) Glutamate as a candidate retrograde messenger at interneurone-Purkinje cell synapses of rat cerebellum. J Physiol (Lond) 497:531–537.

Goda Y, Stevens CF (1994) Two components of transmitter release at a central synapse. Proc Natl Acad Sci USA 91:12942–12946.

Hayashi Y, Momiyama A, Takahashi T, Ohishi H, Ogawa-Meguro R, Shigemoto R, Mizuno N, Nakanishi S (1993) Role of a metabotropic glutamate receptor in synaptic modulation in the accessory olfactory bulb. Nature 366:687–690.

Hille B (1992) Ionic Channels of Excitable Membranes, Second Edition, Sunderland, MA: Sinauer Associates, Inc.

Hoffman DA, Magee JC, Colbert CM, Johnston D (1997) K^+ channel regulation of signal propagation in dendrites of hippocampal pyramidal neurons. Nature 387:869–875.

Huerta PT, Lisman JE (1995) Bidirectional synaptic plasticity induced by a single burst during cholinergic theta oscillation in CA1 *in vitro*. Neuron 15:1053–1063.

Inoue M, Oomura Y, Yakushiji T, Akaike N (1986) Intracellular calcium ions decrease the affinity of the GABA receptor. Nature 324:156–158.

Jessell TM, Kandel ER (1993) Synaptic transmission: a bidirectional and self-modifiable form of cell-cell communication. Neuron 10(Suppl.):1–30.

Johnston D, Magee JC, Colbert CM, Christie BR (1996) Active properties of neuronal dendrites. Ann Rev Neurosci 19:165–186.

Kandel ER, Spencer WA (1961) Electrophysiology of hippocampal neurons II. After-potentials and repetitive firing. J Neurophysiol 24:243–259.

Lambert NA, Wilson WA (1994) Temporally distinct mechanisms of use-dependent depression at inhibitory synapses in the rat hippocampus *in vitro*. J Neurophysiol 72:121–130.

Le Beau FEN, Alger BE (1998) Transient suppression of $GABA_A$-receptor-mediated IPSPs after epileptiform burst discharges in CA1 pyramidal cells. J Neurophysiol 79:659–669.

Lenz RA, Wagner JJ, Alger BE (1998) N- and L-type calcium channel involvement in depolarization-induced suppression of inhibition in rat hippocampal CA1 cells. J Physiol (in press).

Lipscombe D, Madison DV, Poenie M, Reuter H, Tsien RW, Tsien RY (1988) Imaging of cytosolic Ca^{++} transients arising from Ca^{2+} stores and Ca^{2+} channels in sympathetic neurons. Neuron 1:355–365.

Littman L, Robinson MB (1994) The effects of L-glutamate and trans-(+)-1-amino-1,3-cycloentanedicarboxylate on phosphoinositide hydrolysis can be pharmacologically differentiated. J. Neurochem 63:1291–1302.

Llano I, Leresche N, Marty A (1991) Calcium entry increases the sensitivity of cerebellar Purkinje cells to applied GABA and decreases inhibitory synaptic currents. Neuron 6:565–574.

Llano I, DiPolo R, Marty A (1994) Calcium-induced calcium release in cerebellar Purkinje cells. Neuron 12:663–673.

Llano I, Marty A (1995) Presynaptic metabotropic glutamatergic regulation of inhibitory synapses in rat cerebellar slices. J Physiol (Lond) 486:163–176.

Macdonald RL, Olsen RW (1994) $GABA_A$ receptor channels. Ann Rev Neurosci 17:569–602.

Magee JC, Johnston D (1997) A synaptically controlled, associative signal for Hebbian plasticity in hippocampal neurons. Science 275:209–213.

Markram H, Lubke J, Frotscher M, Sakmann B (1997) Regulation of synaptic efficacy by coincidence of postsynaptic APs and EPSPs. Science 275:213–215.

Markram H, Tsodyks M (1996) Redistribution of synaptic efficacy between neocortical pyramidal neurons. Nature 382:807–810.

Martin AR (1977) Junctional transmission II. Presynaptic mechanisms. In: Handbook of Physiology. Section 1: The Nervous System, Volume I, Part 1 Kandel ER, ed), pp 329–355. Bethesda, MD: American Physiological Society.

Martin LA, Pitler TA, Lenz RA, Wagner JJ, Alger BE (1995) Muscarinic enhancement of depolarization-induced suppression of inhibition in rat hippocampal pyramidal cells. Soc Neurosci Abstr 21:1095

Martin LA, Alger BE (1996) Muscarinic receptor activation of interneurons susceptible to depolarization-induced suppression of inhibition (DSI). Soc Neurosci Abstr 22:1989.

McBurney RN, Neering IR (1987) Neuronal calcium homeostasis. Trends Neurosci 10:164–169.

Meldrum BS (1975) Epilepsy and gamma-aminobutyric acid-mediated inhibition. Int Rev Neurobiol 17:1–35.

Messing RO, Sneade AB, Savidge B (1990) Protein kinase C participates in up-regulation of dihydropyridine-sensitive calcium channels by ethanol. J Neurochem 55:1383–1389.

Miledi R (1966) Strontium as a substitute for calcium in the process of transmitter release at the neuromuscular junction. Nature 5067:1233–1234.

Miles R, Toth K, Gulyas AI, Hajos N, Freund TF (1996) Differences between somatic and dendritic inhibition in the hippocampus. Neuron 16:815–823.

Mody I, Lambert JDC, Heinemann U (1987) Low extracellular magnesium induces epileptiform activity and spreading depression in rat hippocampal slices. J Neurophysiol 57:869–888.

Morishita W, Alger BE (1997) Sr^{2+} supports depolarization-induced suppression of inhibition and provides new evidence for a presynaptic expression mechanism in rat hippocampal slices. J Physiol (Lond) 505:307–317.

Morishita W, Kirov SA, Alger BE (1998) Evidence for metabotropic glutamate receptor activation in induction of depolarization-induced suppression of inhibition in hippocampal CA1. J Neurosci 18:4870–4882

Morishita W, Kirov SA, Pitler TA, Martin LA, Lenz RA, Alger BE (1997) N-Ethylmaleimide blocks depolarization-induced suppression of inhibition and enhances GABA release in the rat hippocampal slice *in vitro*. J Neurosci 17:941–950.

Nathan T, Jensen MS, Lambert JDC (1990) The slow inhibitory postsynaptic potential in rat hippocampal CA1 neurones is blocked by intracellular injection of QX-314. Neurosci Lett 110:309–313.

Nicoll RA, Alger BE (1981) A simple chamber for recording from submerged brain slices. J Neurosci Meth 4:153–156.

Ohno-Shosaku T, Sawada S, Yamamoto C (1998) Properties of depolarization-induced suppression of inhibitory transmission in cultured rat hippocampal neurons. Pflugers Arch 435:273–279.

Perkins KL, Wong RKS (1995) Intracellular QX-314 blocks the hyperpolarization-activated inward current Iq in hippocampal CA1 pyramidal cells. J Neurophysiol 73:911–915.

Pin J-P, Duvoisin R (1995) The metabotropic glutamate receptors: structure and function. Neuropharmacology 34:1–26.

Pitler TA, Alger BE (1992a) Postsynaptic spike firing reduces synaptic $GABA_A$ responses in hippocampal pyramidal cells. J Neurosci 12:4122–4132.

Pitler TA, Alger BE (1992b) Cholinergic excitation of GABAergic interneurons in the rat hippocampal slice. J Physiol (Lond) 450:127–142.

Pitler TA, Alger BE (1994) Depolarization-induced suppression of GABAergic inhibition in rat hippocampal pyramidal cells: G protein involvement in a presynaptic mechanism. Neuron 13:1447–1455.

Poncer J-C, Shinozaki H, Miles R (1995) Dual modulation of synaptic inhibition by distinct metabotropic glutamate receptors in the rat hippocampus. J Physiol (Lond) 485:121–134.

Poncer J-C, McKinney RA, Gahwiler BH, Thompson SM (1997) Either N- or P-type calcium channels mediate GABA release at distinct hippocampal inhibitory synapses. Neuron 18:463–472.

Potier B, Dutar P, Lamour Y (1993) Different effects of omega-conotoxin GVIA at excitatory and inhibitory synapses in rat CA1 hippocampal neurons. Brain Res 616:236–241.

Rodriguez-Moreno A, Herreras O, Lerma J (1997) Kainate receptors presynaptically downregulate GABAergic inhibition in the rat hippocampus. Neuron 19:893–901.

Shapiro MS, Wollmuth LP, Hille B (1994) Modulation of Ca^{2+} channels by PTX-sensitive G-proteins is blocked by N-ethylmaleimide in rat sympathetic neurons. J Neurosci 14:7109–7116.

Shigemoto R, Kinoshita A, Wada E, Nomura S, Ohishi H, Takada M, Flor PJ, Neki A, Abe T, Nakanishi S, Mizuno N (1997) Differential presynaptic localization of metabotropic glutamate receptor subtypes in the rat hippocampus. J Neurosci 17:7503–7522.

Staley KJ, Soldo BL, Proctor WR (1995) Ionic mechanisms of neuronal excitation by inhibitory $GABA_A$ receptors. Science 269:977–981.

Stelzer A, Kay AR, Wong RKS (1988) $GABA_A$-receptor function in hippocampal cells is maintained by phosphorylation factors. Science 241:339–341.

Stelzer A (1992) $GABA_A$ receptors control the excitability of neuronal populations. Int Rev Neurobiol 33:195–287.

Taira T, Lamsa K, Kaila K (1997) Posttetanic excitation mediated by $GABA_A$ receptors in rat CA1 pyramidal neurons. J Neurophysiol 77:2213–2218.

Tancredi V, Hwa GGC, Zona C, Brancati A, Avoli M (1990) Low magnesium epileptogenesis in the rat hippocampal slice: electrophysiological and pharmacological features. Brain Res 511:280–290.

Thompson SM (1994) Modulation of inhibitory synaptic transmission in the hippocampus. Prog Neurobiol 42:575–609.

Thompson SM, Fortunato C, McKinney RA, Muller M, Gahwiler BH (1996) Mechanisms underlying the neuropathological consequences of epileptic activity in the rat hippocampus *in vitro*. J Comp Neurol 372:515–528.

Tsubokawa H, Ross WN (1996) IPSPs modulate spike backpropagation and associated $[Ca^{2+}]_i$ changes in the dendrites of hippocampal CA1 pyramidal neurons. J. Neurophysiol. 76:2896–2906.

Tsubokawa H, Ross WN (1997) Muscarinic modulation of spike backpropagation in the apical dendrites of hippocampal CA1 pyramidal neurons. J Neurosci 17:5782–5791.

Vincent P, Armstrong CM, Marty A (1992) Inhibitory synaptic currents in rat cerebellar Purkinje cells: modulation by postsynaptic depolarization. J Physiol (Lond) 456:453–471.

Vincent P, Marty A (1993) Neighboring cerebellar Purkinje cells communicate via retrograde inhibition of common presynaptic interneurons. Neuron 11:885–893.

Wagner JJ, Alger BE (1995) GABAergic and developmental influences on homosynaptic LTD and depotentiation in rat hippocampus. J Neurosci 15:1577–1586.

Wagner JJ, Alger BE (1996) Increased neuronal excitability during depolarization-induced suppression of inhibition in rat hippocampus. J Physiol (Lond) 495:107–112.

Walker DW, King MA, Hunter BE (1993) Alterations in the structure of the hippocampus after long-term ethanol administration. In: Alcohol-induced Brain Damage, Hunt WA, Nixon SJ, eds), pp 231–247. Rockville, MD: U.S. Department of Health and Human Services.

Wang X, Wang G, Lemos JR, Treistman SN (1994) Ethanol directly modulates gating of a dihydropyridine-sensitive Ca^{2+} channel in neurohypophysial terminals. J Neurosci 14:5453–5460.

Westenbroek RE, Ahlijanian MK, Catterall WA (1990) Clustering of L-type Ca2+ channels at the base of major dendrites in hippocampal pyramidal neurons. Nature 347:281–284.

Wigstrom H, Gustafsson B (1983) Facilitated induction of hippocampal long-lasting potentiation during blockade of inhibition. Nature 301:603–604.

Wong RKS, Prince DA (1979) Dendritic mechanisms underlying penicillin-induced epileptiform activity. Science 204:1228–1231.

8

NATIVE $GABA_A$ RECEPTORS GET "DRUNK" BUT NOT THEIR RECOMBINANT COUNTERPARTS

Hermes H. Yeh and Douglas W. Sapp

Departments of Pharmacology and Neurology
Program in Neuroscience
The University of Connecticut Health Center, MC-6125
263 Farmington Avenue, Farmington, Connecticut

1. INTRODUCTION

Ethanol is arguably the most widely used (and abused) among psychoactive substances. The behavioral effects of ethanol consumption are well-acknowledged, as are its psychosocial consequences. Yet, relatively little is known about its mechanism of action in the central nervous system (CNS). At the systems level, it clearly targets a multitude of brain regions. At the cellular level, the prevailing thought is that ethanol exerts relatively specific modulatory effects on a number of neurotransmitter systems (*e.g.*, γ-aminobutyric acid (GABA)ergic, glutamatergic, cholinergic, serotonergic), their corresponding receptors and/or intracellular second messenger intermediaries (for reviews see Deitrich et al., 1989; Grant and Lovinger, 1995; Morrow, 1995). These various effects of ethanol have modified the more traditional notion of a pleiotropic and non-specific action of ethanol on cellular membranes. With specific regard to the $GABA_A$ receptor, acute exposure to ethanol has been shown to exert potentiating effects. This has been implicated to account for sedation/motor incoordination at low ethanol concentrations and anesthetic consequences at higher concentrations.

The diversity of the $GABA_A$ receptor subunits identified to date presents a most formidable problem being faced in elucidating the cellular and molecular bases of ethanol-GABA interactions. Not only does this diversity reflect a complex molecular construct of the $GABA_A$ receptor, the regionally-specific distribution of the subunits in the brain indicates further that functionally-distinct $GABA_A$ receptor isoforms exist (for reviews see De Blas, 1996; Macdonald and Olsen, 1994; Yeh and Grigorenko, 1995). Taking the view that subunit composition is key in determining receptor properties, we and others have postulated that sensitivity to modulation by ethanol can be expected to vary with different $GABA_A$ receptor isoforms and that this may underlie the differential modulatory effects of ethanol not only among brain regions but even among neurons within a given brain region.

The "Drunken" Synapse, edited by Liu and Hunt.
Kluwer Academic / Plenum Publishers, New York, 1999.

A promising and logical experimental strategy towards resolving the issue of whether sensitivity to ethanol may depend on subunit composition has been to examine recombinant $GABA_A$ receptors of defined heterologous subunit combinations in expression systems. However, while numerous studies have reported subunit-specific dependence of the ethanol effect, others have not. Can information gleaned by studying recombinant $GABA_A$ receptors be applied towards understanding the interaction between ethanol and native $GABA_A$ receptors? In this chapter, we focus on a review of data generated in our laboratory relevant to the state of this question and place them into perspective with work reported in the literature. A comprehensive review of the literature on ethanol-$GABA_A$ receptor interactions is beyond the scope of this treatise.

2. THE $GABA_A$ RECEPTOR

2.1. General Considerations

GABA is the major neurotransmitter in the CNS mediating fast, moment-to-moment inhibitory synaptic transmission, which is accomplished via activation of the $GABA_A$ receptor-chloride ionophore complex. As other members of the superfamily of ligand-gated ion channels, the $GABA_A$ receptor is modeled as a heteromeric protein assembly, except that it forms a central chloride-selective pore. As its molecular construct unveils, its complexity is being appreciated. The $GABA_A$ receptor was initially thought to be a dimer, consisting of a 48–53 KDa α subunit and a 55–57 KDa β subunit (Sigel and Barnard, 1984). To date, at least 18 $GABA_A$ receptor subunits have been identified and classified into six families based on the degree of sequence homology, namely, α(1–6), β(1–4), γ(1–3), δ, ε, and ρ(1–3). These subunits delimit the putative pentameric chloride ionophore in as yet loosely-defined stoichiometric configurations.

The heterogeneity and the differential pattern of distribution of the $GABA_A$ receptor subunits are neither trivial nor arbitrary, as compelling evidence to date point to a rather intricate relationship between $GABA_A$ receptor composition, assembly and function. The heterogeneity is reflected in the diversity in regional, cellular and subcellular localization, as well as in differential sensitivity to functional regulation by endogenous and exogenous modulators of the $GABA_A$ receptor. Indeed, the relationship between $GABA_A$ receptor subunit composition and function transcends the elemental direct action of GABA on the chloride ionophore itself to include the efficacy of allosteric modulators. While an ethanol-selective recognition site on the $GABA_A$ receptor complex has not been firmly established, it has been shown to modulate $GABA_A$ receptor-mediated function and receptor properties in a variety of CNS neurons and preparations (Aguayo, 1990; Allan and Harris, 1986; Celentano et al., 1988; Freund and Palmer, 1997; Mehta and Ticku, 1988; Mihic et al., 1997; Nishio and Narahashi, 1990; Reynolds and Prasad, 1991; Sapp and Yeh, 1998; Suzdak et al., 1986; Wan et al., 1996; Yeh and Kolb, 1997).

2.2. Correlating Native $GABA_A$ Receptor Function and Subunit Expression

While the combination of subunits in engineered $GABA_A$ receptors reconstituted in expression systems is known and can be systematically changed by the experimenter, this is not the case with native $GABA_A$ receptors expressed by neurons *in situ*. Indeed, multiple mRNAs encoding $GABA_A$ receptor subunits can be found in virtually every region of the brain, even in regions with relatively simple and organized cytoarchitecture, such as the cerebellum (Figure 1) (Laurie et al., 1992; Wisden et al., 1992). This consideration, coupled with the cell-type specific expression of GABA$_A$ receptor subunits, calls for devising effective strategies to correlate sensitivity to ethanol with subunit expression in individual cells. This has been aided by the introduction of techniques that combine patch

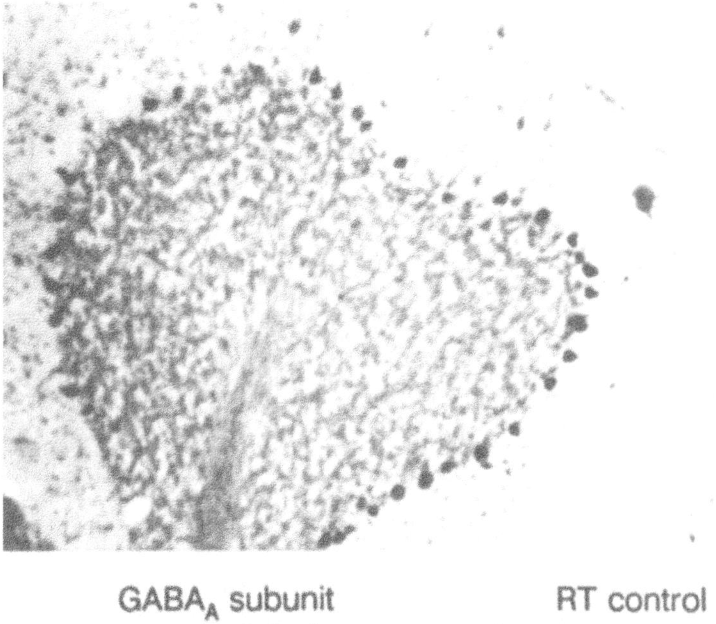

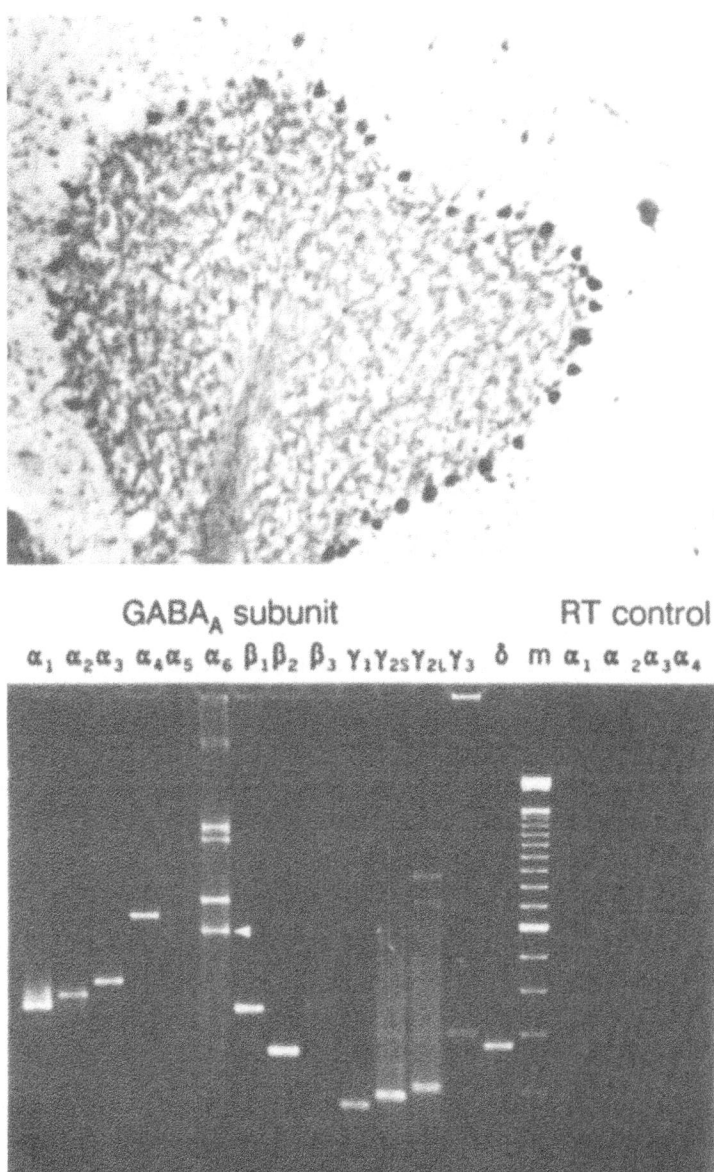

Figure 1. Expression profiling of GABA$_A$ receptor subunit mRNAs in adult rat cerebellum by RT-PCR. Top panel: A 20-μm cryostat section of the rat cerebellum showing the well-delineated laminar cytoarchitecture of the cerebellar cortex and calbindin-immunopositive Purkinje cells. The section was briefly counterstained with toluidine blue. Lower panel: mRNAs were detected encoding the α1–4, α6, β1–3, γ1–3 and δ subunits. The arrowhead denotes PCR-amplified product of size corresponding to the α6 subunit.

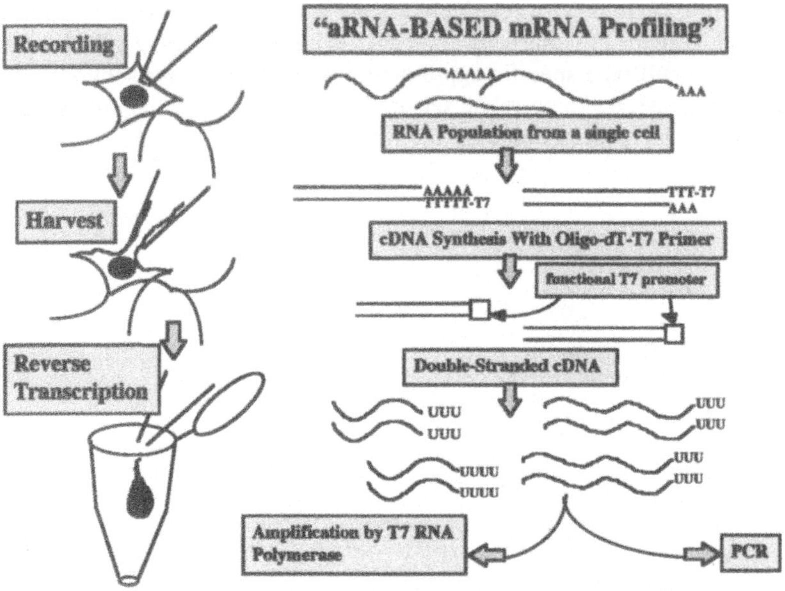

Figure 2. Schematic illustration of combined patch clamp recording and mRNA expression profiling in an individual neuron. Following electrophysiological recording, the neuron examined is harvested, transferred to an Eppendorf tube (left panel). The cellular RNA is reverse transcribed in preparation for aRNA- or PCR-based amplification and subsequent expression profiling (right panel). For details see Eberwine et al. (1992).

clamp electrophysiology with aRNA- or RT/PCR-based amplification, allowing the assessment of receptor function and subunit gene expression in the same cell (Figure 2) (Eberwine et al., 1992; Lambolez et al., 1992; Monyer and Jonas, 1995). Recent demonstrations taking this general approach (Jonas et al., 1994; Monyer et al., 1992; Sapp and Yeh, 1998; Surmeier et al., 1992; Yeh et al., 1996) underscore the need for such analysis at the single cell level. Profiles of subunit mRNA expression could be revealed that would not have been readily uncovered or deduced based on analysis of gene expression in brain tissue or cultured preparations containing heterogeneous populations of neurons and glia (see section 3.2.3).

3. ACUTE EFFECTS OF ETHANOL ON GABA$_A$ RECEPTOR FUNCTION

3.1. Recombinant GABA$_A$ Receptors

Recombinant receptors expressed in a variety of expression systems have been used as models to define the subunit composition of GABA$_A$ receptors that are sensitive to modulation by acute exposure to ethanol. Differences in sensitivity to ethanol have been shown to depend on the α subunit isomer, as recombinant GABA$_A$ receptors expressing either the $\alpha 1$ or $\alpha 6$ subunit differed in their rates of desensitization in response to coapplication of GABA and alcohol (Marszalec et al., 1994). Zolpidem, based on its demonstrated high-affinity binding to benzodiazepine type I receptors, also appears to target selectively GABA$_A$ receptors that are sensitive to potentiation by ethanol (Criswell et al., 1993).

Perhaps the most compelling evidence for subunit specificity in the action of ethanol is that the specific inclusion of the γ2L subunit in the $GABA_A$ receptor assembly is required for an ethanol-induced enhancement of the GABA response (Wafford et al., 1991; Wafford and Whiting, 1992). Hybridization of total brain RNA with antisense oligonucleotide to the γ2L subunit yielded $GABA_A$ receptors that were unresponsive to ethanol. Expression in the *Xenopus* oocyte of the α1/β1/γ2L subunit combination, but not that of the α1/β1/γ2S, resulted in enhancement of GABA-activated current by ethanol. Moreover, ethanol enhanced GABA-stimulated chloride flux in a human embryonic kidney cell line (PA-3) stably transfected with the α1/β1/γ2L combination but not in another (X-25) which expressed only the α1/β1 recombinant $GABA_A$ receptor isoform (Harris et al., 1997; Harris et al., 1995). Qualitatively similar results were obtained in electrophysiological experiments, insofar as GABA-activated current responses were enhanced, albeit less dramatically than in the chloride flux assays, in PA-3 cells but not in X-25 cells. The γ2L subunit has been reported to confer a protein kinase C-dependent component of the modulatory action of ethanol (Whiting et al., 1990).

Other studies have reported that the same subunit combinations (α1/β1/γ2L, α1/β1/γ2S) reconstituted $GABA_A$ receptors that were insensitive to modulation by ethanol and, thus, did not appear to adhere to the stringent requirement for the γ2L subunit (Sapp and Yeh, 1998; Sigel et al., 1993; White et al., 1990). In the study by Sigel et al. (1993), $GABA_A$ receptors of various combinations, including the α1/β1/γ2L and α1/β1/γ2S variant, were expressed in *Xenopus* oocyte. Ethanol (20 - 100 mM) exerted no modulatory effect regardless of whether the recombinant $GABA_A$ receptor contained the γ2L or the γ2S subunit.

The results of our studies using stably transfected cell lines corroborated the finding that recombinant $GABA_A$ receptors are insensitive to modulation by ethanol (Sapp and Yeh, 1998) (Figure 3). GABA responses elicited in the PA-3, X-25 and WSS-1 cells were unaltered by exposure to physiologically-relevant concentrations of ethanol (10–100 mM). The PA-3 and WSS-1 cell lines expressed recombinant α1/β1/γ2L and α1/β2/γ2S $GABA_A$ receptors, respectively, and the presence of these subunits was confirmed by RT-PCR profiling of the individual electrophysiologically studied cells (Sapp and Yeh, 1998). Several immortalized cell lines expressing functional $GABA_A$ receptors were also examined and

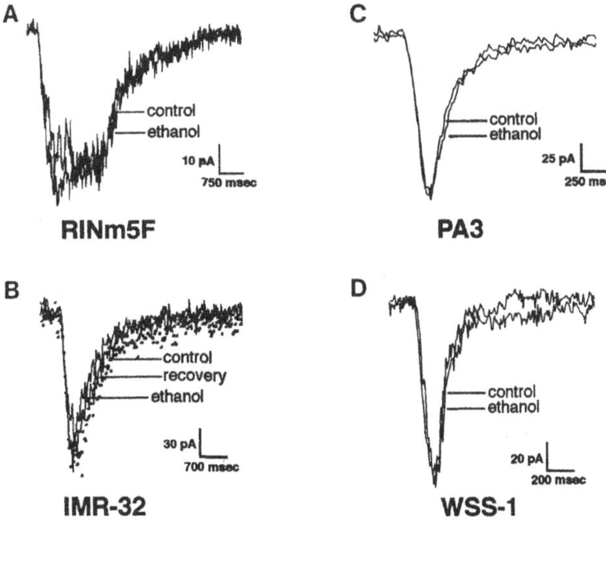

Figure 3. GABA responses elicited in RINm5F (A), IMR-32 (B), PA-3 (C), and WSS-1 (D) cells are insensitive to modulation by ethanol.

found to be insensitive to modulation by ethanol. These cell lines included the IMR-32 (a human neuroblastoma line), RINm5F (a mouse pancreatic tumor line) and more recently, P19 (a mouse teratocarcinoma; Wang and Yeh, unpublished observation). Of these, a subpopulation of the IMR-32 and P19 cells expressed the γ2L $GABA_A$ receptor subunit mRNA and, as expected, the GABA responses were potentiated by diazepam.

3.2. Native $GABA_A$ Receptors

Our primary motivation in undertaking studies involving cell lines expressing functional $GABA_A$ receptors was the hope that such studies would serve as reference for ongoing investigations of ethanol effects on the functional properties of native $GABA_A$ receptors. To this end, electrophysiological recording conditions and protocols identical to those used to examine recombinant $GABA_A$ receptors were employed to assess ethanol-GABA interaction in acutely-dissociated neurons.

3.2.1. Retinal Bipolar Cells. In rod bipolar cells obtained from the adult rat retina following acute dissociation, ethanol (10–50 mM) consistently enhanced GABA responses (Yeh and Kolb, 1997; see also Figure 4A). The GABA response elicited in rod bipolar cells reflects a net activation of two components, namely, a $GABA_A$ receptor-mediated component as well as a bicuculline-resistant, diazepam-insensitive component resembling the $GABA_C$ receptor subtype (Feigenspan et al., 1993; Frumkes and Nelson, 1995;

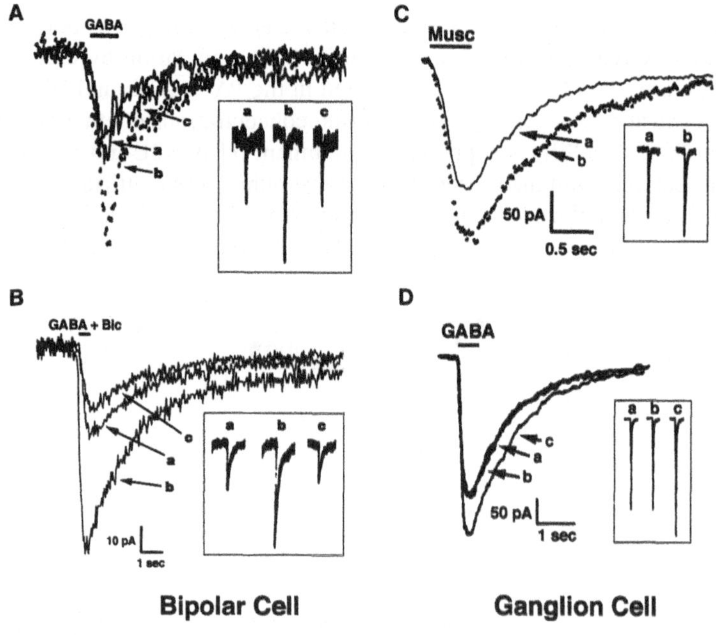

Figure 4. Examples of ethanol-GABA interactions assessed in acutely-dissociated retinal bipolar cells (A, B) and ganglion cells (C, D). Digitized computer-generated traces of responses to GABA (A, B) monitored before (a), during (b), and after (c) acute exposure to ethanol were averaged and superimposed to facilitate comparison. Inset in A and B illustrate representative penwriter records taken during epochs before (a), during (b), and after (c) ethanol exposure. In C and D, the $GABA_A$ receptor agonist muscimol was used in lieu of GABA. The control GABA response (a) was unaltered upon exposure to ethanol (b) but was potentiated during concomitant application of diazepam (c).

Frumkes et al., 1995; Lukasiewicz and Werblin, 1994; Yeh et al., 1996). The latter component is most likely due to the expression of ρ subunit-containing $GABA_A$ receptors in retinal rod bipolar cells. Indeed, under pharmacological conditions that completely blocked activation of the $GABA_A$ receptor, ethanol potentiated the $GABA_C$ receptor-mediated component (Figure 4B). Thus, the acutely-dissociated retinal bipolar cells offered an advantageous model in which GABA receptor isoforms could be distinguished and their interaction with ethanol examined.

3.2.2. Retinal Ganglion Cells. The same study carried out using acutely-dissociated retinal ganglion cells indicated the existence of at least two subpopulations; one with GABA responses that was positively modulated by acute exposure to ethanol (Figure 4C), and the other with GABA responses that was insensitive to modulation by ethanol but sensitive to potentiation by diazepam (Figure 4D). This disparity in the outcome of an ethanol-GABA interaction was perhaps to be expected, given the morphological and functional heterogeneity inherent in retinal ganglion cells. Indeed, it is in line with the body of literature reporting enhanced GABA responses to acute ethanol exposure in some CNS neurons but not others (Leidenheimer and Harris, 1992). Overall, these findings suggest that $GABA_A$ receptor sensitivity to modulation by ethanol may be critically dependent on the brain region examined, on the type of neuron investigated in a given region, as well as on the profile of $GABA_A$ receptor subunits in the neurons studied.

3.2.3. Cerebellar Purkinje Cells. In the adult state, Purkinje cells possess a relatively simple profile of α1/β2/3/γ2L/S $GABA_A$ receptor subunit mRNAs (Laurie et al., 1992), resembling those used to express recombinant receptors in the PA-3 and WSS-1 cell lines. During the neonatal period, the expression of γ2L and γ2S splice variants is developmentally regulated (Figure 5A), whereas mRNA encoding the γ2S subunit is detectable at birth and remains at a relatively steady level of expression, the γ2L message increases progressively over the first two postnatal weeks. Throughout this period, ethanol potentiated GABA responses to Purkinje cells acutely isolated from neonatal rat cerebellum (Figure 5B). Single-cell mRNA profiling of the electrophysiologically-recorded cells indicated prominent expression of the γ2S but not the γ2L message prior to postnatal day-7 (Figure 5C). Thus, the Purkinje cell presents itself as a unique opportunity to examine in native $GABA_A$ receptors the relationship between the γ2 subunit splice variants and sensitivity to ethanol.

Overall, since ethanol-mediated potentiation of GABA could be observed in the absence of the γ2L message, we are left to conclude that the long splice variant is not an absolute requirement for conferring ethanol sensitivity to native $GABA_A$ receptors. An important caveat here is that the γ2L subunit mRNA may have been present in individual neonatal Purkinje cells at levels below detection by expression profiling yet have the capacity to translate subunit protein for incorporation into a functional receptor. In this light, a challenge for future development will be the ability to incorporate detection of subunit proteins into the combined electrophysiological and molecular analysis of interaction between ethanol and native $GABA_A$ receptors.

4. CONCLUSIONS

Can information gleaned by studying recombinant $GABA_A$ receptors be applied towards understanding the interaction between ethanol and native $GABA_A$ receptors?

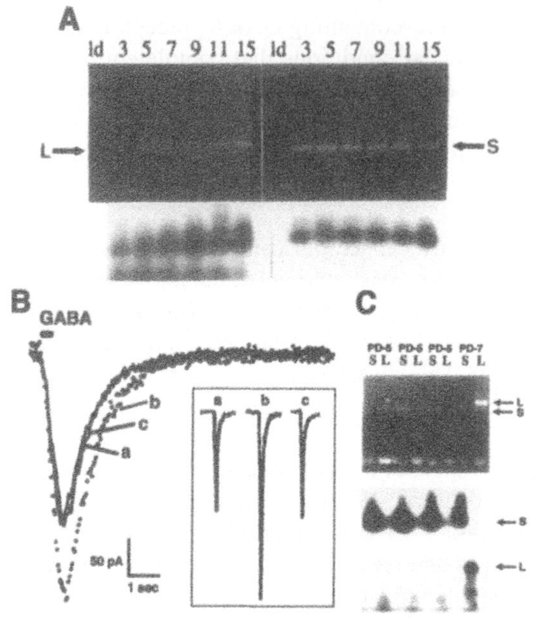

Figure 5. Correlation between ethanol-GABA interaction and γ2L/S subunit mRNA expression in developing cerebellum. (A): Agarose gel (top) and Southern blot (bottom) showing the profile of γ2L and γ2S subunit mRNA expression over the first two postnatal weeks of cerebellar development. (B): Digitized, averaged traces and representative penwriter traces (inset) taken before (a), during (b), and after (c) exposure to ethanol illustrate an ethanol-induced potentiation of the GABA-activated current response in Purkinje cells acutely dissociated from the postnatal rat cerebellum. (C): Agarose gel and corresponding Southern blot from 4 electrophysiologically-recorded Purkinje cells whose GABA responses were potentiated by ethanol. Three postnatal day-5 (PD-5) and 1 PD-7 Purkinje cells are shown. The γ2L subunit mRNA is not detected until PD-7 while the γ2S subunit mRNA is present throughout this postnatal period. In A and C, arrows denote PCR-amplified products corresponding to sizes predicted for γ2l (L) and γ2S (S).

Clearly, in our studies comparing recombinant and native receptors, the hoped for, clear-cut answers related to the relationship between $GABA_A$ receptor subunit expression and sensitivity to modulation by ethanol did not emerge. Nonetheless, even discrepant data are instructive reminders not only that neurons and cells in expression systems are different but also that differences in subcellular regulatory processes may participate in shaping the outcome of an ethanol-$GABA_A$ receptor interaction. In the final analysis, we are continually guided by insights gained from recombinant receptors and challenged by the task of fitting them into the complexities inherent *in situ*.

ACKNOWLEDGMENT

The work described in this chapter was supported by PHS grants AA09861 and AA03510. DWS was supported in part by ARC T32 AA07209.

REFERENCES

Aguayo LG (1990) Ethanol potentiates the $GABA_A$-activated Cl^- current in mouse hippocampal and cortical neurons. Eur J Pharmacol 187:127–130.

Allan AM, Harris RA (1986) Gamma-aminobutyric acid and alcohol actions: neurochemical studies of long sleep and short sleep mice. Life Sci 39:2005–2015.

Celentano JJ, Gibbs TT, Farb DH (1988) Ethanol potentiates GABA- and glycine-induced chloride currents in chick spinal cord neurons. Brain Res 455:377–380.

Criswell HE, Simson PE, Duncan GE, McCown TJ, Herbert JS, Morrow AL, Breese GR (1993) Molecular basis for regionally specific action of ethanol on gamma-aminobutyric acidA receptors: generalization to other ligand-gated ion channels. J Pharmacol Exp Ther 267:522–537.

De Blas AL (1996) Brain $GABA_A$ receptors studied with subunit-specific antibodies. Mol Neurobiol 12:55–71.

Deitrich RA, Dunwiddie TV, Harris RA, Erwin VG (1989) Mechanism of action of ethanol: initial central nervous system actions. Pharmacol Rev 41:489–537.

Eberwine J, Yeh H, Miyashiro K, Cao Y, Nair S, Finnell R, Zettel M, Coleman P (1992) Analysis of gene expression in single live neurons. Proc Natl Acad Sci (USA) 89:3010–3014.

Feigenspan A, Wassle H, Bormann J (1993) Pharmacology of GABA receptor Cl^- channels in rat retinal bipolar cells. Nature 361:159–162.

Freund RK, Palmer MR (1997) Beta adrenergic sensitization of gamma-aminobutyric acid receptors to ethanol involves a cyclic AMP/protein kinase A second-messenger mechanism. J Pharmacol Exp Ther 280:1192–1200.

Frumkes TE, Nelson R (1995) Functional role of GABA in cat retina: I. Effects of $GABA_A$ agonists. Vis Neurosci 12:641–650.

Frumkes TE, Nelson R, Pflug R (1995) Functional role of GABA in cat retina: II. Effects of $GABA_A$ antagonists. Vis Neurosci 12:651–661.

Grant KA, Lovinger DM (1995) Cellular and behavioral neurobiology of alcohol: receptor-mediated neuronal processes. Clin Neurosci 3:155–164.

Harris RA, Mihic SJ, Brozowski S, Hadingham K, Whiting PJ (1997) Ethanol, flunitrazepam, and pentobarbital modulation of $GABA_A$ receptors expressed in mammalian cells and Xenopus oocytes. Alcohol Clin Exp Res 21:444–451.

Harris RA, Proctor WR, McQuilkin SJ, Klein RL, Mascia MP, Whatley V, Whiting PJ, Dunwiddie TV (1995) Ethanol increases $GABA_A$ responses in cells stably transfected with receptor subunits. Alcohol Clin Exp Res 19:226–232.

Jonas P, Racca C, Sakmann B, Seeburg PH, Monyer H (1994) Differences in Ca2+ permeability of AMPA-type glutamate receptor channels in neocortical neurons caused by differential GluR-B subunit expression. Neuron 12:1281–1289.

Lambolez B, Audinat E, Bochet P, Crepel F, Rossier J (1992) AMPA receptor subunits expressed by single Purkinje cells. Neuron 9:247–258.

Laurie DJ, Seeburg PH, Wisden W (1992) The distribution of 13 $GABA_A$ receptor subunit mRNAs in the rat brain. II. Olfactory bulb and cerebellum. J Neurosci 12:1063–1076.

Leidenheimer NJ, Harris RA (1992) Acute effects of ethanol on $GABA_A$ receptor function: molecular and physiological determinants. Adv Biochem Psychopharmacol 47:269–279.

Lukasiewicz PD, Werblin FS (1994) A novel GABA receptor modulates synaptic transmission from bipolar to ganglion and amacrine cells in the tiger salamander retina. J Neurosci 14:1213–1223.

Macdonald RL, Olsen RW (1994) $GABA_A$ receptor channels. Ann Rev Neurosci 17:569–602.

Marszalec W, Kurata Y, Hamilton BJ, Carter DB, Narahashi T (1994) Selective effects of alcohols on gamma-aminobutyric acid A receptor subunits expressed in human embryonic kidney cells. J Pharmacol Exp Ther 269:157–163.

Mehta AK, Ticku MK (1988) Ethanol potentiation of GABAergic transmission in cultured spinal cord neurons involves -aminobutyric $acid_A$-gated chloride channels. J Pharmacol Exp Ther 246:558–564.

Mihic SJ, Ye Q, Wick MJ, Koltchine VV, Krasowski MD, Finn SE, Mascia MP, Valenzuela CF, Hanson KK, Greenblatt EP, Harris RA, Harrison NL (1997) Sites of alcohol and volatile anaesthetic action on GABA(A) and glycine receptors [see comments]. Nature 389:385–389.

Monyer H, Jonas P. (1995) Polymerase chain reaction analysis of ion channel expression in single neurons of brain slices. In: Single-Channel Recording (Sakmann B, Neher E, eds), pp. 357–374. New York, Plenum Press.

Monyer H, Sprengel R, Schoepfer R, Herb A, Higuchi M, Lomeli H, Burnashev N, Sakmann B, Seeburg PH (1992) Heteromeric NMDA receptors: molecular and functional distinction of subtypes. Science 256:1217–1221.

Morrow AL (1995) Regulation of $GABA_A$ receptor function and gene expression in the central nervous system. Int Rev Neurobiol 38:1–41.

Nishio M, Narahashi T (1990) Ethanol enhancement of GABA-activated chloride current in rat dorsal root ganglion neurons. Brain Res 518:283–286.

Reynolds JN, Prasad A (1991) Ethanol enhances $GABA_A$ receptor-activated chloride currents in chick cerebral cortical neurons. Brain Res 564:138–142.

Sapp DW, Yeh HH (1998) Ethanol-GABA(A) Receptor Interactions - a Comparison Between Cell Lines and Cerebellar Purkinje Cells. J Pharmacol Exp Ther 284:768–776.

Sigel E, Barnard EA (1984) A γ-aminobutyric acid/benzodiazepine receptor complex from bovine cerebral cortex. J Biol Chem 259:7219–7223.

Sigel E, Baur R, Malherbe P (1993) Recombinant $GABA_A$ receptor function and ethanol. FEBS Lett 324:140–142.

Surmeier D J, Eberwine J, Wilson C J, Cao Y, Stefani A, Kitai ST (1992) Dopamine receptor subtypes colocalize in rat striatonigral neurons. Proc Natl Acad Sci (USA) 89:10178–10182.

Suzdak PD, Schwartz RD, Skolnick P, Paul SM (1986) Ethanol stimulates gamma-aminobutyric acid receptor-mediated chloride transport in rat brain synaptoneurosomes. Proc Natl Acad Sci (USA) 83:4071–4075.

Wafford KA, Burnett DM, Leidenheimer NJ, Burt DR, Wang JB, Kofuji P, Dunwiddie TV, Harris RA, Sikela JM (1991) Ethanol sensitivity of the $GABA_A$ receptor expressed in Xenopus oocytes requires 8 amino acids contained in the gamma 2L subunit. Neuron 7, 27–33.

Wafford KA, Whiting PJ (1992) Ethanol potentiation of $GABA_A$ receptors requires phosphorylation of the alternatively spliced variant of the gamma 2 subunit. FEBS Lett 313, 113–117.

Wan FJ, Berton F, Madamba SG, Francesconi W, Siggins GR (1996) Low ethanol concentrations enhance GABAergic inhibitory postsynaptic potentials in hippocampal pyramidal neurons only after block of $GABA_B$ receptors. Proc Natl Acad Sci USA 93, 5049–5054.

White G, Lovinger DM, Weight FF (1990) Ethanol inhibits NMDA-activated current but does not alter GABA-activated current in an isolated adult mammalian neuron. Brain Res. 507, 332–336.

Whiting P, McKernan RM, Iversen LL (1990) Another mechanism for creating diversity in gamma-aminobutyrate type A receptors: RNA splicing directs expression of two forms of gamma2 subunit, one of which contains a protein kinase C phosphorylation site. Proceedings of the National Academy of Sciences USA 87, 9966–9970.

Wisden W, Laurie DJ, Monyer H, Seeburg PH (1992) The distribution of 13 $GABA_A$ receptor subunit mRNAs in the rat brain. I. Telencephalon, diencephalon, mesencephalon. J Neurosci 12, 1040–1062.

Yeh HH, Grigorenko EV (1995) Deciphering the native $GABA_A$ receptor: is there hope? J Neurosci Res 41, 567–571.

Yeh HH, Grigorenko EV, Veruki ML (1996) Correlation between a bicuculline-resistant response to GABA and $GABA_A$ receptor rho 1 subunit expression in single rat retinal bipolar cells. Vis Neurosci 13, 283–292.

Yeh HH, Kolb JE (1997) Ethanol modulation of GABA-activated current responses in acutely dissociated retinal bipolar cells and ganglion cells. Alcohol Clin Exp Res 21, 647–655.

9

ADENOSINE AND ETHANOL

Is There a Caffeine Connection in the Actions of Ethanol?

Thomas V. Dunwiddie

Department of Pharmacology and Program in Neuroscience
University of Colorado Health Sciences Center
4200 East 9th Avenue
Denver, Colorado 80262
Veterans Administration Medical Research Service
Denver VA Medical Center
1055 Clermont Street
Denver, Colorado 80220

1. INTRODUCTION

The study of the actions of ethanol on the brain has generated a wide range of hypotheses concerning the mechanisms by which it alters neuronal activity. These include relatively non-specific actions (e.g., increases in membrane fluidity), that might ultimately alter the function of proteins embedded in the lipid bilayer. However more recent studies have focused on specific interactions between ethanol with membrane proteins, such as neurotransmitter receptors and voltage-gated ion channels. It has been hypothesized that the interaction is between ethanol and a hydrophobic pocket in the protein molecule (Figure 1). The enhancement or antagonism of the function of such membrane proteins has been posited to underlie the alterations in neural activity that ultimately result in the intoxicating effects of ethanol.

A third target of the effects of ethanol that has been proposed is the adenosine receptor in brain. However, rather than specifically interacting with these receptors, ethanol has been hypothesized to act in ways that increase the extracellular concentrations of adenosine, and that the subsequent increased activation of adenosine receptors provides a mechanistic basis for ethanol action. It is unlikely that such a mechanism could possibly explain *all* the actions of ethanol, because although adenosine receptor antagonists, such as caffeine, antagonize some of the behavioral effects of ethanol (Stephenson, 1977), many ethanol effects are unaffected by these antagonists. Nevertheless, a variety of evidence makes this an attractive hypothesis concerning ethanol action.

The "Drunken" Synapse, edited by Liu and Hunt.
Kluwer Academic / Plenum Publishers, New York, 1999.

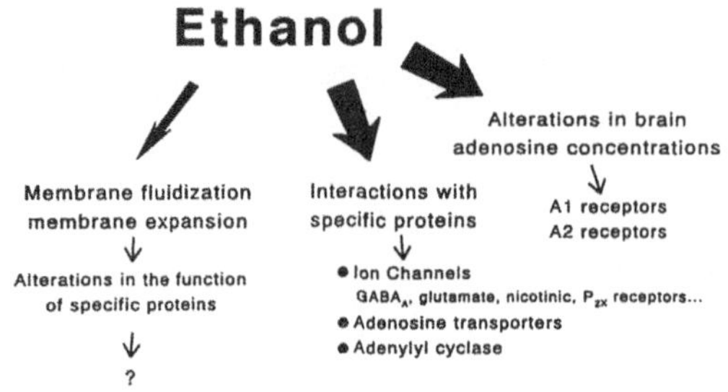

Figure 1. Proposed mechanisms of ethanol action. Although early theories of ethanol action suggested effects based upon the disordering of lipid membranes, most recent work favors more specific interactions with proteins, particularly neurotransmitter receptors. In addition, adenosine can be formed and released as a result of the metabolism of acetate derived from ethanol, resulting in increases in extracellular adenosine.

Adenosine and ethanol elicit many similar types of physiological responses, such as hypothermia, sedation, anticonvulsant effects, and vasodilation. There is evidence that some of the actions of ethanol (e.g., vasodilation in the portal circulation; Carmichael et al., 1988) can be completely antagonized by adenosine receptor antagonists. Thus, the key issue to be resolved is the extent to which "purinergic" mechanisms involving adenosine receptors underlie the actions of ethanol. The evidence both for and against such mechanisms is the subject of this review.

2. ADENOSINE RECEPTORS AND THEIR FUNCTION IN BRAIN

Adenosine is a generally inhibitory modulator of the activity of the central nervous system that exerts its actions via a family of related, GTP-binding protein (G protein)-coupled adenosine receptors (Fredholm et al., 1994; Olah and Stiles, 1995). Adenosine A_1 receptors are linked to the inhibition of adenylyl cyclase, but also to a hyperpolarizing effect on neurons (an action mediated via a pertussis toxin-sensitive G protein and the subsequent activation of an inwardly rectifying K^+ channel), and an inhibition of neurotransmitter release, which is thought to reflect the inhibitory modulation of calcium (Ca^{++}) channels via a pertussis toxin-insensitive G protein (Linden, 1991; Brundege and Dunwiddie, 1997). The inhibition of excitatory glutamatergic transmission by adenosine is particularly striking, and in some instances excitatory postsynaptic responses can be inhibited by >95% via this mechanism (Dunwiddie and Hoffer, 1980). The physiological consequences of activation of A_{2A} and A_{2B} receptors, which both activate adenylyl cyclase, and of A_3 receptors, which can activate phospholipase C, are less well understood, but in many cases do not appear to be as profound as the effects induced by A_1 receptor activation (Mogul et al., 1993; Cunha et al., 1994; Cunha et al., 1995; Fleming and Mogul, 1996; Kessey and Mogul, 1997; Dunwiddie et al., 1997a). In any case, the cellular consequences of A_1 receptor activation (neuronal hyperpolarization and inhibition of excitatory transmission) result in a profound suppression of the electrical activity of the brain (Dunwiddie, 1985).

Behavioral studies that have characterized the effects of adenosine receptor activation are consistent with this basic idea, in that agonists induce a high degree of sedation (Dunwiddie and Worth, 1982; Katims et al., 1983; Nikodijevic et al., 1991). Although this effect initially was ascribed to activation of A_1 receptors, more recent data favors a primary role for A_{2A} receptors, which are found primarily in the caudate nucleus and olfactory tubercle (Ongini and Fredholm, 1996; Wolfgang and Münkle, 1997). Of greater relevance to human pharmacology is the fact that the almost ubiquitous drug caffeine is a relatively non-selective antagonist of adenosine receptors. In the concentrations that are typically achieved in the blood of coffee and tea drinkers, antagonism of adenosine receptors is likely to be its primary pharmacological mechanism of action. Animal studies have demonstrated that caffeine, as well as more potent and selective antagonists of adenosine receptors, generally elicit increases in spontaneous motor activity that correspond well with the affinity of these agents for adenosine receptors (Snyder et al., 1981). However, other effects (release of intracellular Ca^{++}, or phosphodiesterase inhibition) are not observed at behaviorally active concentrations.

What remains something of an enigma is the origin of the adenosine that normally interacts with adenosine receptors. Despite the profound and widespread actions of adenosine, most current evidence does not support the idea that adenosine is a neurotransmitter *per se*, because in general it is not released in a Ca^{++}-dependent fashion from brain tissue upon stimulation (see Dunwiddie, 1985 for review). There is also no compelling evidence for the existence of "purinergic" neurons that release adenosine as their major transmitter, either in brain, or in the peripheral nervous system. However, there are several fairly well characterized mechanisms by which adenosine is either released from tissue, or is formed in the extracellular space.

Adenosine transporters facilitate the passive equilibration of adenosine across cellular membranes (Geiger and Fyda, 1991). In general, intracellular metabolism (phosphorylation by adenosine kinase, deamination by adenosine deaminase, or incorporation into S-adenosyl-homocysteine) keeps intracellular adenosine concentrations low, and the net flux of adenosine is into the cell. Under conditions of metabolic stress (ischemia, seizures, etc.), increased adenosine that is formed by the breakdown of ATP may be released via these transporters into the extracellular space.

Another primary mechanism that affects adenosine concentrations in the extracellular space is the formation of adenosine from adenine nucleotides via the actions of ectoenzymes. Adenosine triphosphate (ATP) is co-released into the extracellular space along with conventional transmitters such as NE and ACh (Silinsky and Hubbard, 1973; Fredholm et al., 1982; Richardson and Brown, 1987). Adenine nucleotides can also be released as a result of N-methyl D-aspartate (NMDA) receptor activation (Hoehn and White, 1990; Craig and White, 1993), and via the release of cyclic adenosine monophosphate (cAMP) via a probenecid-sensitive transporter (Rosenberg and Dichter, 1989; Rosenberg and Li, 1995). Virtually all adenine nucleotides are rapidly converted extracellularly to adenosine (with the exception of cAMP, this conversion probably takes place in less than 1 second; Dunwiddie et al., 1997b), so this provides a mechanism by which quite rapid changes in extracellular adenosine can take place.

The relative importance of each of these mechanisms is still not clear, and may differ significantly in different brain regions. Nevertheless, it seems likely that extracellular concentrations generally "track" intracellular concentrations, although localized processes may alter these levels significantly. Regardless of the source, it is clear that adenosine is found in the extracellular space of normal brain in sufficient concentrations to exert a tonic inhibitory effect on neural activity. The increases in behavioral and electrophysiological activity that are observed upon the administration of adenosine antagonists, such as caffeine or theophylline (Dunwiddie et al., 1981; Snyder et al., 1981; Motley and Collins, 1983; Haas and Greene, 1988), are thought to reflect the antagonism of the tonic inhibitory effects of endogenous adenosine.

3. ADENOSINE-ETHANOL INTERACTIONS

As discussed in the Introduction, it has been proposed that some of the effects of ethanol are mediated via adenosine receptors. However, unlike the situation with gamma-aminobutyric acid ($GABA_A$) or NMDA receptors, where ethanol has been suggested to interact directly with the receptors, most of the evidence with respect to adenosine suggests that ethanol may act by altering extracellular adenosine concentrations, either by affecting adenosine formation or adenosine uptake. One such mechanism for increased adenosine formation involves the acetate that is formed as a consequence of ethanol metabolism in the liver. When ethanol is metabolized, blood acetate concentrations rise to the 1–2 mM range, even with relatively low blood concentrations of ethanol (Carmichael et al., 1991). When acetate is incorporated into acetyl-CoA, significant amounts of 5'-adenosine monophosphate (5'-AMP) are formed, and the action of 5'-nucleotidases can lead to the subsequent formation of adenosine.

A second potential site of adenosine-ethanol interaction involves adenosine transporters, and in particular, an adenosine transporter that is inhibited by nitrobenzylthioinosine (NBTI). The other major class of facilitatory transporters, which is relatively insensitive to NBTI, but is potently inhibited by dipyridamole (DIPY), appears to be unaffected by ethanol (Krauss et al., 1993). Although these transporters can mediate transport of adenosine in either direction across the membrane, under most conditions the inhibition of transport leads to increases in the interstitial concentration of adenosine. Furthermore, although conventional inhibitors of transport such as NBTI and DIPY block transport of adenosine in both directions, it has been reported that ethanol only inhibits the transport of adenosine *into* cells (see below). The inhibition would act to further increase extracellular levels of adenosine, because efflux via this transporter is unaffected. Work from a number of laboratories has indicated that even relatively low concentrations of ethanol can significantly inhibit adenosine transport, which in turn could increase extracellular brain concentrations of adenosine.

A third proposed mechanism by which adenosine and ethanol have been proposed to interact is at the adenosine receptor itself, or more likely, via the G-protein coupled effector mechanisms through which adenosine receptors act. There is considerable evidence to suggest that ethanol can facilitate the receptor mediated activation of adenylyl cyclase by various hormones and neurotransmitters (Rabin and Molinoff, 1981; Luthin and Tabakoff, 1984; Hoffman and Tabakoff, 1990; Rabin, 1990). Thus, the actions of adenosine at A_{2A} and A_{2B} receptors, which both activate adenylyl cyclase, might be enhanced by ethanol. Because ethanol enhances basal and receptor stimulated adenylyl cyclase activity, and disrupts the inhibitory regulation of cyclase (Hynie et al., 1980; Rabin and Molinoff, 1981; Luthin and Tabakoff, 1984; Gordon et al., 1986; Bauche et al., 1987; Rabe et al., 1990), this provides yet another point at which ethanol could modulate effects mediated via adenosine receptors.

Thus, ethanol could potentially enhance adenosine actions on the brain via increases in adenosine formation, inhibition of adenosine transport, and by means of specific interactions with GTP-binding proteins and the associated adenosine receptors. Support for these hypothetical mechanisms of action have come from behavioral studies in animals and in man, from direct biochemical measurements of adenosine concentrations in brain following ethanol administration, and from studies at the cellular level that implicate adenosine as a mediator of some of the effects of ethanol. Behavioral studies relevant to these hypotheses are generally supportive of a role for adenosine, although they could not be said to be compelling in this regard. Because much of the behavioral evidence that relates to this putative role for adenosine has been reviewed and critiqued elsewhere (Dunwiddie, 1995), this work will not be discussed here. However, the evidence from biochemical and physiological studies relating to adenosine-ethanol interactions will be discussed in detail in the following sections.

3.1. Effects of Ethanol Mediated via Adenosine Formed during Acetate Metabolism

Following ethanol administration, low millimolar concentrations of acetate are formed from the metabolism of ethanol, and circulate through the blood. The subsequent metabolic conversion of acetate to acetyl CoA, a process which utilizes ATP, results in the formation of 5'-AMP, which can then be converted to adenosine by the action of 5'-nucleotidase. It has been proposed that some portion of the effects of ethanol are mediated via adenosine formed by this mechanism (Orrego et al., 1988; Carmichael et al., 1993). Some ethanol-induced responses, such as the vasodilation in the portal vasculature, appear to be entirely mediated via adenosine (Orrego et al., 1988; Carmichael et al., 1988). If acetate formed from hepatic metabolism enters the brain, a similar process could occur in the brain, which then would result in the local generation of adenosine. This would provide a relatively direct mechanism by which ethanol could elevate the extracellular concentrations of adenosine in brain.

Carmichael et al (1991) demonstrated that both acetate and ethanol potentiate the effects of general anesthetics, and that the adenosine receptor antagonist 8-phenyltheophylline completely reverses the effects of acetate, but only partially antagonizes the effects of ethanol. In a more recent study, Campisi et al (1997) demonstrated that the effects of systemically administered acetate and ethanol were centrally mediated, since they could be antagonized by direct administration of adenosine receptor antagonists into the brain. The effect of acetate was completely blocked, whereas the effects of ethanol were only partially reversed by antagonists. In addition, a highly selective A_1 agonist reduced the anesthetic requirement as well.

Taken together, these results suggest that a part of the centrally mediated effects of ethanol may occur via increases in the extracellular adenosine concentration in brain acting upon A_1 receptors. Because the metabolic pathways for acetate production are saturated by relatively low doses of ethanol (approximately 0.5 g/kg of ethanol in rat), acetate levels never rise above approximately 1 mM regardless of the dose of ethanol (Carmichael et al., 1991). For this reason, it would appear that this mechanism may be of particular importance in mediating low-dose effects of ethanol, but are unlikely to account for those effects that have a threshold dose that is above 0.5 g/kg.

Electrophysiological studies have attempted to further define this putative role for acetate formed from ethanol metabolism as a mediator of the central effects of ethanol. In a study in anesthetized rats (Phillis et al., 1992), intraperitoneal injections of ethanol were shown to inhibit spontaneous firing of cerebral cortical neurons, and ethanol prolonged the depression in firing induced by local iontophoretic application of adenosine. Local application of acetate also inhibited firing, but this response was not antagonized by the adenosine receptor antagonist 8-*p*-sulfophenyltheophylline. These authors concluded that the ability of ethanol to potentiate adenosine responses was most likely the result of an inhibition of adenosine uptake (see below), and that their evidence did not support the hypothesis that acetate could lead to increased adenosine formation in the brain. However, at this point relatively little is known about the dynamics of adenosine formation from acetate, and it is conceivable that local administration of acetate does not permit the generation of sufficient adenosine to affect cellular activity.

Investigations of a putative role for acetate in ethanol action using brain slices have not provided much support for this hypothesis, although there have been some differences in the findings of various groups. In an initial study in the dentate gyrus (Cullen and Carlen, 1992), ethanol, acetate, and adenosine were found to produce a variety of effects that could be antagonized by 8-phenyltheophylline, which included a hyperpolarization of the resting membrane potential and an enhancement of a Ca^{++}-dependent afterhyperpolarization following depolarizing current injection. What was surprising, however, was that acetate did not inhibit the amplitude of evoked excitatory postsynaptic potentials (EPSPs), a response that is probably the best characterized effect of adenosine on hippocampal electrophysiological activity. This might suggest that bath superfusion with acetate leads to the formation of adenosine in a very restricted region surrounding the cell body, a region that does not include the dendrites of these cells where excitatory inputs terminate.

In a more recent study from our own laboratory (Brundege and Dunwiddie, 1995), we confirmed that acetate had no effect on EPSPs either in the dentate gyrus, or in the CA1 region, where the effects of adenosine have also been extensively characterized. Unlike the dentate gyrus, bath superfusion with acetate was found to have no effect on the resting membrane potential, input resistance, number of spikes evoked by a depolarizing current pulse, or the afterhyperpolarization elicited by the current pulse in CA1 pyramidal neurons, although all of these parameters are affected significantly by adenosine. We also compared the physiological properties of CA1 neurons recorded with KCl and K-methylsulfate electrodes to neurons recorded with K-acetate electrodes (which would result in leakage of substantial amounts of acetate into the cell), and did not find differences on any parameter that would be affected by altered concentrations of adenosine. Thus, it is difficult to find strong support in these electrophysiological studies for an effect of acetate mediated via adenosine.

Further evidence that mitigates against such a role comes from a biochemical study of hippocampal brain slices (Fredholm and Wallman-Johansson, 1996), in which it was found that neither ethanol nor acetate altered the basal or evoked release of ATP metabolites from hippocampal brain slices. This study reported a slight increase in the release of adenosine *per se*, but based upon the changes in the pattern of metabolites that were observed, the authors concluded that the effects were most consistent with an action of ethanol and acetate on adenosine *uptake*, rather than release. The concentrations of acetate that were tested in this study (5 and 20 mM) were above the pharmacologically relevant range (0.5–1 mM), so it is difficult to determine the relevance of these results to what would be observed following ethanol administration *in vivo*. However, the general conclusions reached in this study echo those of earlier studies from Phillis' group (Phillis et al., 1980; Phillis et al., 1992), who found no enhancement in the efflux of either adenosine or its primary metabolite inosine from cerebral cortex following administration, i.p., of acetate or ethanol, but did report results that would be consistent with an ethanol inhibition of adenosine transport.

At this point, it is difficult to evaluate the relative role of adenosine receptor-mediated actions in the central effects of acetate derived from the metabolism of ethanol. One possibility is that the adenosine, which is formed via this mechanism, is very localized, so that some adenosine receptors (e.g., somatic receptors) are activated, and others (e.g., presynaptic modulatory receptors) are not. This might explain why little or no adenosine seems to be generated in brain slices from acetate, and why there is no evidence for increased activation of presynaptic adenosine receptors in brain slices superfused with acetate. However, the studies from our laboratory do not support even this very limited view of adenosine action. Some of the differences outlined above in the results obtained from different laboratories may relate to the strains of animals tested, the conditions under which the effects of acetate are examined, the brain region tested, or to other unknown variables.

3.2. Ethanol Effects Mediated via Inhibition of Adenosine Transport

The increased formation of adenosine from ethanol-derived acetate is one mechanism by which brain adenosine concentrations could be elevated. However, an entirely different mechanism has been suggested by biochemical studies that have demonstrated that ethanol can inhibit the uptake of adenosine into cells. Adenosine uptake via non-energy-requiring (also called *facilitatory* or *equilibrative*) transporters is mediated primarily by two major subtypes of transporters, the NBTI-sensitive transporters (now most commonly termed the *es*, or *e*quilibrative *s*ensitive) and the NBTI-insensitive (*ei*, or *e*quilibrative *in*sensitive) transporters (Geiger and Fyda, 1991). The *ei* transporters are also sometimes referred to as DIPY-sensitive transporters, since DIPY is one of the better inhibitors of these transporters. Several of these transporters have been recently cloned (Griffiths et al., 1997; Yao et al., 1997).

Work from several laboratories has indicated that even relatively low concentrations of ethanol (10 mM) can significantly inhibit adenosine transport (Clark and Dar, 1989b; Gordon et al., 1990; Nagy et al., 1990), and this effect appears to be restricted to one of the *es* transporters (Krauss et al., 1993). A novel aspect of this inhibition is that unlike any of the other known inhibitors of adenosine transporter function, ethanol only inhibits adenosine uptake, and does not reduce adenosine efflux (Nagy et al., 1990). Studies conducted primarily on the NG108–15 neuroblastoma-glioma, S49 lymphoma, and related cell lines (Gordon et al., 1990; Krauss et al., 1993) have established that ethanol produces increases in cAMP that are mediated by activation of A_2 adenosine receptors (Gordon et al., 1986), and can be blocked either by treatment with adenosine deaminase (which establishes the involvement of extracellular adenosine), or by adenosine receptor antagonists. Thus, the primary effect of ethanol is to inhibit adenosine transport. That inhibition subsequently leads to the accumulation of extracellular adenosine, activation of A_2 receptors, and increased cAMP levels.

With chronic ethanol treatment, the corresponding persistent increase in cAMP concentrations leads to a heterologous desensitization of adenylyl cyclase-coupled receptor systems (Gordon et al., 1986). The cellular mechanism underlying this effect is a reduction in the amount of functional $G_{\alpha s}$ (Mochly-Rosen et al., 1988). Similar types of effects have been observed in brain or cultured neurons as well (Saito et al., 1987; Rabin, 1990). This heterologous desensitization appears to have several consequences. First, with chronic ethanol treatment, the continued presence of ethanol is required to see normal sensitivity to adenosine (Gordon et al., 1986). In the absence of ethanol, the response to activation of adenosine receptors is approximately 50% of that observed in control cells. Thus, in a sense the cells are "dependent" upon ethanol, because ethanol is required in order to elicit the normal response to adenosine. A second effect occurs as a consequence of the regulation of adenosine transporter activity by protein kinase A. Phosphorylation of the transporter does not appear to directly affect its activity, but has a permissive effect on the ethanol-mediated inhibition of transport, such that in the dephosphorylated state, the transporter is relatively unaffected by ethanol. In cells that have been chronically treated with ethanol, it appears that most of the transporter reverts to a dephosphorylated state, and ethanol is no longer able to inhibit transporter activity (Sapru et al., 1994; Coe et al., 1996a). This may come about at least in part because of a reduction in cAMP-dependent phosphorylation associated with chronic ethanol, but the activation of a phosphatase in a protein kinase C (PKC)-dependent manner also appears to play a role (Coe et al., 1996b). Regardless, dephosphorylation of the transporter essentially produces a cellular analog of ethanol tolerance, *i.e.*, a reduced responsiveness to previously effective concentrations of a drug. Ethanol is no longer able to inhibit adenosine transport, and thus the acute effects of ethanol that are mediated via increases in extracellular adenosine are effectively blocked.

An interesting variant to these responses occurs in cells that express primarily A_1 receptors, which inhibit adenylyl cyclase and reduce cAMP levels. In cultured hepatocytes, for example, chronic exposure to ethanol has an opposite effect to what is seen in cell lines such as the NG-108. There is a decrease in $G_{\alpha i}$ protein (rather than $G_{\alpha s}$), and an *increased* sensitivity in systems that are coupled to increases in cAMP (Nagy and DeSilva, 1994). Nevertheless, these effects are still mediated via increases in extracellular adenosine that come about as a result of inhibition of the adenosine transporter. Thus, although the primary response to ethanol is the same (inhibition of transport, and increased extracellular adenosine), both the acute and chronic effects are completely reversed in cells that have predominantly A_1 rather than A_2 receptors. Given the complexities of A_1 and A_2 receptor localization in brain, this would suggest that the effects of acute and chronic ethanol on cAMP and adenylyl cyclase in brain might be very difficult to predict.

These studies have provided compelling evidence for the inhibition of an adenosine transporter playing a central role in the cellular actions of ethanol. However, this model of ethanol action has been developed primarily based upon studies of cells in culture, where the regulation of extracellular adenosine concentrations might be expected to be very different from intact tissue. Is there evidence that similar types of processes occur in intact systems? Along these lines, some striking differences have been reported in adenosine receptor mediated responses in lymphocytes and red blood cell ghosts obtained from heavy alcohol drinkers. Both basal and adenosine stimulated cAMP formation was found to be markedly reduced in alcoholics when compared with appropriate controls (Diamond et al., 1987; Diamond and Gordon, 1997). The ability of ethanol to enhance stimulated cAMP production was blunted as well, whereas a normal response to ethanol was seen in lymphocytes from abstinent alcoholics (Gordon et al., 1990). In chronic ethanol feeding studies in rats, similar types of changes occur. A reduction in basal levels of adenosine transport in isolated hepatocytes obtained from chronically fed rats has been described, as well as a near complete loss of the ethanol sensitivity of adenosine transport (Wannamaker and Nagy, 1995). There is also an *increase* in $G_{\alpha s}$ (Iles and Nagy, 1995), which is the opposite of the effect seen in NG108 cells following chronic ethanol (Mochly-Rosen et al., 1988). Thus, the common elements that are observed following chronic ethanol are a reduction in adenosine transport and a loss of sensitivity of the adenosine transporter to inhibition by ethanol. However, the effects on transduction mechanisms linked to adenylyl cyclase appear to be dependent upon the system that is examined, and may be determined by the complement of adenosine receptors on the cells that are tested.

Although these effects of ethanol are quite striking, they leave open the issue as to whether similar responses occur in brain, and the extent to which they can account for pharmacological responses to ethanol. Other cell lines that have some neuronal characteristics (e.g., PC12 cells) do not show the same regulation of transporter and cyclase function as do the NG108 cells. Rabin et al (1993) have shown that in PC12 cells, chronic ethanol induces a similar increase in extracellular adenosine, and a similar desensitization of cAMP production, but there is no relationship between the two effects, *i.e.*, desensitization is still observed under conditions where extracellular adenosine does not accumulate. Thus, although the results obtained with lymphocytes, NG108 cells and S49 cells are straightforward, the generality of these effects is not yet clear. In this context, it is perhaps not entirely surprising that the studies cited in the previous section (Phillis et al., 1980; Phillis et al., 1992; Fredholm and Wallman-Johansson, 1996) suggest that ethanol has little if any effect on adenosine concentrations in brain.

To address this issue directly, we have investigated whether superfusion of hippocampal brain slices with ethanol inhibits adenosine transport, and elevates extracellular adenosine concentrations (Diao and Dunwiddie, 1996). Superfusing hippocampal brain slices with DIPY, an inhibitor of the *ei* transporter, markedly inhibited field excitatory postsynaptic potential (fEPSP) responses from hippocampal slices in a theophylline-reversible fashion, which is consistent with accumulation of substantial amounts of extracellular adenosine. On the other hand, superfusion with NBTI, which is selective for the *es* transporter, had no effect (Figure 2A). Superfusion with ethanol also had no statistically significant effect on the fEPSP (Figure 2B). When slices had been pretreated with DIPY, NBTI had a small but statistically significant effect, but even under these conditions (i.e., DIPY pretreatment), superfusion with either 20 mM or 100 mM ethanol did not. These results suggest that a) in hippocampus, the *es* transporter does not play a major role in regulating extracellular adenosine concentrations, and b) that the inhibition of this transporter by ethanol in hippocampus is not robust enough to be detected using this paradigm.

An intriguing interaction was observed between ethanol and theophylline in terms of the fEPSP response. Even though ethanol had no effect on the average amplitude of fEPSP responses in individual slices, fairly significant increases and/or decreases were observed, and adenosine receptor antagonists apparently blocked both types of responses. Thus, there was a highly significant reduction in the *variability* in the response following superfusion with either 20 mM or 100 mM ethanol in slices that had been pretreated with either the non-selective adenosine receptor antagonist theophylline (250 μM), or the selective A_1 antagonist cyclopentyltheophylline (1 μM; Diao and Dunwiddie, 1996). This reduction in variability was significant at both concentrations of ethanol tested, and with both antago-

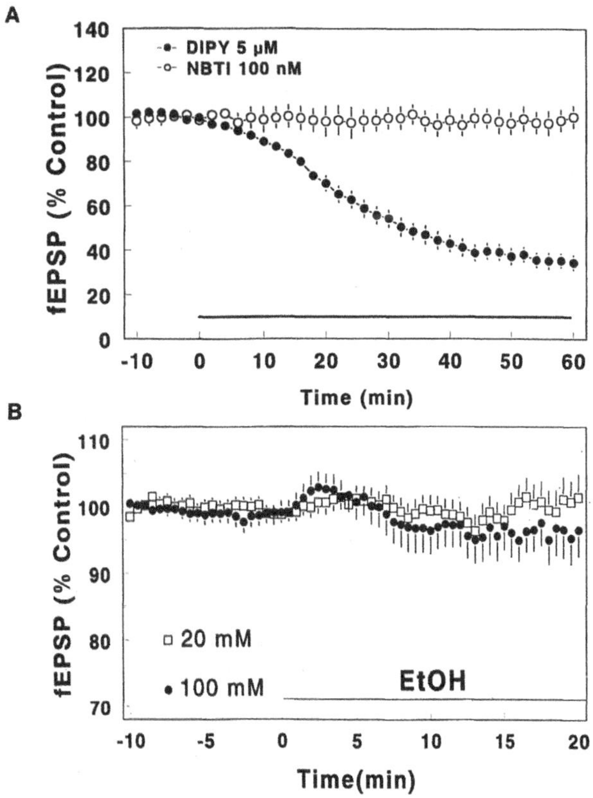

Figure 2. Effects of uptake inhibitors and ethanol on hippocampal fEPSP responses. In A, the average effect on the fEPSP response of superfusion with DIPY (5 μM) and NBTI (0.1 μM) is shown; the duration of drug superfusion is denoted by the horizontal line above the abscissa. The NBTI concentration was chosen to selectively inhibit the *es* (IC_{50} 0.11 nM) adenosine transporter while having minimal effect on other transporters (Geiger et al., 1988). Each point represents the mean ± SEM response from 14 slices treated with an identical protocol. The inhibition of the fEPSP response by DIPY was highly significant ($p<0.01$), but NBTI had no significant effect on the response. In B, the mean ± SEM response to superfusion with ethanol are illustrated for groups of 32 (20 mM) and 26 (100 mM) slices. At no time during the ethanol superfusion were mean responses significantly different from those evoked during the control period ($p<0.05$). (Data from Diao and Dunwiddie, 1996, with permission).

nists. What this suggests is that ethanol induces two different competing responses, both of which involve A_1 receptors. Possibilities in this regard might be an increase in extracellular adenosine and an antagonism of the effector pathways by which A_1 receptors inhibit transmitter release. Alternatively, an enhancement of the interaction of adenosine with A_1 receptors and an uncoupling of A_1 receptors from effector mechanisms would be another possible combination of antagonistic effects. Further studies will be needed to resolve these possible mechanisms. In any case, it is clear that ethanol does not produce a simple increase in extracellular adenosine that is manifested by an increased activation of hippocampal A_1 receptors.

Why is it that extracellular adenosine concentrations in brain appear to be relatively insensitive to ethanol (Phillis et al., 1980; Clark and Dar, 1988b; Phillis et al., 1992), despite the fact that ethanol clearly affects adenosine transport? Probably the most important factor in this respect is that the inhibition of adenosine uptake by ethanol that has been described in brain is not particularly profound. For example, Clark and Dar (1989b) found that the maximal inhibition that could be achieved in cerebellar synaptosomes was about 15%, and that this occurred at an ethanol concentration of about 25 mM, and even 200 mM ethanol inhibits the *es* transporter by only 50% (Nagy et al., 1990). Second, there appear to be fewer *es* transporters, as defined by the binding of NBTI, than *ei* transporters, as defined by DIPY binding, in brain (Bisserbe et al., 1985; Bisserbe et al., 1986). The relatively small magnitude of the ethanol inhibition may explain the negative reports, which suggest that ethanol, does not elevate extracellular brain adenosine levels at all. We have demonstrated using electrophysiological techniques that in hippocampus, where there are relatively few NBTI-sensitive transporters, inhibition of uptake by NBTI does not appear to increase extracellular adenosine (Dunwiddie and Diao, 1994), whereas DIPY produces a marked increase in extracellular adenosine (see also Figure 2A). In the olfactory cortex, where the proportion of NBTI-sensitive sites is higher (Geiger and Nagy, 1984; Bisserbe et al., 1985), NBTI produces electrophysiological responses that are consistent with elevations in extracellular adenosine (Sanderson and Scholfield, 1986). On this basis, we hypothesize that ethanol inhibition of adenosine uptake might be functionally significant only in brain regions that have relatively high levels of *es* transport relative to *ei* transport (olfactory cortex, locus coeruleus, superior colliculus), whereas in brain regions that have low ratios (hippocampus, cerebellum), this particular interaction may be functionally unimportant. These issues remain to be resolved by future studies.

3.3. Indirect Studies of Ethanol-Adenosine Interactions

The studies discussed previously have suggested that adenosine concentrations in brain might be altered both by an increase in the formation of adenosine, as well as by an inhibition of its uptake, although direct evidence for such changes in intact brain is lacking. However, there are also a number of studies that involve indirect methods for investigating adenosine-ethanol interactions that have provided suggestive evidence for increases in adenosine. Clark and Dar (1989c) indirectly characterized changes in adenosine receptor-mediated actions by studying glutamate release from cerebellar synaptosomes. Adenosine inhibits glutamate release in this system via an action at A_1 receptors. Ethanol itself appeared to inhibit release, although this was not statistically significant. The adenosine receptor antagonist theophylline facilitated glutamate release, presumably by antagonizing the effects of endogenous adenosine, and this effect was antagonized by ethanol. The significance of this observation is unclear. If ethanol had either enhanced or decreased adenosine levels by itself, this should have been reflected in changes in glutamate release with ethanol alone, which was not observed. Thus, these results suggest that ethanol is somehow able to interact with an adenosine receptor-mediated response, but presumably not by altering the basal level of adenosine.

In another study from this same group, the efflux of endogenous adenosine from cerebellar synaptosomes was characterized directly (Clark and Dar, 1989a), and ethanol was found to produce a highly significant increase in adenosine release. The adenosine overflow was enhanced by nearly 50% by 25 mM ethanol, and was more than doubled by 100 mM ethanol. As has been reported previously in other systems, the adenosine transport inhibitor dilazep *inhibited* endogenous adenosine release, which is consistent with a bidirectional inhibition of transport, *i.e.*, both efflux as well as influx are inhibited, and the effects of dilazep and ethanol were completely independent. Thus, in accordance with what has been reported by others (Nagy et al., 1990), ethanol does not appear to act in the same fashion as other transport inhibitors, because it produced only increases in release. The mechanism underlying this effect is unclear, but evidently does not involve an interaction with the same site as inhibitors such as dilazep.

3.4. Ethanol Effects on Adenosine Receptors

As discussed in the introduction, there are four known subtypes of adenosine receptors, all of which belong to the G-protein coupled family of membrane receptors, and these receptors themselves might be targets for ethanol action. Although there have been occasional reports of enhancement of adenosine receptor-mediated effects (e.g., Hynie et al., 1980), these effects have generally not been specific for adenosine receptors *per se*, but are shared by other G-protein coupled receptors as well. Although early studies have suggested that there might be differences in adenosine receptors in lines of animals selected for ethanol sensitivity (Fredholm et al., 1985), subsequent studies have failed to find significant differences in either A_1 or A_{2A} receptors (Fredholm, unpublished; Smolen and Smolen, 1993a; Smolen et al., 1993b). Clark and Dar (1988a) reported that acute ethanol administration (1.5 g/kg, *in vivo*) produces a 40% increase in A_1 receptors as determined by subsequent radioligand binding studies *in vitro*. However, it appears that this effect is unlikely to be mediated via an activation of adenosine receptors, because theophylline enhanced this effect of ethanol. A possible alternative is that ethanol alters in some way the interaction between the adenosine receptor and G proteins, and that this indirectly causes changes in post-mortem adenosine receptor binding. However, more direct evidence that this is the case is lacking.

4. CONCLUSIONS

Current research in this field clearly suggests some intriguing interactions between ethanol and purinergic mechanisms in brain, although the nature of these interactions is sometimes obscure. However, based upon this work, a number of conclusions can be made, and hypotheses developed that clearly merit further investigation:

- Adenosine is likely to make a significant contribution to the behavioral effects of ethanol. Although the behavioral work has not been explicitly reviewed here, the observation that adenosine receptor antagonists reverse some (but clearly not all) of the effects of ethanol, suggests that certain behavioral responses to ethanol may have a purinergic component.

- Some ethanol responses may be completely mediated via purinergic mechanisms. Although this has not been well established for biochemical or electrophysiological responses, there are certain physiological effects of ethanol (e.g., vasodilation in the portal circulation) that can be completely antagonized by adenosine receptor antagonists. A more careful consideration of the range of responses to ethanol may point to other instances where this is the case.
- The cellular mechanisms by which ethanol affects adenosine systems are not well understood, but there are multiple candidate mechanisms, and it is quite likely that these may be brain region dependent. The ethanol-mediated inhibition of inwardly directed adenosine flux via the *es* transporter and facilitation of adenosine receptor activation have been clearly established as ethanol actions at the cellular level. What are lacking are functional studies on more intact systems that can establish the relative importance of these effects as cellular mechanisms underlying ethanol actions.

Finally, it is important to recognize that these potential effects of ethanol on adenosine systems may interact in a synergistic fashion. Acetate-mediated increases in adenosine and the ethanol inhibition of adenosine transport may each be marginally effective in altering the level of activation of adenosine receptors in isolation, but might produce substantial alterations in physiology or behavior when combined. In experiments on isolated systems, such as brain slices, which cannot metabolize ethanol to acetate, it may be quite important to characterize the effects of the combination of ethanol + acetate in order to fully understand the effects that may be occurring in an intact system. Thus, a challenge for the future will be to determine if and where these actions are sufficient to produce physiologically significant events.

ACKNOWLEDGMENT

This work was supported by grant AA 03527 from the National Institute on Alcohol Abuse and Alcoholism, and by the Veterans Administration Medical Research Service.

REFERENCES

Bauche F, Bourdeaux-Jaubert AM, Giudicelli Y, Nordmann R (1987) Ethanol alters the adenosine receptor-Ni-mediated adenylate cyclase inhibitory response in rat brain cortex *in vitro*. FEBS Lett 219:296–300.

Bisserbe JC, Deckert J, Marangos P (1986) Autoradiographic localization of adenosine uptake sites in guinea pig brain using [3H]dipyridamole. Neurosci Lett 66:341–345.

Bisserbe JC, Patel J, Marangos PJ (1985) Autoradiographic localization of adenosine uptake sites in rat brain using [^{3}H]nitrobenzylthioinosine. J Neurosci 5:544–550.

Brundege JM, Dunwiddie TV (1995) The role of acetate as a potential mediator of the effects of ethanol in the brain. Neurosci Lett 186:214–218.

Brundege JM, Dunwiddie TV (1997) Role of adenosine as a modulator of synaptic activity in the central nervous system. Adv Pharmacol 39:353–391.

Campisi P, Carmichael FJL, Crawford M, Orrego H, Khanna JM (1997) Role of adenosine in the ethanol-induced potentiation of the effects of general anesthetics in rats. Eur J Pharmacol. 325:165–172.

Carmichael FJ, Israel Y, Crawford M, Minhas K, Saldivia V, Sandrin S, Campisi P, Orrego H (1991) Central nervous system effects of acetate: contribution to the central effects of ethanol. J Pharmacol Exp Ther 259:403–408.

Carmichael FJ, Orrego H, Israel Y (1993) Acetate-induced adenosine mediated effects of ethanol. Alcohol Alcohol (Supplement) 2:411–418.

Carmichael FJ, Saldivia V, Varghese GA, Israel Y, Orrego H (1988) Ethanol-induced increase in portal blood flow: role of acetate and A1- and A2-adenosine receptors. Amer J Physiol 255:G417–23.
Clark M, Dar MS (1988a) Mediation of acute ethanol-induced motor disturbances by cerebellar adenosine in rats. Pharmacol Biochem Behav 30:155–161.
Clark M, Dar MS (1988b) The effects of various methods of sacrifice and of ethanol on adenosine levels in selected areas of rat brain. J Neurosci Meth 25:243–249.
Clark M, Dar MS (1989a) Effect of acute ethanol on release of endogenous adenosine from rat cerebellar synaptosomes. J Neurochem 52:1859–1865.
Clark M, Dar MS (1989b) Effect of acute ethanol on uptake of [^{3}H]adenosine by rat cerebellar synaptosomes. Alcohol Clin Exp Res 13:371–377.
Clark M, Dar MS (1989c) Release of endogenous glutamate from rat cerebellar synaptosomes: interactions with adenosine and ethanol. Life Sci 44:1625–1635.
Coe IR, Dohrman DP, Constantinescu A, Diamond I, Gordon, AS (1996a) Activation of cyclic AMP-dependent protein kinase reverses tolerance of a nucleoside transporter to ethanol. J Pharmacol Exp Ther 276:365–369.
Coe IR, Yao L, Diamond I, Gordon AS (1996b) The role of protein kinase C in cellular tolerance to ethanol. J Biol Chem 271:29468–29472.
Craig CG, White TD (1993) N-methyl-D-aspartate- and non-N-methyl-D-aspartate-evoked adenosine release from rat cortical slices: distinct purinergic sources and mechanisms of release. J Neurochem 60:1073–1080.
Cullen N, Carlen PL (1992) Electrophysiological action of acetate, a metabolite of ethanol, on hippocampal dentate granule neurons: interaction with adenosine. Brain Res 588:49–57.
Cunha RA, Johansson B, Fredholm BB, Ribeiro JA, Sebastiáo AM (1995) Adenosine A_{2A} receptors stimulate acetylcholine release from nerve terminals of the rat hippocampus. Neurosci Lett 196:41–44.
Cunha RA, Johansson B, Van der Ploeg I, Sebastiao AM, Ribeiro JA, Fredholm BB (1994) Evidence for functionally important adenosine A_{2A} receptors in the rat hippocampus. Brain Res 649:208–216.
Diamond I, Gordon AS (1997) Cellular and molecular neuroscience of alcoholism. Physiol Rev 77:1–20.
Diamond I, Wrubel B, Estrin W, Gordon A (1987) Basal and adenosine receptor-stimulated levels of cAMP are reduced in lymphocytes from alcoholic patients. Proc Natl Acad Sci (USA) 84:1413–1416.
Diao LH, Dunwiddie TV (1996) Interactions between ethanol, endogenous adenosine and adenosine uptake in hippocampal brain slices. J Pharmacol Exp Ther 278:542–546.
Dunwiddie TV (1985) The physiological role of adenosine in the central nervous system. Int Rev Neurobiol 27:63–139.
Dunwiddie TV (1995) Acute and chronic effects of ethanol on the brain: interactions of ethanol with adenosine, adenosine transporters, and adenosine receptors. In: Pharmacological Effects of Ethanol on the Nervous System (Deitrich RA, Erwin VG, eds), pp. 147–161. Boca Raton: CRC Press.
Dunwiddie TV, Diao LH (1994) Extracellular adenosine concentrations in hippocampal brain slices and the tonic inhibitory modulation of evoked excitatory responses. J Pharmacol Exp Ther 268:537–545.
Dunwiddie TV, Diao LH, Kim HO, Jiang JL, Jacobson KA (1997a) Activation of hippocampal adenosine A_3 receptors produces a desensitization of A_1 receptor-mediated responses in rat hippocampus. J Neurosci 17:607–614.
Dunwiddie TV, Diao LH, Proctor WR (1997b) Adenine nucleotides undergo rapid, quantitative conversion to adenosine in the extracellular space in rat hippocampus. J Neurosci 17:7673–7682.
Dunwiddie TV, Hoffer BJ (1980) Adenine nucleotides and synaptic transmission in the *in vitro* rat hippocampus. Br J Pharmacol 69:59–68.
Dunwiddie TV, Hoffer BJ, Fredholm BB (1981) Alkylxanthines elevate hippocampal excitability: Evidence for a role of endogenous adenosine. Naunyn Schmiedebergs Arch Pharmacol 316:326–330.
Dunwiddie TV, Worth TS (1982) Sedative and anticonvulsant effects of adenosine analogs in mouse and rat. J Pharmacol Exp Ther 220:70–76.
Fleming KM, Mogul DJ (1996) Adenosine A_3 receptors potentiate hippocampal calcium current by a PKA-dependent/PKC-independent pathway. Drug Dev Res 37:121.
Fredholm BB, Abbracchio MP, Burnstock G, Daly JW, Harden TK, Jacobson KA, Leff P, Williams M (1994) VI. Nomenclature and classification of purinoceptors. Pharm Rev 46:143–156.
Fredholm BB, Fried G, Hedqvist P (1982) Origin of adenosine released from rat vas deferens by nerve stimulation. Eur J Pharmacol 79:233–243.
Fredholm BB, Wallman-Johansson A (1996) Effects of ethanol and acetate on adenosine production in rat hippocampal slices. Pharmacol Toxicol 79:120–123.
Fredholm BB, Zahniser NR, Weiner GR, Proctor WR, Dunwiddie TV (1985) Behavioral sensitivity to PIA in selectively bred mice is related to a number of A_1 adenosine receptors but not to cyclic AMP accumulation in brain slices. Eur J Pharmacol 111:133–136.

Geiger JD, Fyda DM (1991) Adenosine transport in nervous system tissues. In: Adenosine in the Nervous System (Stone TW ed), pp 1–20. London: Academic Press.
Geiger JD, Johnston ME, Yago V (1988) Pharmacological characterization of rapidly accumulated adenosine by dissociated brain cells from adult rat. J Neurochem 51:283–291.
Geiger JD, Nagy JI (1984) Heterogeneous distribution of adenosine transport sites labeled by [^{3}H]Nitrobenzylthioinosine in rat brain: An autoradiographic and membrane binding study. Brain Res Bull 13:657–666.
Gordon AS, Collier K, Diamond I (1986) Ethanol regulation of adenosine receptor-stimulated cAMP levels in a clonal neural cell line: an *in vitro* model of cellular tolerance to ethanol. Proc Natl Acad Sci (USA) 83:2105–2108.
Gordon AS, Nagy L, Mochly-Rosen D, Diamond I (1990) Chronic ethanol-induced heterologous desensitization is mediated by changes in adenosine transport. Biochem Soc Symp 56:117–136.
Griffiths M, Beaumont N, Yao SYM, Sundaram M, Boumah CE, Davies A, Kwong FYP, Coe I, Cass CE, Young JD, Baldwin SA (1997) Cloning of a human nucleoside transporter implicated in the cellular uptake of adenosine and chemotherapeutic drugs. Nature Med 3:89–93.
Haas HL, Greene RW (1988) Endogenous adenosine inhibits hippocampal CA1 neurones: further evidence from extra- and intracellular recording. Naunyn Schmiedebergs Arch Pharmacol 337:561–565.
Hoehn K, White TD (1990) N-methyl-D-aspartate, kainate and quisqualate release endogenous adenosine from rat cortical slices. Neurosci 39;441–450.
Hoffman PL, Tabakoff B (1990) Ethanol and guanine nucleotide binding proteins: A selective interaction. FASEB J. 4:2612–2622.
Hynie S, Lanefelt F, Fredholm BB (1980) Effects of ethanol on human lymphocyte levels of cyclic AMP *in vitro*: potentiation of the response to isoproterenol, prostaglandin E2 or adenosine stimulation. Acta Pharmacol Toxicol 47:58–65.
Iles KE, Nagy LE (1995) Chronic ethanol feeding increases the quantity of $G_{\alpha s}$-protein in rat liver plasma membranes. Hepatology 21:1154–1160.
Katims JJ, Annau Z, Snyder SH (1983) Interactions in the behavioral effects of methylxanthines and adenosine derivatives. J Pharmacol Exp Ther 227:167–173.
Kessey K, Mogul DJ (1997) NMDA-independent LTP by adenosine A_2 receptor-mediated postsynaptic AMPA potentiation in hippocampus. J Neurophysiology 78:1965–1972.
Krauss SW, Ghirnikar RB, Diamond I, Gordon AS (1993) Inhibition of adenosine uptake by ethanol is specific for one class of nucleoside transporters. Mol Pharmacol 44:1021–1026.
Linden J (1991) Structure and function of A1 adenosine receptors. FASEB J. 5:2668–2676.
Luthin GR, Tabakoff B (1984) Activation of adenylate cyclase by alcohol requires the nucleotide-binding protein. J Pharmacol Exp Ther 228:579–587.
Mochly-Rosen D, Chang FH, Cheever L, Kim M, Diamond I, Gordon AS (1988) Chronic ethanol causes heterologous desensitization of receptors by reducing alpha-s messenger RNA. Nature 333:848–850.
Mogul DJ, Adams ME, Fox AP (1993) Differential activation of adenosine receptors decreases N-type but potentiates P-type Ca^{2+} current in hippocampal CA3 neurons. Neuron 10:327–334.
Motley SJ, Collins GGS (1983) Endogenous adenosine inhibits excitatory transmission in the rat olfactory cortex slice. Neuropharmacol 22:1081–1086.
Nagy LE, DeSilva SEF (1994) Adenosine A_1 receptors mediate chronic ethanol-induced increases in receptor-stimulated cyclic AMP in cultured hepatocytes. Biochem J 304:205–210.
Nagy LE, Diamond I, Casso DJ, Franklin C, Gordon AS (1990) Ethanol increases extracellular adenosine by inhibiting adenosine uptake via the nucleoside transporter. J Biol Chem 265:1946–1951.
Nikodijevic O, Sarges R, Daly JW, Jacobson KA (1991) Behavioral effects of A_1- and A_2-selective adenosine agonists and antagonists: Evidence for synergism and antagonism. J Pharmacol Exp Ther 259:286–294.
Olah ME, Stiles GL (1995) Adenosine receptor subtypes: characterization and therapeutic regulation. Ann Rev Pharmacol Toxicol 35:581–606.
Ongini E, Fredholm BB (1996) Pharmacology of adenosine A_{2A} receptors. Trends Pharmacol 17:364–372.
Orrego H, Carmichael FJ, Saldivia V, Giles HG, Sandrin S, Israel Y (1988) Ethanol-induced increase in portal blood flow: role of adenosine. Amer J Physiol 254:G495-G501
Phillis JW, O'Regan MH, Perkins LM (1992) Actions of ethanol and acetate on rat cortical neurons: ethanol/adenosine interactions. Alcohol 9:541–546.
Phillis JW, Ziang ZG, Chelack BJ (1980) Effects of ethanol on acetylcholine and adenosine efflux from *in vivo* rat cerebral cortex. J Pharm Pharmacol 32:871–872.
Rabe CS, Giri PR, Hoffman PL, Tabakoff B (1990) Effect of ethanol on cyclic AMP levels in intact PC12 cells. Biochem Pharmacol 40:565–571.
Rabin RA (1990) Direct effects of chronic ethanol exposure on beta-adrenergic and adenosine-sensitive adenylate cyclase activities and cyclic AMP content in primary cerebellar cultures. J Neurochem 55:122–128.

Rabin RA, Fiorella D, Van Wylen DG (1993) Role of extracellular adenosine in ethanol-induced desensitization of cyclic AMP production. J Neurochem 60:1012–1017.

Rabin RA, Molinoff PB (1981) Activation of adenylate cyclase by ethanol in mouse striatal tissue. J Pharmacol ExpTher 216:129–134.

Richardson PJ, Brown SJ (1987) ATP release from affinity-purified rat cholinergic nerve terminals. J Neurochem 48:622–630.

Rosenberg PA, Dichter MA (1989) Extracellular cAMP accumulation and degradation in rat cerebral cortex in dissociated cell culture. J Neurosci 9:2654–2663.

Rosenberg PA, Li Y (1995) Adenylyl cyclase activation underlies intracellular cyclic AMP accumulation, cyclic AMP transport, and extracellular adenosine accumulation evoked by β-adrenergic receptor stimulation in mixed cultures of neurons and astrocytes derived from rat cerebral cortex. Brain Res 692:227–232.

Saito T, Lee JM, Hoffman PL, Tabakoff B (1987) Effects of chronic ethanol treatment on the beta-adrenergic receptor-coupled adenylate cyclase system of mouse cerebral cortex. J Neurochem 48:1817–1822.

Sanderson G, Scholfield CN (1986) Effects of adenosine uptake blockers and adenosine on evoked potentials of guinea-pig olfactory cortex. Pflügers Arch 406:25–30.

Sapru MK, Diamond I, Gordon AS (1994) Adenosine receptors mediate cellular adaptation to ethanol in NG108–15 cells. J Pharmacol Exp Ther 271:542–548.

Silinsky EM, Hubbard JI (1973) Release of ATP from rat motor nerve terminals. Nature 243:404–405.

Smolen TN, Smolen A (1993a) Down-regulation of adenosine A_2 receptors in long-sleep mice following chronic ethanol administration. Alcohol Clin Exp Res 17:498(Abstract)

Smolen TN, Smolen A, Han PC (1993b) Upregulation of adenosine A_1 receptors following chronic purinergic agonist and antagonist administration in mice. FASEB J 7:A255(Abstract)

Snyder SH, Katims JJ, Annau Z, Bruns RF, Daly JW (1981) Adenosine receptors and behavioral actions of methylxanthines. Proc Natl Acad Sci (USA) 78:3260–3264.

Stephenson PE (1977) Physiologic and psychotropic effects of caffeine on man. A review. J Amer Dietetic Ass 71:240–247.

Wannamaker VL, Nagy LE (1995) Equilibrative adenosine transport in rat hepatocytes after chronic ethanol feeding. Alcohol Clin Exp Res 19:735–740.

Wolfgang H, Münkle M (1997) Motor depressant effects mediated by dopamine D_2 and adenosine A_{2A} receptors in the nucleus accumbens and the caudate-putamen. Eur J Pharmacol 323:127–131.

Yao SY, Ng AM, Muzyka WR, Griffiths M, Cass CE, Baldwin, SA, Young JD (1997) Molecular cloning and functional characterization of nitrobenzylthioinosine (NBMPR)-sensitive (es) and NBMPR-insensitive (ei) equilibrative nucleoside transporter proteins (rENT1 and rENT2) from rat tissues. J Biol Chem 272:28423–28430.

10

A METABOTROPIC HYPOTHESIS FOR ETHANOL SENSITIVITY OF GABAergic AND GLUTAMATERGIC CENTRAL SYNAPSES

George R. Siggins, Zhiguo Nie, and Samuel G. Madamba

The Scripps Research Institute
Department of Neuropharmacology and Alcohol Research Center
10550 North Torrey Pines Road
La Jolla, California 92037

1. INTRODUCTION

This chapter will address the sensitivity of ligand-gated ion channels to ethanol and show that such ethanol sensitivity (or lack of sensitivity) is not invariable for any given neuron. In fact, bringing together several pieces of electrophysiological and pharmacological data leads us to put forward a hypothesis that the ethanol sensitivity of ligand-gated ion channels is regulated by 'metabotropic' systems defined in the generic sense: that is, by receptor-activated, G-protein-linked, non-ionotropic, energetic mechanisms. This hypothesis might provide an explanation for the variability of ethanol-transmitter interactions seen in various laboratories.

For this discussion we will draw on data obtained from two kinds of ligand-gated systems—glutamatergic (primarily N-methyl-D-aspartate (NMDA) receptor-mediated) synaptic transmission and gamma-aminobutyric acid $(GABA)_A$ergic synaptic transmission—studied in two different brain regions: the nucleus accumbens (NAcc) and the hippocampus. These data will be used to posit a 'metabotropic' regulation of ethanol sensitivity that may play a role at both pre- and postsynaptic sites in these models. We chose the NAcc in part because recent studies have suggested that alcoholism, alcohol-seeking behavior, or alcohol dependence might involve the NAcc. As part of the limbic system, the hippocampus may also play a role in these phenomena. The hippocampus also has the advantage of a considerable database of functional data on synaptic mechanisms and the effects of alcohol.

The "Drunken" Synapse, edited by Liu and Hunt.
Kluwer Academic / Plenum Publishers, New York, 1999.

2. METHODS

We use male Sprague Dawley rats to prepare standard hippocampal transverse slices (Siggins et al., 1987) for intracellular current- and voltage-clamp recording from CA1 pyramidal neurons. The NAcc slice was cut as a coronal section on a vibratome, and we recorded primarily from the core region (see Yuan et al., 1992). Both types of slices were completely submerged and continuously superfused with artificial cerebrospinal fluid (ACSF) at a constant rate (2–4 ml/min) for the duration of the experiment. We also used two relatively new methods to look at the details of alcohol effects on the isolated $GABA_A$, $GABA_B$, non-NMDA and NMDA glutamate receptor-mediated synaptic components of the GABA and glutamate systems. These involved local stimulation to obtain monosynaptic activation of the synaptic potentials or synaptic currents, combined with pharmacological isolation of the subtypes of receptors of interest. For example, we used 6-cyano-7-nitroquinoxaline-2,3-dione (CNQX) to block the (R,S)-α-amino-3-hydroxy-5-methylisoxazole-4-propionic acid (AMPA)/kainate glutamate receptors, DL-2-amino-5-phophonovaleric acid (APV) to block NMDA glutamate receptors, bicuculline to block $GABA_A$ receptors, and CGP 35348 and CGP 55845A to block $GABA_B$ receptors. CGP 55845A is about 500 times more potent than CGP 35348 as an antagonist of $GABA_B$ receptors, although both are quite selective. In most cases, application of all the antagonists combined would completely abolish synaptic responses evoked by local stimulation. We also tried to mimic synaptic release of transmitter, by applying from pipettes and by rapid superfusion, the ligands for the receptors under study. These include GABA and baclofen for the $GABA_A$ and $GABA_B$ receptors, respectively, and glutamate, NMDA, AMPA and kainate for the glutamate receptor subtypes. In these cases of local ligand application, tetrodotoxin (TTX; 0.5–1 μM) was added to block Na^+-dependent action potentials and synaptic transmission so that a postsynaptic site of ethanol action could be tested.

The usual drug- or alcohol-testing protocol was: recording of membrane potentials and currents for 10–15 min during superfusion of ACSF alone ("control"), followed by switching to ACSF with drug and repeating these current measures after 3–15 min of drug, then followed by switching again to ACSF alone for 10–35 min with subsequent current measures ("washout").

3. NUCLEUS ACCUMBENS AND NMDA-EPSPs

Our previous studies of ethanol effects on NMDA receptors in the NAcc slice preparation (Nie et al., 1994) used either local application or rapid superfusion of NMDA with CNQX and bicuculline in the bath, to generate a mean dose-response curve of grouped data pooled from multiple NAcc core neurons. These dose-response curves essentially replicated the earlier results originally obtained almost a decade ago by others, but showed that ethanol inhibits NMDA receptors in the NAcc, as in other neuronal preparations. However, the novel aspect of neuronal NMDA receptors is their high sensitivity to ethanol. Thus, we found an apparent IC_{50} of 13 mM ethanol, with virtually every cell showing such an antagonism by ethanol at reasonable (11–66 mM) concentrations. In addition, the extent of that inhibition was large, averaging about a 70% decrease at maximal concentrations of 100 mM ethanol (Nie et al., 1994).

We also found the same kind of ethanol inhibition of NMDA receptor-mediated excitatory postsynaptic potentials (NMDA-EPSPs) evoked by stimulating remotely (Nie et al., 1993a,b; Nie et al., 1994) or locally in NAcc near the recorded neuron (Martin et al.,

1997). Ethanol blunted these NMDA-EPSPs with subsequent reversal of the effect on washout. To obtain clear NMDA-EPSPs, we depolarized the NAcc neurons slightly to about –60 mV from their normally large resting membrane potentials (about –82 mV).

One of our more intriguing but perplexing findings arises from recent *in vivo* research. Our results (see below and Wan et al., 1996) showed a $GABA_B$ receptor influence on ethanol interactions with $GABA_A$- inhibitory postsynaptic potentials (IPSPs). This research centered on the question as to a possible influence of $GABA_B$ receptors on the ethanol–NMDA interaction described above. The *in vivo* data (Nie et al., 1996) were derived from extracellular recordings of excitatory responses to iontophoretically-applied NMDA in spontaneously firing hilar interneurons in the hippocampus and from evoked activity in the NAcc. In these two brain regions, both baclofen and ethanol blunted the excitatory effects of NMDA. Interestingly, the effects of both these agents were blocked by CGP 35348, suggesting a possible $GABA_B$ influence on the ethanol sensitivity of NMDA receptors there.

To help unravel the mechanisms behind this *in vivo* ethanol-NMDA receptor–$GABA_B$ receptor interaction, we isolated NMDA-EPSPs in the NAcc slice by recording in the presence of CNQX and bicuculline. The involvement of NMDA receptors in these EPSPs was subsequently verified by their total block with APV, and by their voltage dependence. Under these conditions, 44–66 mM ethanol markedly inhibited the NMDA-EPSPs (Figure 1), as previously shown (Nie et al., 1994). Interestingly, CGP 55845A blocked or blunted this effect of ethanol, in accord with the *in vivo* findings. Ethanol's inhibition of non-NMDA EPSPs was not altered by CGP 55845A, suggesting a postsynaptic (probably metabotropic) $GABA_B$ receptor effect. However, subsequent studies have been unable to pinpoint a pre- or postsynaptic locus for this action.

4. HIPPOCAMPUS AND GABAergic SYSTEMS

In his chapter, Yeh has covered the considerable variability seen from study to study with respect to whether ethanol enhances, or even inhibits, GABAergic responses. Our early studies of cerebellum *in vivo* were negative (Bloom et al., 1984), as were those in hippocampus with local application of GABA either *in vivo* (Mancillas et al., 1986) or *in vitro* in a slice preparation (Siggins et al., 1987). In the hippocampal slice, with either GABA superfusion, local application or evoked IPSPs, we saw little or no influence (or an actual depression) of 10–200 mM ethanol on the GABA system.

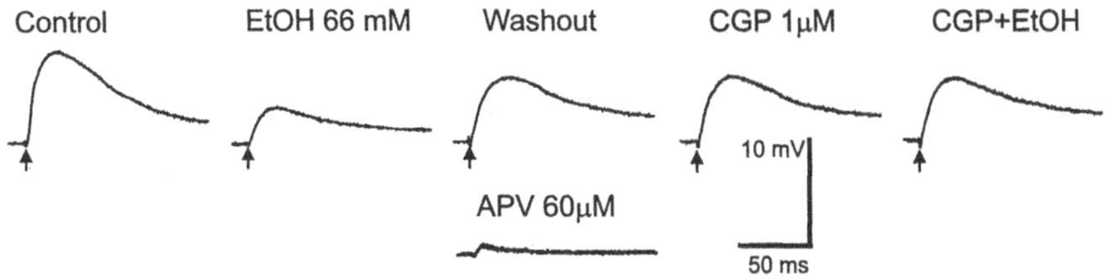

Figure 1. A $GABA_B$ antagonist blocks the ethanol inhibition of NMDA-EPSPs in the NAcc *in vitro*. Voltage recordings of synaptic responses to near maximal (subthreshold) local stimulation (at arrows) of NAcc slices superfused with 10 μM CNQX and 30 μM bicuculline to isolate NMDA-EPSPs. Superfusion of 66 mM ethanol reduced the NMDA-EPSPs by 40% with partial recovery on washout (15 min). The $GABA_B$ antagonist CGP 55845A (1 μM) alone slightly decreased NMDA-EPSP size, but blocked the ethanol inhibition of these EPSPs (right panel). Subsequent superfusion of the NMDA receptor antagonist APV (60 μM) markedly reduced the EPSPs, verifying their mediation by NMDA receptors. (Resting membrane potential (RMP) = holding potential (Vh) = -60 mV; 3 M KCl-filled electrode.)

In our early negative *in vitro* studies, we stimulated Schaffer collaterals in the stratum radiatum, and thus evoked compound synaptic inputs to CA1 pyramidal neurons mediated by combined glutamatergic EPSPs as well as $GABA_A$- and $GABA_B$-IPSPs. Therefore, to re-evaluate these earlier negative data, we used the more modern pharmacological methods described above for isolating synaptic components mediated by the subtypes of GABA and glutamate receptors (Wan et al., 1996). Considering that the glutamate system might interfere with an ethanol-GABA interaction, we first superfused CNQX and APV, and locally stimulated evoked monosynaptic compound IPSPs or inhibitory postsynaptic currents (IPSP/IPSCs) containing both the $GABA_A$ and the $GABA_B$ components. However, even in these conditions superfusion of ethanol had little reproducible effect on the compound IPSP.

We then superfused CGP 35348 (0.5 mM, together with CNQX and APV) to fully block $GABA_B$ receptors in hippocampus, leaving a pure monophasic $GABA_A$-IPSP, as shown by its subsequent elimination with bicuculline superfusion. Under these conditions, ethanol (in the presence of CGP 35348) caused a small but significant increase in the peak amplitude of the $GABA_A$-IPSP (Wan et al., 1996). More pronounced was the prolongation of the response, in parallel with findings described in chapters by Narahashi and Lovinger, for 5-HT, GABA and nicotinic receptor activation in other systems.

In the data pooled from all cells studied, with $GABA_B$ receptor antagonism the ethanol effect on the IPSP/IPSC peak amplitude was statistically significant at about a 20% increase, but the increase in the IPSP/IPSC area (that would factor in the prolonged duration of the $GABA_A$ IPSP/IPSC) was much more pronounced. The apparent EC_{50} for this effect was about 10 mM ethanol (Wan et al., 1996).

Although we initially thought this must be a postsynaptic effect because ethanol did not have any effect on the isolated $GABA_B$-IPSP/IPSCs, we were concerned that such IPSP/IPSCs were obtained without blocking the presynaptic $GABA_B$ receptors. Therefore, we tested exogenous GABA, applied close to the recording electrode from a pipette filled with GABA. In voltage clamp mode, the resulting GABA currents were unaffected by superfusion of 44–66 mM ethanol, even in the presence of the $GABA_B$ antagonist CGP 35348 (0.5–1 mM; Figure 2). In 7 neurons studied to date, there was little change in the peak or duration (area) of the GABA response, as we previously reported (Siggins et al., 1987). Subsequent total block of the GABA currents by bicuculline verified their mediation by $GABA_A$ receptors.

Our working hypothesis is that there is a presynaptic effect of the $GABA_B$ receptors in preventing the interaction of ethanol and the $GABA_A$-IPSP in CA1 hippocampus. To our knowledge, there are few if any convincing reports of ethanol enhancement of responses evoked by exogenous GABA in hippocampus. However, as we (and others) used relatively prolonged applications of GABA (several seconds, compared to tens of a millisecond for IPSP durations), receptor desensitization could be a confound, as discussed elsewhere in this volume. Thus, we still have not completely ruled out a postsynaptic site of $GABA_B$-ethanol interaction. Other methods (e.g., analysis of spontaneous and miniature IPSCs) will be required to further examine this issue.

5. ETHANOL AND GABA IN THE NUCLEUS ACCUMBENS

We examined a similar phenomenon in NAcc. However, unlike the situation in hippocampus, in NAcc we found that some neurons (40–50%) did show a clear enhancement of currents evoked by exogenously applied GABA, by ethanol superfusion in reasonable

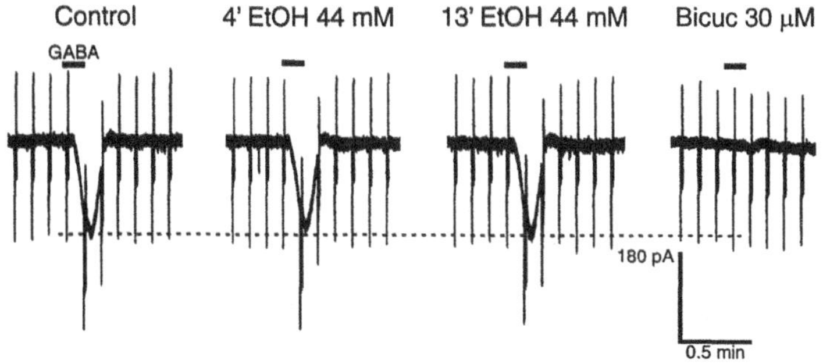

Figure 2. Ethanol has little effect on currents evoked by exogenous GABA in hippocampal CA1 pyramidal neurons, even after block of $GABA_B$ receptors. Voltage clamp: representative current tracings from a CA1 pyramidal neuron recorded in the presence of 0.5 mM CGP 35348, 20 μM CNQX, 30 μM APV and 1 μM TTX. Pressure application of GABA (2 mM in pipette; 8 sec; 7 psi; bars above records) near the recorded neuron. Note that at 4 minutes ethanol superfusion, the peak of the GABA current is even slightly smaller than control peak current (dashed line). The total block of the GABA current by bicuculline (Bicuc, right panel) shows that the current results from activation of $GABA_A$ receptors. (RMP = Vh = -70 mV; 3 M KCl-filled electrode.)

concentrations (22–100 mM). As in the hippocampus, these studies were done in the presence of TTX to block presynaptic effects and CNQX to block, at the high resting membrane potentials, most of the glutamate system.

Interestingly, ethanol dose-response plots for these NAcc cells using just the responsive cells showed a large peak enhancement of about 40%. However, at 200 mM ethanol, the frequency of responding cells decreased, as did the magnitude of the ethanol enhancement of the GABA current. The resulting inverted U-shaped curve (data not shown) could suggest a sort of short-term tolerance to alcohol, and it contradicts the generalization that higher concentrations should cause a better interaction.

Turning to evoked IPSCs in NAcc, we have now essentially repeated the observation of the phenomenon seen in hippocampus, that $GABA_B$ receptors could condition the sensitivity of $GABA_A$ergic IPSP/IPSCs to ethanol in NAcc. Despite clear evidence for ethanol enhancement of exogenous GABA in some NAcc neurons, we have seen no enhancement of GABAergic IPSPs without the use of $GABA_B$ antagonism. However, after addition of CGP 55845A, we now find the expected enhancement of the IPSPs by ethanol, as in hippocampus (Figure 3). CGP 55845A had no effect by itself. Although these studies are still ongoing, a presynaptic role for $GABA_B$ receptors is suspected.

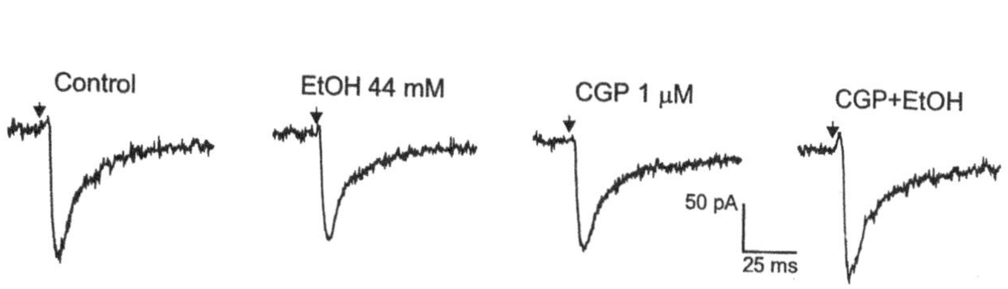

Figure 3. Ethanol enhances $GABA_A$-IPSCs in NAcc but only when $GABA_B$ receptors are antagonized; in the presence of 10 μM CNQX and 30 μM APV, local stimulation (arrows; 4 V) evokes pharmacologically isolated $GABA_A$-IPSCs. In this NAcc core neuron, 44 mM ethanol superfusion (13 min) actually reduced IPSC amplitude with partial recovery on washout with 1μM CGP 55845A (CGP) in the bath (15 min). Subsequent superfusion of 44 mM ethanol together with 1 μM CGP (CGP+EtOH) for 13 min increased IPSC amplitude by 30%. (RMP = Vh = -89 mV; 3 M KCl-filled electrode.)

At the postsynaptic level, we were puzzled why GABA currents in some cells responded to ethanol but in others they did not. We also were struck by an older study by Stelzer and Wong, (1989), showing that low concentrations of glutamate that do not alter resting membrane properties can enhance GABA responses in hippocampal neurons. Therefore, we repeated these studies, now in NAcc, and saw essentially the same thing: glutamate at low doses also can enhance GABA currents in NAcc neurons. Interestingly, in those cells that were responsive to ethanol, applying ethanol and glutamate together resulted in an even greater enhancement of the GABA current.

To test the possibility that ethanol enhancement of GABA could involve endogenous glutamate release or activation of some glutamate receptor subtype, we applied several glutamate receptor antagonists. In brief, neither of the ionotropic antagonists, CNQX or APV, had any influence on the ethanol or glutamate sensitivity of GABA currents in the NAcc, verifying that the ionotropic receptors were not involved.

We then examined the possible role of metabotropic glutamate receptors in the ethanol-GABA interaction. Interestingly, trans-1-aminocyclopentane-1,3-decarboxylic acid (trans-ACPD), the group 1 and 2 metabotropic receptor agonist, mimicked the glutamate enhancement of GABA currents (data not shown). The effects of both glutamate and trans-ACPD were blocked by the group 1 and 2 metabotropic receptor antagonist, (+)-α-methyl-4-carboxyphenylglycine (MCPG). Even more interesting is that in cells showing ethanol enhancement of GABA currents, such an effect could be antagonized by the same metabotropic antagonist (Figure 4) in five of five cells studied.

We believe this effect is a postsynaptic interaction, as the studies were performed in the presence of TTX. More recently, reasoning that such a metabotropic effect could be mediated by G-protein-linked kinase activation, we first examined protein kinase C (PKC) activators (see also Weiner et al., 1994b). One such compound, phorbol 12-myristate 13-acetate (PMA), a phorbol ester, was superfused onto NAcc neurons. PMA by itself had no effect on GABA currents in these cells. However, it is interesting that PMA enhanced the ethanol effect on these GABA currents (data not shown), in effect converting non-responding cells to responding cells. Thus, instead of the usual 40–50% of ethanol-responding cells as in our original study (see above), in the presence of PMA 88% of cells showed a clear enhancement by ethanol of the GABA current. In addition, with PMA superfusion there was a much greater total potentiation of the GABA currents by ethanol. Averaged over all 8 cells studied, 44 mM ethanol alone enhanced GABA currents to only 120 ± 3% of control, whereas in the presence of 5 μM PMA, 44 mM ethanol enhanced the GABA currents to 141 ± 10% of control. Furthermore, these effects of PMA were blocked by the

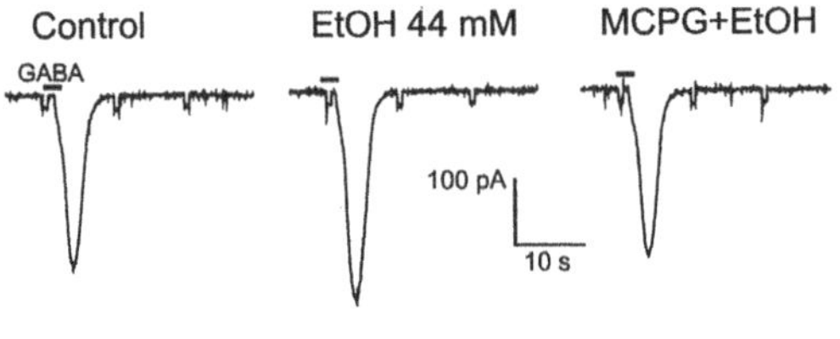

Figure 4. The ethanol enhancement of GABA currents in the NAcc can be blocked by inhibition of metabotropic glutamate receptors. GABA currents evoked in the presence of CNQX (10 μM), APV (30 μM) and TTX (1 μM). Pressure application of GABA (2 mM in pipette; 2 sec; 7 psi; bars above records) near the recorded neuron. Ethanol (44 mM) superfusion for 9 min enhanced the GABA currents, and the group 1 and 2 metabotropic receptor antagonist MCPG (1 mM) blocked the ethanol enhancement of GABA currents. (RMP = Vh = -90 mV; 3 M KCl-filled electrode.)

kinase inhibitor sphingosine. Interestingly, phorbol 12,13-diacetate (PDAc), another PKC activator, had no influence on the ethanol-GABA interaction, perhaps because it acts on a different isozyme of PKC.

6. DISCUSSION: A METABOTROPIC HYPOTHESIS

In this chapter we have presented an admittedly complicated but potentially interrelated set of data on the influence of two different metabotropic receptors on the ethanol sensitivity of two transmitter systems obtained from two brain regions. Figure 5 attempts to summarize how we believe these data begin to point to a hypothesis of metabotropic influences on the ethanol sensitivity of ligand-gated ion channels. In this figure, we have combined all of the ligand-gated ion channels that we have studied, as shown on the left side. On the right are the G-protein-linked metabotropic receptors. On the postsynaptic side (bottom), there is good evidence for metabotropic receptors of the glutamate type in NAcc, but also $GABA_B$ receptors (at least in hippocampus), and these receptors may link to a PKC via a G-protein. This and other kinases may somehow condition, perhaps by phosphorylation or dephosphorylation, the sensitivity of these ligand-gated channels to ethanol.

Similar phenomena may operate presynaptically (top of figure), with respect to glutamate and GABA release. Whether or not the metabotropic receptors are directly linked or not to release mechanisms (*e.g.*, Ca^{++} influx), or whether a G-protein is involved through a protein kinase linkage, is open to discussion. Our NAcc studies on opiates (Yuan et al., 1992; Martin et al., 1997) suggest that there is also an opiate receptor link here, at least presynaptically. Such receptors are also metabotropic in the generic sense and could condition ethanol sensitivity, as some of our previous studies have suggested (Nie et al., 1993b).

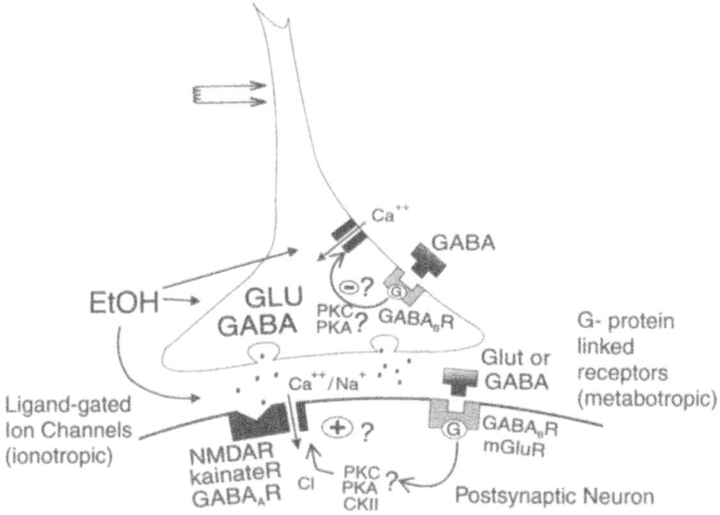

Figure 5. Schematic of postulated synaptic sites of metabotropic modulation of ethanol action on the transmitter release or ligand-gated ion channels for GABA and glutamate (glu: NMDA, AMPA and kainate) receptors. See the text for explanation. (PKA = protein kinase A; CKII = calcium/calmodulin kinase II; G = G-protein. Other abbreviations as in the text.)

The tabulation below essentially summarizes our evidence for a post-translational metabotropic regulation of ethanol sensitivity. Whether the final result is a change in the receptor/channel subunit composition (see the chapter by Yeh), that is, its stoichiometry, or whether something is changed post-translationally in the subunits or channels themselves (*e.g.*, via phosphorylation), remains for further research and discussion.

A summary of our evidence for metabotropic regulation of ethanol sensitivity of synapses is:

- $GABA_B$ receptors 'permit' ethanol-NMDA-EPSP interactions in NAcc and other central neurons.
- $GABA_B$ receptors prevent ethanol-$GABA_A$-IPSP interactions in hippocampus and NAcc, possibly at presynaptic sites.
- Postsynaptic ethanol potentiation of locally-applied GABA in NAcc is mimicked by glutamate and trans-ACPD and blocked by a metabotropic glutamate receptor antagonist.
- The PKC activator PMA augments ethanol potentiation of GABA currents in NAcc and increases the percentage of neurons showing ethanol-GABA interactions.

These data favor the idea that $GABA_B$ receptors permit or potentiate ethanol interactions with NMDA receptors, perhaps presynaptically by reducing glutamate release, at least in NAcc and perhaps other brain regions as well (Nie et al., 1996). By contrast, $GABA_B$ receptors prevent ethanol-$GABA_A$ receptor interactions in terms of the IPSPs studied in both hippocampus and NAcc, so the vector here is in the opposite direction. At this point, our data favor a presynaptic $GABA_B$ receptor locus of action. Interestingly, these two $GABA_B$ receptor effects would tend to act in opposite directions in terms of ethanol's effects on neuronal excitability, by assisting in ethanol's reduction of glutamate release (to act on NMDA receptors) but preventing ethanol's enhancement of GABA release.

However, we have also shown clear postsynaptic metabotropic influences, in terms of the response to locally-applied GABA in NAcc and perhaps other brain regions as well, that appear to involve a protein kinase step. Of course, data from other laboratories (Lovinger, 1993; Weiner et al., 1994a; Weiner et al., 1994b) support the hypothesis as well. There also are other related ancillary data suggesting that $GABA_B$ receptors, for example, might be involved in ethanol's actions. At least two anti-alcoholism drugs, acamprosate and γ-hydroxybutyrate, can act through $GABA_B$ receptors (Madamba et al., 1996; Berton et al., 1998; Madamba et al., 1996). As noted above, μ-opiate receptors also provide another metabotropic system that can postsynaptically alter glutamate receptor efficacy in NAcc neurons (Martin et al., 1997). Interestingly, we have recently found that chronic morphine treatment dramatically alters the depressant effect of presynaptic group 2 metabotropic receptors on NMDA-EPSPs.

Finally, our data lead us to build a testable hypothesis, as follows: Metabotropic systems (in the generic sense) play a regulatory role in the acute actions of alcohol on transmitter release or $GABA_A$ and NMDA receptors, at both pre- and postsynaptic levels. This regulation may involve energetic G-protein-coupled transduction systems that alter protein kinases and/or Ca^{++} channels or NMDA/GABA receptor phosphorylation/dephosphorylation. By extension of this hypothesis, we also must consider the possibility that these metabotropic systems will play a role in the effects (or the neuroadaptation) following chronic ethanol. We would predict from the data that other ligand-gated channels might be conditioned similarly by these types of metabotropic systems, and therefore be state-dependent, as Narahashi has pointed out in his chapter for some nicotinic channels and Lovinger has reported for AMPA glutamate receptors (Lovinger, 1993). Perhaps the metabotropic systems are involved in the ethanol-sensitivity of those channels as well. Obviously, extensive additional studies, including those delineating subunit stoichiometry and phosphorylation states, will be required to test these related hypotheses.

ACKNOWLEDGMENTS

We thank Drs. F.J. Wan, F. Berton, W. Francesconi, S. Steffenson and S.J. Henriksen for extracts of their published or submitted data, W. Fröstl and A. Suter (Novartis Pharma) for the gift of CGP 35348 and CGP 55845A, and P.L. Herrling (Novartis Pharma) for the gift of various drugs. This work was supported by grants from NIH (AA-06420 and DA-03665) and Groupe Lipha.

REFERENCES

Berton F, Francesconi W, Madamba SG, Zieglgänsberger W, Siggins GR (1998) Acamprosate enhances N-Methyl-D-Aspartate receptor-mediated neurotransmission but inhibits presynaptic $GABA_B$ receptors in nucleus accumbens neurons. Alcohol Clin Exp Res 22:183–191.

Bloom FE, Siggins GR, Foote SL, Gruol D, Aston-Jones G, Rogers J, Pittman Q, Staunton D (1984) Noradrenergic involvement in the cellular actions of ethanol. In: Catecholamines, Neurology and Neurobiology, Volume 13. (Usdin E, ed.), pp. 159–168, New York: Alan R. Liss, Inc.

Lovinger DM (1993) High ethanol sensitivity of recombinant AMPA-type glutamate receptors expressed in mammalian cells. Neurosci Lett 159:83–87.

Madamba SG, Schweitzer P, Zieglgaensberger W, Siggins GR (1996) Acamprosate (calcium acetylhomotaurinate) enhances the NMDA component of excitatory neurotransmission in rat hippocampal CA1 neurons *in vitro*. Alcohol Clin Exp Res 20:651–658.

Mancillas JR, Siggins GR, Bloom FE (1986) Ethanol selectively enhances responses to acetylcholine and somatostatin in the rat hippocampus. Science 231:161–163.

Martin G, Nie Z, Siggins GR (1997) Mu-opioid receptors modulate NMDA receptor-mediated responses in nucleus accumbens neurons. J Neurosci 17:11–22.

Nie Z, Madamba SG, Siggins GR (1993a) Low ethanol concentrations reduce NMDA currents in rat nucleus accumbens neurons. Soc Neurosci Abst 19:377.

Nie Z, Madamba SG, Siggins GR (1994) Ethanol inhibits glutamatergic neurotransmission in nucleus accumbens neurons by multiple mechanisms. J Pharmacol Exp Ther 271:1566–1573.

Nie Z, Yuan X, Madamba SG, Siggins GR (1993b) Ethanol decreases glutamatergic synaptic transmission in rat nucleus accumbens *in vitro*: naloxone reversal. J Pharmacol Exp Ther 266:1705–1712.

Nie Z, Steffensen SC, Criado JR, Henriksen SJ, Siggins GR (1996) Ethanol inhibition of NMDA responses involves presynaptic $GABA_B$ receptors. Soc Neurosci Abst 22:2074

Siggins GR, Pittman Q, French E (1987) Effects of ethanol on CA1 and CA3 pyramidal cells in the hippocampal slice preparation: An intracellular study. Brain Res, 414:22–34.

Stelzer A, Wong RK (1989) $GABA_A$ responses in hippocampal neurons are potentiated by glutamate. Nature 337:170–173.

Wan FJ, Berton F, Madamba SG, Francesconi W, Siggins GR (1996) Low ethanol concentrations enhance GABAergic inhibitory postsynaptic potentials in hippocampal pyramidal neurons only after block of $GABA_B$ receptors. Proc Natl Acad Sci (USA) 93:5049–5054.

Weiner JL, Zhang L, Carlen PL (1994a) Guanosine phosphate analogs modulate ethanol potentiation of $GABA_A$-mediated synaptic currents in hippocampal CA1 neurons. Brain Res 665:307–310.

Weiner JL, Zhang L, Carlen PL (1994b) Potentiation of $GABA_A$-mediated synaptic current by ethanol in hippocampal CA1 neurons: Possible role of protein kinase C. J Pharmacol Exp Ther 268:1388–1395.

Yuan X, Madamba SG, Siggins GR (1992) Opioid peptides reduce synaptic transmission in the nucleus accumbens. Neurosci Lett 134:223–228.

11

QUESTIONS AND ANSWERS OF SESSION II

Synaptic Modulation

1. Q&As BETWEEN AUDIENCE AND INDIVIDUAL SPEAKERS

1.1. Q&As between Audience and Dr. Alger

1.1.1. How Is Glutamate Released in the DSI Model?

DR. HARRISON: (Neil Harrison from University of Chicago). Dr. Alger, if this is glutamate—it certainly looks like the pharmacology fits quite well and the cell that you were actually recording from is blocked with QX-314. My question is: How does the glutamate get out of the pyramidal cell and onto terminals of interneurons?

DR. ALGER: That's an excellent question. At this point we have no idea how it's released. We don't think that there are synapses from the pyramidal cell back to the interneuron. So perhaps some more unusual mechanisms are involved.

1.1.2. Calcium Dependence of DSI

DR. TSIEN: Can you block DSI with EGTA instead of BAPTA?

DR. ALGER: Yes, 10 mM EGTA will block DSI, as will 10 mM BAPTA. However, we have not yet done a careful comparison of EGTA and BAPTA at a variety of concentrations, and so it is not clear whether or not they are equipotent in blocking DSI.

DR. TSIEN: You very carefully mapped the voltage-dependence of the induction. Did you also check the duration-dependence, in order to try to grade the amount of calcium signal? Do you know how high the calcium has to rise and for how long?

DR. ALGER: We haven't done that. In the initial experiments, we showed that DSI can be induced with a 250 ms train of action potentials, so it is clear that the 1- or 2-sec long voltage steps that we usually use are not mandatory. However, the dependence of DSI on voltage step duration, and hence its dependence on intracellular calcium concentration, will probably be affected by the internal calcium buffering conditions. We need to optimize these conditions before some of the other measurements will be meaningful. Measuring the calcium dependence of DSI directly using calcium imaging techniques is something we want to do as well.

1.1.3. Metabotropic Receptors and DSI

AUDIENCE MEMBER: I was interested in the lack of change in the paired-pulse response, presumably with the metabotropic receptor activation. That looks a little bit different than what you see when you activate a lot of presumed presynaptic metabotropic receptors. I wonder if you know much about the metabotropic receptor distribution, specifically the group I subtype on those interneurons, and their mechanism of inhibiting release.

DR. ALGER: The short answer is no. We have not yet investigated the localization of the mGluR subtypes, and we do not understand the mechanism by which mGluR activation inhibits transmitter release. We agree that the failure of DSI or presynaptic mGluR activation to alter paired-pulse release is very interesting and will no doubt provide an important clue as to how DSI does suppress GABA release. I should point out that in other systems, I am thinking of the work of Ian Forsythe's on the calyx of Held, ACPD also reduces the EPSC without changing the EPSC paired-pulse ratio, although it acts by reducing the presynaptic calcium current group (Barnes-Davies et al., 1995). Regarding the localization of different mGluRs, there has been some controversy concerning the localization of group I mGluRs. Evidence for and against their localization near presynaptic nerve terminals has been reported. It is clearly going to be important to resolve this issue.

AUDIENCE MEMBER: Does NEM block the effect of ACPD?

DR. ALGER: Yes, NEM very effectively blocks its suppressive effects on IPSCs.

1.1.4. Glutamate Transporter and DSI

DR. DUNWIDDIE: Since neurotransmitter transporters show some interesting voltage-dependent behavior, is it possible that a glutamate transporter is involved in this effect?

DR. ALGER: It could be, but this seems unlikely. We have tried a number of glutamate transporter blockers and none of them block DSI. We also tried loading pyramidal cells with high concentrations of glutamate salts in the recording pipette to see if that would enhance DSI, perhaps by increasing the export of glutamate through a reverse transporter action, but saw no effects on DSI.

1.1.5. Induction of DSI

DR. MORRISETT: Dr. Alger, you should be in the plasticity section. Actually, along those lines, I'm talking about Wyllie's depolarization-induced potentiation of mEPSPs in the next section. What is your induction paradigm and how does that compare with others?

DR. ALGER: The mEPSC potentiation studied by Wyllie and co-workers was induced by similar amplitude voltage steps as ours, but their protocol utilized 3 s long voltage steps given once every 6 s for a total of 5 min. The potentiation became maximal 5 min after the end of the train (Wyllie et al., 1994). Strowbridge and Schwartzkroin showed that trains of voltage pulses lasting from 25 to 225 s caused a potentiation of spontaneous EPSPs in mossy cells (Strowbridge & Schwartzkroin, 1996). Our protocol uses a single voltage step to 0 mV.

DR. MORRISETT: But your period of induction there would most likely intuitively require a lower level of calcium rise than the depolarization potentiation?

DR. ALGER: That seems likely. But we have not determined the calcium dependence of DSI yet.

DR. MORRISETT: What is your induction protocol?

DR. ALGER: In the experiments I reported here, we induced it by a 60- to 70 mV voltage step, lasting one or two seconds.

DR. MORRISETT: Once?

DR. ALGER: Once.

1.2. Q&As between Audience and Dr. Yeh

1.2.1. Subunit Composition of GABA$_A$ Receptors in Native Cells

AUDIENCE MEMBER: There is an important question that's always begged by these studies. We talk about transfecting subunits into oocytes or HEK-293 cells and saying we're reproducing what's in the native cell. And yet when you describe the simplest cell in the brain probably, the Purkinje cell, you're talking at least a β_2 and a β_3 in the same cell. Now, could that be important? Could it be that you have to have two different betas? Could it be that once in a while you'll see an α_1 and an α_3 together? Might it not be the system itself, but the particular set of subunits that's actually there, that maybe no one has looked at the right set?

DR. YEH: As simple as you would like these subunit combinations to be, in the real cells they're not. Let me try to answer your question in a couple of ways. First of all, when one looks at the β subunits and based on what we know about the contribution of the β subunits to the function of the GABA$_A$ receptor, it's not really quite clear whether there is, in fact, a difference between β_2 and β_3, especially with regards to second messenger mediated modulation. There are consensus sites on the β_2 and β_3 subunits, either by PKA or tyrosine kinase-mediated phosphorylation. From what I can recall from Steve Moss' work (Moss et al., 1992, 1995), I don't think there is a big difference. The cell does make β_2 and β_3 subunit mRNAs. Are there different subpopulations of Purkinje cells that only make the β_2 protein versus the β_3 protein? I think that's something we'll have to get to the protein level to find out.

DR. BREESE: (George Breese from North Carolina). I'd just like to ask one question. I assume from your slides that you normally see ethanol enhancing GABA from Purkinje neurons after it dissociated. Is that correct?

DR. YEH: Yes. We see it after dissociation. We see it after maintenance in long-term culture conditions.

DR. BRESSE: The reason I ask that question is, if you look *in vivo*—and you can obviously recognize Purkinje cells when you're doing extracellular recording—you do not see this. In fact, a large number of the neurons do not respond to ethanol initially. We've got studies showing that you can add GABA, etc., and then they become sensitive to ethanol. But it's not at all clear, and maybe it has something to do with this symposium. There's something about the *in vivo* study that seems to separate out, and there appears to be another controlling system there that we're not quite certain of what it is. And these cells, by the way, almost invariably are sensitive to zolpidem; all of them are. So they do have $\alpha_1\beta_2\gamma_2$ in them.

DR. YEH: Yes.

DR. BREESE: And Hugh Criswell has recently shown in dissociated neurons, just to back up your statement about the γ_2 short subunit, that they are ethanol-sensitive and have only the γ_2 short in them, with $\alpha_1\beta_2$ (Criswell et al., 1997).

DR. YEH: Yes. I don't know now exactly how to explain the apparent lack of sensitivity of adult Purkinje cells *in vivo*. I don't think Purkinje cells, especially in the adult, in the entire cerebellum, are entirely insensitive. I don't think the sensitivity is as dramatic as we see with acutely dissociated Purkinje cells from the neonatal cerebellum. This is just a hunch, we don't have enough data, especially with the acutely dissociated cells recorded that were also profiled for mRNA expression to make the statement. But I would be willing to bet that when we do have enough cells and we're able to sort this out statistically, that we will see a difference in the sensitivity to modulation by ethanol between young rats and older rats. I'll bet you that the difference will be that the older rats are less sensitive.

1.2.2. Distributions of GABA$_A$ Receptor Subtypes in Vivo

AUDIENCE MEMBER: I'm very impressed by your experiment showing a variety of different GABA$_A$ subtypes in the cerebellum. My question is: You see such a variety distributed in the cerebellum, do you see this kind of distribution as constant over all the different animals? The second question is: Does such a distribution have some biological significance or biological meanings?

DR. YEH: I guess the first question is whether this kind of heterogeneous pattern of distribution is seen across brain regions, or across cerebella of different animals. I believe it is seen across different brain regions within any given animal. I don't know about whether other kinds of species apply with regards to heterogeneity. I would not be surprised if they are. I think the heterogeneity is something that's there.

As to why there should be this kind of heterogeneity, I don't know. But I do know that this kind of heterogeneity could account for some of the specificity or selectivity of modulatory drugs or agents in the brain.

1.2.3. Post-Translational Modification of GABA$_A$ Receptors and Sensitivity to Ethanol.

DR. VALENZUELA: (Fernando Valenzuela from Colorado). Hermes, have you determined which factors essentially determine why the native receptors are more sensitive than the recombinant receptors? Can you reduce the sensitivity of the GABA receptors by using kinase inhibitors or phosphatase inhibitors? Have you done anything like that?

DR. YEH: We've not done that experiment. That certainly is in the works. The question that we're grappling with is whether indeed applying kinase inhibitors is really the best and most effective strategy with regard to addressing these issues. Rather than artificially activating or inhibiting kinases, my bias would be to try to find a physiological model that we know phosphorylation plays a role in, and then try to manipulate that system. But your question is well taken. It's something that we're thinking about.

1.2.4. Sensitivity to Ethanol of $GABA_A$ Receptors in Different Preparations

DR. HARRISON: (Neil Harrison from University of Chicago). Hermes, my reading of the recombinant literature and my own experimental experience is a little bit different from yours. There are many papers that show that high concentrations of ethanol will be effective in modulating the $GABA_A$ receptor, and that seems to be rather subunit-independent. And, indeed, that seems to be relatively robust, at least in the oocyte expression system, insofar as we were recently able to completely remove alcohol sensitivity of various recombinant $GABA_A$ receptors. And if the effect hadn't been fairly robust, we probably would have had very little success in removing it by mutagenesis. It seems to me that the big difference in the field is between the neuronal expression, which is highly variable and in the oocyte expression system, where high concentrations of ethanol always seem to work. But in mammalian cell expression, like fibroblasts, kidney cells, and so on, where you have a very hard time in getting any effects at all, as what you showed today.

So to follow Fernando Valenzuela's question, it seems really as though we may have evidence for what may be critical or direct sites of action of ethanol within the $GABA_A$ receptor from the mutagenesis work. What we need to look at very closely now is post-translational modifications that may lead to differences in ethanol sensitivity, both in its potency and efficacy, as a regulator of the receptor. I think that's where the action's going to be in the next two or three years.

DR. YEH: Your point's well taken. The recent paper you published on chimeric receptors certainly supports what you've said. I think in reading the literature, though, high concentrations of ethanol clearly works, but those are often concentrations that exceed physiological or even lethal levels of ethanol. There are other studies, for example, with the CA3 cells where, in fact, high concentrations of alcohol does not work. It's at the really low concentrations of alcohol, like 3 mM and 10 mM, where things work. So there you have it. But I agree with you, the general trend is toward the high concentrations of alcohol.

DR. TSIEN: So Neil Harrison is actually proposing that the post-translational modification mechanisms are different in a cultured mammalian cell than in a mammalian neuron. This suggests someone could take a Purkinje cell and introduce an engineered GABA receptor to look for ethanol sensitivity of that particular channel. Otherwise, all these ideas need reevaluation.

DR. YEH: Yes. That study is best done in Purkinje cells that are maintained long-term in culture, because of the difficulties involved with expressing or isolating engineered GABA receptors in an acutely dissociated neuron. Even so, we've had a lot of problems trying to either transfect or to knock out subunits or change subunits in primary neurons. But I think that's an experiment to do.

1.3. Q&As between Audience and Dr. Dunwiddie

1.3.1. Locations of Adenosine Receptors

DR. TSIEN: You stressed the point that the adenosine receptors are absent from the presynaptic GABAergic terminals. Are they also absent from the excitatory glutamatergic terminals that make synapses onto the inhibitory GABAergic neurons?

DR. DUNWIDDIE: No. Adenosine receptors do seem to be present on those nerve terminals. So if you activate that pathway, i.e., if you antidromically fire pyramidal neurons and then the collaterals to the GABAergic interneurons, adenosine will reduce the IPSPs that you get by stimulating in that fashion. Thus, the excitatory synapse that drives the interneuron is a potential target for ethanol modulation. However, if you use DNQX and APV to block excitatory transmission, and directly stimulate the GABAergic neurons, there's no modulation of the directly evoked IPSP.

1.3.2. Paired-Pulse Facilitation and Adenosine Receptors

AUDIENCE MEMBER: You showed some field potential recordings. Have you looked at any paired-pulse effects, with and without alcohol, to see whether or not they are modulated by adenosine?

DR. DUNWIDDIE: No, we haven't looked at that specifically. We know that adenosine, like most other presynaptic modulators, does change paired-pulse facilitation, but we haven't looked at that specifically with alcohol.

1.3.3. A_2 Receptor and Effects of Ethanol

AUDIENCE MEMBER: Have you looked at all at A_2 receptor agonists and whether they might interact with ethanol's effect?

DR. DUNWIDDIE: No, but that's certainly an important question. There aren't a lot of A_2 receptors in the hippocampal formation, so that might not be the right place to look for such an effect. A_2 receptors are located primarily in the striatum, and those are probably the receptors that are involved in the locomotor stimulant effects of caffeine. Therefore, the striatum might be the appropriate place to look for an electrophysiological substrate for the differences between the long-sleep and short-sleep mice in terms of their locomotor responses to adenosinergic drugs.

1.3.4. Caffeine and Intracellular Calcium

DR. ALGER: Caffeine and some of the other things you used are well-known to affect internal calcium stores. Calcium is something that's sort of big where I come from, University of Maryland, but I haven't heard much about it in your talk. Is it possible that caffeine's effects on internal calcium stores play any role in the effects that you reported?

DR. DUNWIDDIE: Yes. We've heard of calcium even as far west as Colorado. The ability of caffeine to release calcium from intracellular stores is fairly well established, but the concentrations that you need to get that kind of effect is basically in the 1–10 mM range.

Caffeine's affinity for the adenosine receptor is on the order of 15 μM. That's actually about the concentration of circulating caffeine that you would expect to find in someone who has had a cup or two of coffee. Therefore, we think that that's the pharmacologically relevant range for caffeine concentrations. For those reasons, I guess the bottom line is that we don't think calcium mobilization has much to do with the effects that you would normally see at pharmacologically relevant caffeine concentrations.

DR. SIGGINS: And phosphodiesterase either?

DR. DUNWIDDIE: You begin to see phosphodiesterase antagonism at concentrations of about 100 μM caffeine, and that's starting to get into the clinically dangerous range, i.e., where you start to see seizures and things like that.

1.4. Q&As between Audience and Dr. Siggins

1.4.1. Location of Effects of Ethanol: Pre- vs. Postsynaptic

DR. LOVINGER: When you showed the metabotropic or the $GABA_B$ blockers antagonize the ethanol effects on NMDA receptors, which were all in intact synapses. That were all stimulating presynaptic fibers and recording EPSPs. Right?

DR. SIGGINS: Right.

DR. LOVINGER: Have you done that same experiment with applied NMDA?

DR. SIGGINS: Yes, we have. CGP55845 doesn't alter the ethanol inhibition of the nucleus accumbens cell's response to exogenous NMDA. In other words, there seems to be no effect of CGP there, so that's why we thought it was a presynaptic effect. However, if it is presynaptic, you'd also expect the non-NMDA EPSP effect of ethanol to be reduced by CGP55845. But, in fact, it isn't. And that's why we're in this sort of no-man's land of not knowing whether it's pre- or postsynaptic. We're in the process of doing paired-pulse facilitation studies, and someday we hope to be able to record spontaneous and miniature NMDA EPSPs. But right now that experiment is rather difficult in these cells, to see whether the effect is pre- or postsynaptic. So that's why I went over it rather quickly, because we really don't know where the locus of that action is yet.

DR. LOVINGER: I guess one possible interpretation is that the receptors you activate synaptically are different from the ones you activate with local application.

DR. SIGGINS: It could be. That's another good interpretation. We've seen that sort of thing with the opiates, where μ receptor agonists, in fact, reduce the NMDA EPSP, like ethanol. But when you apply NMDA locally by pipette, μ receptor agonists enhance the NMDA current (Martin et al., 1997). So we postulated that we were dealing with extrasynaptic versus subsynaptic NMDA receptors in that system, in that paradigm. So what you suggested is quite possible. In the slice preparation, I think we're always up against that kind of problem.

DR. LOVINGER: Yes. Maybe we need a system more like what Dick Tsien showed this morning.

DR. SIGGINS: Right, exactly.

DR. LOVINGER: Activating a single synapse with puffing or local application.

DR. SIGGINS: It would be helpful to look at something like that.

1.4.2. Role of $GABA_B$ Receptor

DR. MORRISETT: If there's anybody that probably should have done the $GABA_B$-ethanol-NMDA reproduction of your work, it's probably me, and I never got around to it.

DR. SIGGINS: In hippocampus, in dentate, you mean?

DR. MORRISETT: Right, either one, CA1 or dentate. What's the situation with that? Have other people seen similar effects as well?

DR. SIGGINS: Do you mean the NMDA-CGP interaction?

DR. MORRISETT: Right, the role of $GABA_B$ receptor.

DR. SIGGINS: Yes, we hope it's $GABA_B$. Since we don't find much in the way of postsynaptic $GABA_B$ receptors in terms of inotropic effects in nucleus accumbens core neurons, like you see with a $GABA_B$-IPSP in hippocampus, we're assuming that maybe those receptors aren't there. That's just an assumption that again leads me to believe we're dealing with a presynaptic effective of the CGP compounds in some way, in enhancing or "allowing" the ethanol effect. However, we can not rule out a non-ionotropic postsynaptic $GABA_B$ receptor.

DR. MORRISETT: I'm asking if other labs also showing the CGP reversal of the ethanol inhibition of NMDA.

DR. SIGGINS: Oh, *in vivo*, you mean?

DR. MORRISETT: No. In any other system are there other people who have looked at this?

DR. SIGGINS: I don't know of any other than Scott Steffensen's *in vivo* data.

DR. MORRISETT: Yes, right. I was curious as to whether or not other slice people have been more diligent than I have.

DR. SIGGINS: It's fairly new and quite unexpected. When Scott Steffensen came to us with that observation, I didn't think it was real. However, it turned out to happen in the slice as well.

1.4.3. Involvement of Phosphorylation in Effects of Ethanol

DR. TSIEN: I have a suggestion to make to George Siggins. In your summary you didn't mention the work of Ken Swartz, Bruce Bean and David concerning the effects of PKC in interrupting G-protein modulation (Swartz et al., 1993). Some of your fine studies on ethanol and blockade of $GABA_B$ receptors have that faint ring about them. Let's suppose that the GABA that is released from presynaptic inhibitory terminals also feeds back via $GABA_B$ receptors to inhibit its own release in an autoreceptor kind of way. Now, if you put on CGP and block the $GABA_B$ receptors, you increase transmitter release. If you put on a metabotropic agonist, like t-ACPD, you also inhibit increased GABA currents, presumably by an inhibitory effect on signaling from the presynaptic $GABA_B$ receptors. Is it possible that ethanol might in some way modify the $GABA_B$ feedback?

DR. LOVINGER: The thing that seems unclear to me at this point is the relationship between glutamate release and ethanol effects on the NMDA receptor-mediated synaptic response. And to tell you the truth, I was trying to think this over even before Dick Tsien brought this up—I can't think of any experiment that addresses that. The only kind of experiments we've done is to do different stimulation strengths and look at the inhibition, and it always looked the same. But that just is really the number of fibers we're activating; that doesn't say anything about the amount of glutamate in the cleft and the relationship to that. And when we look at NMDA-activated currents—that is, activated by applied NMDA in cultured neurons or in, dare I say, recombinant systems—there the ethanol inhibition seemed not to vary with receptor occupancy. But that doesn't mean that it's the same at the synapse. So I think that scenario could well be true, and the important thing would be to get at that relationship perhaps between glutamate release and ethanol inhibition, at least at one part of that.

DR. SIGGINS: Right. I'm sorry that because of time constraints, I neglected to bring up the autoreceptor idea. In fact, enhancement of the duration of the $GABA_A$ IPSP—since we're only stimulating it once every ten seconds, it doesn't seem likely that the synapses will accumulate a lot of GABA to feed back on the terminals. We are not sure what's happening with respect to a single stimulus that releases GABA that could feed back and prevent more GABA release with the same shock. I just don't know if that's going to enhance the duration or not.

DR. TSIEN: You can test it all just by putting pertussis toxin on, since this is likely to interrupt the G-protein signaling involved in the putative auto feedback system.

DR. SIGGINS: Right. Actually we're working with N-ethylmaleimide (NEM) now, with Brad Alger's help, we're going to try that. Pertussis toxin has been real difficult in our hands in the hippocampal slice preparation. But NEM, in fact, does work, thanks to Brad Alger telling us exactly how to do that experiment.

DR. HENRIKSEN: (Steve Henriksen from Scripps). George, I want to remind you—I don't know if Scott Steffensen is going to mention this when he talks—*in vivo*, when one records from inhibitory interneurons that are actually releasing GABA, one of the most sensitive effects of alcohol is increasing the excitability of those GABAergic interneurons. So following stimulation, you get bursts of discharge on those GABAergic neurons. Therefore, if any feedback that's going to take place, it works the opposite way. Alcohol is actually increasing the sensitivity to afferent stimulation in those interneurons. So if there's feedback, it's overcome by some other mechanism.

DR. SIGGINS: Actually we tried to interest another person in our department, Dr. Bert Weiss, to do microdialysis studies of GABA, and we're still trying to push him in that direction with ethanol, and CGP for that matter.

DR. LIU: Dr. Alger, do you have any comment on this?

DR. ALGER: Dr. Siggins, you mentioned that the lipophilic phorbol ester, PMA, had different effects than the comparatively hydrophilic phorbol ester, PDAC. Have you tried other phorbol esters, such as PDBu, which have intermediate degrees of lipophilicity? Have you tried internal application of pseudosubstrate, inhibitory peptiedes to confirm that the phorbol esters are actually acting via PKC?

DR. SIGGINS: No. We're just starting those studies really. The reason I don't think it's a false negative in that case is because we got the reverse order of potency (PDAc >> PMA) with enhancement of NMDA currents in the same accumbens slice (Martin & Siggins, unpublished). With NMDA currents, PDAc works at that same concentration, but PMA didn't. So we think there is an effect. Maybe it's not penetrating the right part of the cell or something—I don't know—but there is a positive effect at that concentration with PDAc, and that effect, in fact, is also blocked by sphingosine. So we're beginning to feel that there is a kinase there. Something else unrelated to PKC could be involved, of course, like calcium currents, calcium levels, and that sort of thing. But these are just early parts of the study, beginnings.

DR. DUNWIDDIE: I think that the importance of phosphorylation has been demonstrated in a number of different labs. Certainly Jeff Weiner and Fernando Valenzuela at the University of Colorado have shown that manipulations that change the level of PKC-dependent phosphorylation of proteins in hippocampal brain slices will change GABAergic sensitivity to ethanol as well (Weiner et al., 1997).

DR. ALGER: Referring to the difference between the different kinds of phorbol esters?

DR. DUNWIDDIE: Yes. I'm not aware of anything other than George's data, though, where those two specific phorbol esters seem to differ in terms of their effects.

DR. ALGER: Did you have a comment, Hermes?

DR. YEH: I was just going to say basically the same thing. George, I think you have a particularly unique system to look at these things. Although not directly related to alcohol research, I've introduced a variety of different kinds of modulators of kinases and catalytic subunits into cells and looked at GABA modulation. One of the potentially complicating factors there is that once you do that, usually the GABA response itself—

DR. SIGGINS: Goes away. Yes.

DR. YEH: In my favorite Purkinje cells, dialysis of catalytic PKA actually increases GABA current. Those are fairly complicated experiments. If you had to add another effect of ethanol onto this paradigm, you will really titer your initial potentiation first. In your case, at least with PKC, it looks like you could do that fairly safely and not have any kind of initial change, and that serves as a nice baseline.

DR. SIGGINS: Yes. That's again in the planning stages. These cells are easy to patch and clamp than hippocampal pyramidal neurons in some way.

2. DISCUSSION BETWEEN AUDIENCE AND SPEAKERS OF SESSION II

2.1. Future Direction of Research on Synaptic Modulation

DR. LIU: Dr. Alger, since you're one of the pioneers of research on synaptic modulation, after hearing this session as an "outsider", a non-alcohol researcher, do you have any comments or suggestions for the future direction on synaptic modulation in alcohol research?

DR. ALGER: The suggestions from Dr. Tsien's work that the variability in quantal event size in the central nervous system might reflect, in part, variability in the extent of receptor saturation at individual receptor patches seems very important. It raises the possibility that changes of quantal size, which is usually taken to reflect postsynaptic factors, may actually be determined by presynaptic factors. This is extremely important for the interpretation of electrophysiological experiments on synaptic function and the effects of drugs, such as ethanol, on it. Dick will correct me on this.

DR. TSIEN: Yes, Brad, we must be careful to acknowledge that while the postsynaptic response is going to depend on the postsynaptic receptors and their state of modulation, there are serious dangers in the classical assumption that a change in the unitary size can be absolutely interpreted in a postsynaptic way. If you allow for the fact that the variability exists and the receptors are not saturated, then it stands to reason that anything that modifies the filling of a vesicle, the degree to which the transmitter is totally dumped, or the properties of transporters near the cleft—any and all of these factors could modify what is traditionally taken as an ironclad measure of postsynaptic function.

DR. ALGER: The point is well taken.

DR. LIU: Since the bell already rang, we're going to close this session. I just want to make one more comment. In our question list, we asked many questions, such as: Are the effect of ethanol on the synapses pre- or postsynaptic? How can we identify them? Are the effects of ethanol on synaptic functions direct or indirect? If the effects are direct, what are the molecular targets at the synaptic sites? What are the best ways to make coherent conclusions of ethanol's effects on molecular targets based on the information derived from recombinant, brain slice, *in vitro* cell culture and *in vivo* preparations? If the effects are indirect, what modulation mechanisms and signaling processes are involved? Is there a common mechanism that may underlie the actions of ethanol on multiple synaptic targets? How can we best approach the identification of subunit composition of native receptors in light of ethanol-induced modulation? What new techniques can be used to more directly measure neurotransmitter release processes at glutamatergic and GABAergic synapses? Are we using the appropriate ethanol concentrations in our experiments? How closely can we relate alcohol effects *in vitro* to those that are relevant *in vivo*, especially in relation to alcohol-induced behaviors? From this morning's first session and this second session, I think we all realized that answers to those questions are extremely complex and we still have a long way to go in synaptic research.

REFERENCES

Barnes-Davies M, Forsythe ID (1995) Pre- and postsynaptic glutamate receptors at a giant excitatory synapse in rat auditory brainstem slices. *J Physiol (Lond)* **488(Pt 2):**387–406

Criswell HE, McCown TJ, Moy SS, Oxford GS, Mueller RA, Morrow AL, Breese GR (1997) Action of zolpidem on responses to GABA in relation to mRNAs for GABA(A) receptor alpha subunits within single cells: evidence for multiple functional GABA(A) isoreceptors on individual neurons. *Neuropharmacology* **36(11–12):**1641–1652

Martin G, Nie Z, Siggins GR (1997) mu-Opioid receptors modulate NMDA receptor-mediated responses in nucleus accumbens neurons. *J Neurosci* 1997 Jan 1;**17(1):**11–22

Moss SJ, Gorrie GH, Amato A, Smart TG (1995) Modulation of $GABA_A$ receptors by tyrosine phosphorylation. *Nature* **377:**344–348

Moss SJ, Smart TG, Blackstone CD, Huganir RL (1992) Functional modulation of $GABA_A$ receptors by cAMP-dependent protein phosphorylation. *Science* **257:**661–665

Strowbridge BW, Schwartzkroin PA (1996) Transient potentiation of spontaneous EPSPs in rat mossy cells induced by depolarization of a single neurone. *J Physiol (Lond)* **494(Pt 2):**493–510

Swartz KJ, Merritt A, Bean BP, Lovinger DM *(1993)* Protein kinase C modulates glutamate receptor inhibition of Ca^{2+} channels and synaptic transmission. *Nature* **361:**165–168

Weiner JL, Valenzuela CF, Watson PL, Frazier CJ, Dunwiddie TV (1997) Elevation of basal protein kinase C activity increases ethanol sensitivity of GABA(A) receptors in rat hippocampal CA1 pyramidal neurons. *J Neurochem* **68(5):**1949–1959

Wyllie DJ, Manabe T, Nicoll RA (1994) A rise in postsynaptic Ca^{2+} potentiates miniature excitatory postsynaptic currents and AMPA responses in hippocampal neurons. *Neuron* 12(1):127–138

Section III

SYNAPTIC PLASTICITY

12

ALCOHOL, MEMORY, AND MOLECULES

Michael Browning,[1,2] James Schummers,[1,3] and Scott Bentz[1,4]

[1]Department of Pharmacology
[2]Program in Neuroscience
University of Colorado Health Sciences Center
4200 East 9th Avenue
Denver, Colorado 80262
[3]Massachusetts Institute of Technology
77 Massachusetts Avenue
Cambridge, Massachusetts 02139
[4]Wright State University School of Medicine
3640 Colonel Glenn Highway
Dayton, Ohio 45435

1. INTRODUCTION

The amnesic effects of acute ingestion of ethanol are well documented in both human and animal studies (Goodwin et al., 1969; Lowy, 1970; Lister et al., 1987; Zimmerberg et al., 1991; Ryback 1971; Castellano and Populin, 1990; Tako et al., 1991). However, the molecular and cellular mechanisms that underlie these effects are unknown. In the past, the vast majority of studies of learning and memory have been conducted under conditions where it was very difficult to perform cellular and molecular analyses—namely in studies of animal behavior. There are, of course, good reasons to begin studying learning and memory under such complex circumstances, perhaps the most obvious being the limitations of simple systems models of memory. Nonetheless, within the last 25 years, there have been a number of important discoveries which indicate that future research may provide extremely important new information about the cellular and molecular substrates of learning and memory deficits. The discovery of long-term potentiation (LTP) provided a cellular mechanism that had many properties thought to be essential for a biological substrate of memorial processes. Subsequent work revealed that both chronic and acute ethanol inhibited this form of neuronal plasticity (Durand and Carlen, 1984; Sinclair and Lo, 1986; Mulkeen et al., 1987; Blitzer et al., 1990).

The "Drunken" Synapse, edited by Liu and Hunt.
Kluwer Academic / Plenum Publishers, New York, 1999.

1.1. Long-Term Potentiation (LTP)

Hebb (1949) suggested in his influential model of the cellular basis of memory that the most likely locus of the physiological substrate of memory would be a synaptic system that exhibited long-lasting changes in efficacy after brief periods of use. The ensuing years saw a substantial amount of research devoted to the search for such a synaptic system. It was not until 1973 that Bliss and colleagues (Bliss and Lomo, 1973) demonstrated that tetanic stimulation in the dentate region of the hippocampus elicited a long-lasting enhancement of synaptic responses. This phenomenon, LTP, can be elicited in all three of the major excitatory synaptic pathways in the hippocampus (Schwartkroin and Wester, 1975; Yamamoto and Chujo, 1978) and has also been seen in visual, motor and piriform cortex (reviewed in Teyler and DiScenna, 1987). In addition, Bliss and Gardner-Medwin (1973), and Douglas and Goddard (1975) demonstrated that with repeated bursts of stimulation, the potentiation would last for months. Thus, LTP appears, intuitively, to be an excellent cellular substrate for memory. More substantive support for such a role has come from pharmacological studies. In these studies, APV, an antagonist of the *N*-methyl-D-aspartate (NMDA) subtype of the glutamate receptor, was used. APV, which has been shown to block LTP in some brain areas (see below), also blocked learning (Davis et al., 1992; Morris et al., 1986; Morris, 1989; Morris et al., 1990). Moreover, LTP-like synaptic potentiation has also been observed during learning (Shors et al., 1989; Weisz et al., 1984; Roman et al., 1987).

1.2. LTP and Ethanol

A number of authors have shown that acute ethanol blocks LTP induction. Sinclair and Lo (1986) reported that 100 mM, but not 50 mM, ethanol produced a significant reduction in LTP induced by high frequency tetanic stimulation. Mulkeen et al., (1987) and Morrisett and Swartzwelder (1993) reported that ethanol inhibited LTP at concentrations of 86 and 75 mM, respectively. In contrast, Blitzer et al., (1990) reported that ethanol concentrations as low as 5 mM produced significant inhibition of tetany-induced LTP. It is essential that we know the concentration of ethanol that inhibits LTP as such information relates both to the molecular mechanisms of ethanol's effects on LTP and to the clinical relevance of ethanol inhibition of LTP. One likely molecular target of ethanol's effect on LTP is the NMDA receptor, which plays a critical role in LTP induction.

1.3. Ethanol and NMDA Receptor

There is virtually unanimous agreement that ethanol inhibits the NMDA receptor. NMDA-activated ion currents in voltage clamped hippocampal neurons were reduced more than 60% by 50 mM ethanol in early reports (Lovinger et al., 1989). Ion currents activated by NMDA in voltage-clamped sensory neurons were inhibited by ethanol with an IC_{50} equal to 10 mM (White et al., 1990). NMDA-stimulated Ca^{++} uptake into cerebellar granule cells was reduced 30% by ethanol concentrations as low as 10 mM (Hoffman et al., 1989). Ethanol also inhibited NMDA stimulated, Ca^{++} dependent, cyclic GMP accumulation in cerebellar granule cells in a dose-dependent manner (Hoffman et al 1989). Single channel currents in cultured hippocampal neurons were shown to be inhibited by rather high concentrations of ethanol (86.5–174 mM), whereas very low concentrations (1.74–8.65 mM) were reported to be stimulatory (Lima-Landman and Albuquerque, 1989). Given that activation of the NMDA receptor is critical for LTP induction, we have been particularly interested in the possibility that ethanol's blockade of LTP and its amnestic effects might be due to effects on the NMDA receptor. In the present report, we review recent results from our lab, which investigate this hypothesis.

2. RESULTS AND DISCUSSION

As indicated in the Introduction, we and others have suggested that the amnestic effects of acute ethanol could be due to ethanol's inhibition of LTP. One difficulty with this interpretation is the controversy concerning the doses of ethanol that are required to inhibit LTP. We tested the effects of various doses of ethanol on LTP in hippocampal slices. When we tested the effects of 25 mM ethanol on LTP we saw no significant effect (Figure 1). In our hands, 50 mM ethanol was required to produce a significant inhibition of LTP (44 % inhibition), whereas 100 mM ethanol produced complete (97 %) inhibition.

Having established the doses of ethanol required to inhibit LTP induction in our system, we next turned to the question of whether ethanol could block the maintenance or expression of LTP as this issue had not previously been addressed. We tested various doses (50, 100 mM) of ethanol applied at various times (5, 10, 15, 30 minutes) after LTP inducing stimulation had been delivered. As shown in Figure 2, we found no evidence for any effect of ethanol on the maintenance phase of LTP. We were also interested in determining

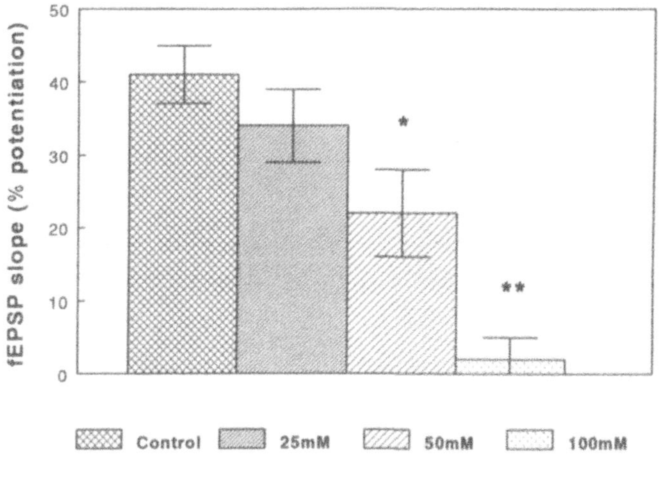

Figure 1. Bar graph showing average data for the potentiation of the extracellular field excitatory postsynaptic potential (fEPSP) in the CA1 region of the hippocampus. Shown are the mean (±SEM) for the slope of the fEPSP 30 min after high frequency stimulation (HFS: 100 stimuli delivered at 100 Hz) delivered in the absence or presence of various concentrations of ethanol. Ethanol when present was perfused onto the slice for 10 min before and during the delivery of the HFS. Ethanol perfusion was then stopped and fEPSPs were monitored 30 min after the cessation of the HFS. Percentage potentiation is determined by comparison of responses taken 30 min after tetanus with basal responses obtained during the 10-minute perfusion with ethanol. In control slices that were not perfused with ethanol, HFS produced a 41.4 ± 4.1% potentiation of the fEPSP (n = 13 slices from 8 rats). In contrast, HFS produced potentiation of 34.3 ± 5.8% in slices perfused with 25 mM ethanol (n = 8 slices from 4 rats), 22.8 ± 6.2% potentiation in slices perfused with 50 mM ethanol (n = 11 slices from 8 rats) and (4.2 ± 1.7%) potentiation in slices perfused with 100 mM ethanol (n = 5 slices from 5 rats). Percentage potentiation was significantly different from control slices in the presence of 50 M and 100 mM ethanol (* = $p < 0.05$; ** = $p < 0.1$, two-tailed, unpaired Student's t test). (Reprinted by permission from Schummers et al., 1997).

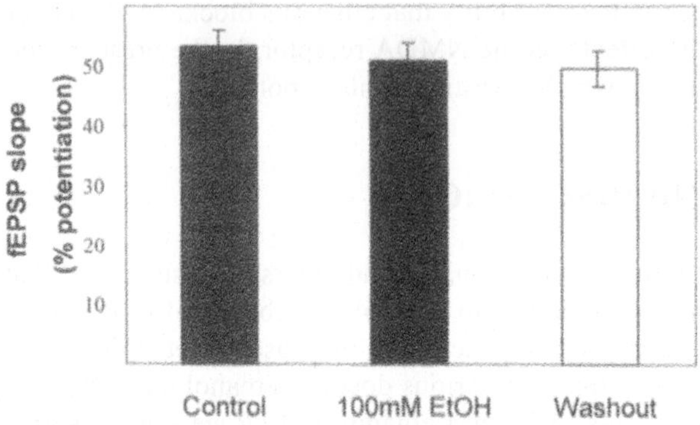

Figure 2. Bar graph showing the lack of effect of 100 mM ethanol on the maintenance of LTP. Shown is the mean (± SEM) for the slope of the fEPSP at various times after LTP induction. In the control condition, average responses taken 30 min after high frequency stimulation are shown. In the 100 mM ethanol condition, average responses taken during a 10-minute ethanol perfusion delivered 35–45 min after HFS are shown. In the washout condition, average responses taken during a 10-minute washout of ethanol 45–55 min after HFS are shown. As can be seen in the graph, ethanol has no significant effect on LTP maintenance. Percentage potentiation is determined by comparison of responses taken at various times after HFS with basal responses obtained during the 10 min period immediately preceding HFS.

whether the effect of ethanol on LTP was reversible. Accordingly, we first perfused the slice with 100 mM ethanol and demonstrated that LTP induction was blocked as shown above. We then washed the slice in buffer without ethanol for 10 minutes and attempted to induce LTP. As shown in Figure 3 the effects of ethanol on LTP induction were readily reversible as a typical LTP was obtained after a 10-minute washout.

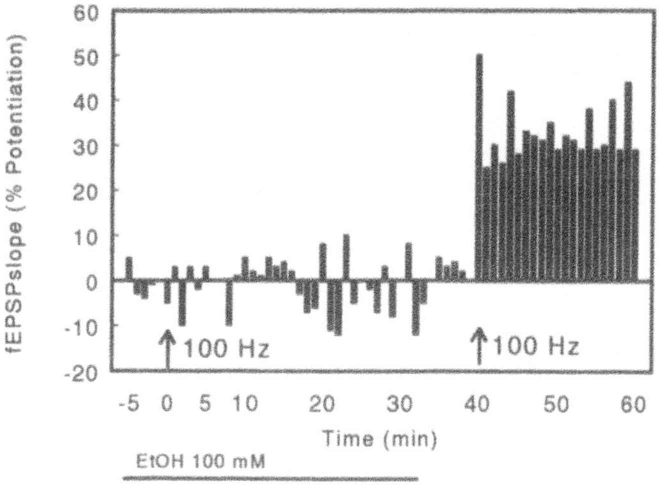

Figure 3. Bar graph demonstrating that ethanol's blockade of LTP is readily reversible. Shown are representative data for the fEPSP slope measurements taken every 60 seconds before and after delivery of HFS (100 Hz HFS was delivered at the time points indicated by the arrows). The first HFS was delivered in the presence of 100 mM ethanol (ethanol was perfused at the times indicated by the horizontal bar at the bottom of the figure) and LTP was not seen following such stimulation. After a 10-minute washout of the ethanol, a second HFS was delivered and this resulted in robust LTP.

Thus, relatively high doses of ethanol are required to block LTP induction, whereas such doses are ineffective in blocking previously established LTP. Moreover, the blockade of LTP by ethanol is readily reversed by a 10-minute washout. We next turned to the issue of the molecular mechanisms that could underlie this effect of ethanol on LTP. As the NMDA receptor plays a critical role in LTP, we were interested in testing the hypothesis that ethanol's inhibition of the NMDA receptor might underlie ethanol's blockade of LTP.

To evaluate the contribution of NMDA receptors in the effect of ethanol on LTP induction, we directly tested the effects of ethanol on NMDA fEPSPs in our hippocampal slice preparation (Schummers et al., 1997). Although a number of authors have previously reported that ethanol inhibits NMDA responses, these experiments have been conducted most often in whole cell analyses and in cultured or recombinant systems, and we wished to characterize this response in the CA1 region of our hippocampal slice preparation. We pharmacologically isolated the NMDA fEPSP using antagonists of the K/AMPA (NBQX), $GABA_A$ (picrotoxin) and $GABA_B$ (CGS 35348) receptors. We then stimulated in the Schaffer/collateral commissural region and recorded in the stratum moleculare of CA1. Under these conditions, we recorded typical NMDA-mediated slow fEPSPs that were totally inhibited by 50 mM APV. In our hands, 100 mM ethanol produced a very modest but statistically significant inhibition (20.5% inhibition) of NMDA receptor fEPSP slope (Figure 4). In agreement with Morrisett and Swartzwelder (1993), who focussed on the dentate region of the hippocampus, and in contrast to Wright and Weight (1992), we saw a pronounced Mg^{++}-dependency of ethanol's effects on NMDA responses in area CA1 of the hippocampus (Schummers et al., 1997). Thus, ethanol had no effect on NMDA receptor responses in slices bathed with buffer containing 0.1 mM or 0.0 mM Mg^{++}.

We next attempted to determine whether such a modest inhibition of NMDA responses would be sufficient to produce a blockade of LTP equivalent to that seen with 100 mM ethanol. To test this, we first determined the concentration of APV that produced a comparable inhibition of NMDA fEPSPs. We found that 5 μM APV produced an inhibition of the NMDA fEPSP slope of 23.7%. We then tested effects of 5 μM APV on LTP induced by HFS. Our results show that 5 μM APV produced a substantial inhibition of LTP induced by HFS (Figure 5). However, this inhibition was significantly less than that seen with 100 mM ethanol. Because APV is a competitive inhibitor of NMDA, it is possible

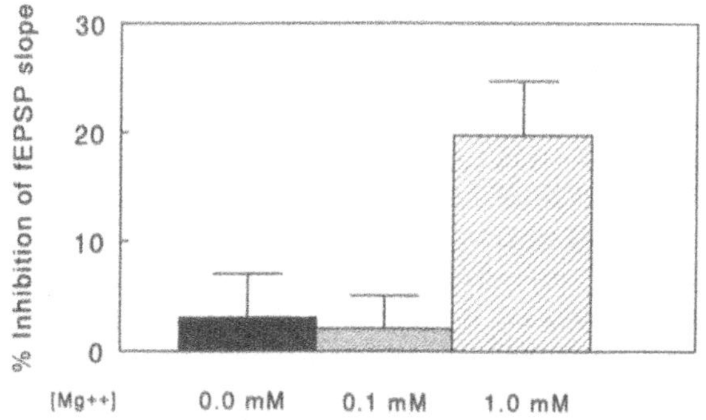

Figure 4. Bargraph showing average data for ethanol's effect on NMDA-mediated fEPSPs in the presence of various concentrations of Mg^{++}. The 100 mM ethanol produced a 22.6 ± 3.6% inhibition of the NMDA-mediated fEPSP when 1.0 mM Mg^{++} was present in the perfusion medium (n = 28 slices from 14 rats). However, when the Mg^{++} was 0.1 or 0.0 mM, the effect of ethanol was 1.7 ± 3.0% (n = 5 slices from 5 rats) and 0.2 ± 2.2% (n = 6 slices from 6 rats) respectively. Shown are the means (± SEM) for the percentage inhibition of the slope of the NMDA-mediated fEPSP. (Modified with permission of Figure 4 from Schummers et al., 1997.)

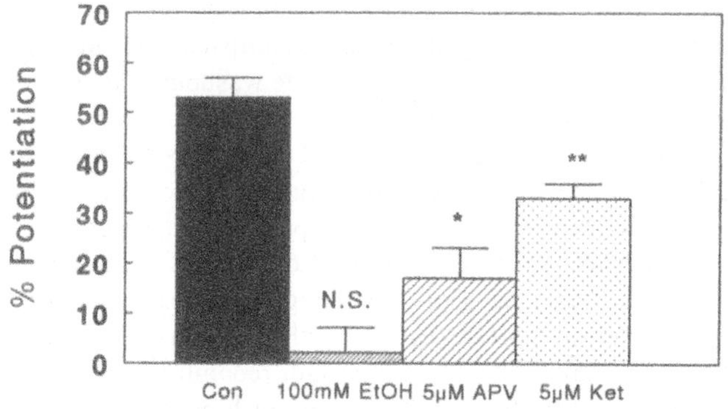

Figure 5. Bargraph comparing the effects of ethanol, APV, and ketamine on LTP induced by HFS. Ethanol and APV, when present, were perfused onto the slice for 10 min before and during the delivery of HFS. After the stimulation perfusion with drug was stopped and fEPSPs were then monitored 30 min after the cessation of the HFS. Percentage potentiation was determined by comparison with basal responses obtained during the 10 to 20 min drug perfusion period before delivery of HFS. In the presence of 100 mM ethanol, HFS did not produce a significant potentiation of the fEPSP slope (4.2 ± 1.7%; n = 5 slices from 5 rats). In the presence of 5 μM APV, HFS produced a significant potentiation of the fEPSP slope (17.3 ± 6.3%; n = 19 slices from 13 rats). In the presence of ketamine, HFS produced a significant potentiation of the fEPSP (32.7 ± 5.3%, n = 12 slices from 9 rats). Two-factor, repeated measures ANOVA indicated significant potentiation of responses taken 27–30 min after HFS, compared with those taken 3–0 min before HFS in the presence of drug (ANOVA, F[2,32]-7.14). Post hoc analysis indicated significant potentiation of responses in the presence of 5 μM APV and 5 μM ketamine, but not in the presence of 100 mM ethanol (paired, two-tailed Student's t test; * = p < 0.02; ** = p < 0.0002; ns [not significant] = p > 0.05). (Modified with permission of Figure 5 from Schummers et al., 1997.)

that the inhibition of the NMDA receptor by APV during tetany was less than that seen in our low frequency stimulation experiments. This might explain the inability of 5μM APV to produce an inhibition of LTP comparable to that seen with 100 mM ethanol. To control for this possibility, we also tested the effects of the non-competitive NMDA antagonist ketamine. We first determined that 5 μM ketamine produced an inhibition (20.1%) of NMDA receptor responses comparable to that of 100 mM ethanol (Figure 5). We then tested the effects of 5 μM ketamine on HFS-induced LTP. Our results show that 5 μM ketamine produced a reduction in LTP. However this inhibition was also significantly less than that seen with 100 mM ethanol (Figure 5).

Since ethanol inhibits NMDA receptor activity (Lovinger et al., 1989; White et al., 1990; Hoffman et al., 1989), which is known to be necessary for LTP induction in area CA1 of the hippocampus, it has been proposed that ethanol's blockade of LTP induction is mediated through the blockade of the NMDA receptor. However, the data we present here demonstrate that the level of NMDA receptor inhibition produced by ethanol in area CA1 of the hippocampal slice is not sufficient to account for ethanol's complete inhibition of LTP induction in this brain region. It is possible that some of the effects of ethanol on LTP could be due to effects on the $GABA_A$ receptor. When we tested the effects of 100 mM ethanol on NMDA responses, GABA responses were blocked. When we tested 100 mM ethanol on LTP (*i.e.*, when GABA responses were not blocked), the effect of ethanol on GABA responses may be potentiating, thus leading to hyperpolarization and NMDA receptor inhibition over and above that due to ethanol's direct effect on the NMDA receptor. This additional NMDA receptor inhibition could thus contribute to the full effect on 100 mM ethanol on LTP. Experiments to address this interesting possibility are currently under way.

In summary, concentrations of ethanol associated with profound intoxication are required to block LTP, a putative cellular substrate of memory. Moreover, the ability of ethanol to inhibit the NMDA receptor can account for some but not all of the effect of ethanol on LTP.

ACKNOWLEDGMENT

The work described in this chapter was supported by grants from the National Institute on Alcohol Abuse and Alcoholism R01AA11428 and R01AA09675.

REFERENCES

Bliss TV, Gardner-Medwin AR (1973) Long-lasting potentiation of synaptic transmission in the dentate area of the unanaesthetized rabbit following stimulation of the perforant path. J Physiol (Lond) 232:357–374.

Bliss TVP, Lomo T (1973) Long-lasting potentiation of synaptic transmission in the dentate area of synaptic transmission in the dentate area of the anaesthetized rabbit following stimulation of the perforant path. J Physiol (Lond) 232:331–356.

Blitzer RD, Gil O, Landau EM (1990) Long-term potentiation in rat hippocampus is inhibited by low concentrations of ethanol. Brain Res 537:203–208.

Castellano C, Populin R (1990) Effect of ethanol on memory consolidation in mice: Antagonism by the imidazobenzodiazepine Ro 15–4513 and decrement by familiarization with the environment. Behav Brain Res 40:67–72.

Davis S, Butcher SP, Morris RGM (1992) The NMDA receptor antagonist D-2-amino-5-phosphonopentanoate (D-AP5) impairs spatial learning and LTP *in vivo* at intracerebral concentrations comparable to those that block LTP *in vitro*. J Neurosci 12:21–34.

Douglas RM, Goddard GV (1975) Long-term potentiation of the perforant path-granule cell synapse in the rat hippocampus. Brain Res. 86:205–15.

Durand D, Carlen PL (1984) Impairment of long-term potentiation in rat hippocampus following chronic ethanol treatment. Brain Res 308:325–332.

Goodwin DW, Crane JB, Guze SB (1969) Alcoholic blackouts: A review and clinical study of 100 alcoholics. Am J Psychiatry 126:191–198.

Hebb DO (1949) The Organization of Behavior. New York: Wiley Publishers.

Hoffman PL, Rabe CS, Moses F, Tabakoff B (1989) N-methyl-D-aspartate receptors and ethanol: Inhibition of calcium flux and cyclic GMP production. J Neurochem 52:1937–1940.

Lima-Landman MT, Albuquerque EX (1989) Ethanol potentiates and blocks NMDA-activated single-channel currents in rat hippocampal pyramidal cells. FEBS Lett 247:61–67.

Lister RG, Eckardt MJ, Weingartner H (1987) Ethanol intoxication and memory. Recent developments and new directions. Recent Dev Alcohol 5:111–126.

Lovinger DM, White G, Weight FF (1989) Ethanol inhibits NMDA-activated ion current in hippocampal neurons. Science 243:1721–1724.

Lowy R (1970) Toxicology of single doses of ethyl alcohol. Int Encycl Pharmacol Therapeut 20:277–299.

Morris RGM (1989) Synaptic plasticity and learning: Selective impairment of learning in rats and blockade of long-term potentiation *in vivo* by the *N*-methyl-D-aspartate receptor antagonist AP5. J Neurosci 9:3040–3057.

Morris RGM, Anderson E, Lynch G, Baudry M (1986) Selective impairment of learning and blockade of long-term potentiation by an N-methyl-D-aspartate receptor antagonist, AP5. Nature 319:774–776.

Morris RGM, Davis S, Butcher SP (1990) Hippocampal synaptic plasticity and NMDA receptors: A role in information storage. Philos Trans R Soc Lond [Biol] 329:187–204.

Morrisett RA, Swartzwelder HS (1993) Attenuation of hippocampal long-term potentiation by ethanol: A patch-clamp analysis of glutamatergic and GABAergic mechanisms. J Neurosci 13:2264–2272.

Mulkeen D, Anwyl R, Rowan MJ (1987) Enhancement of long-term potentiation by the calcium channel agonist Bayer K8644 in CA1 of the rat hippocampus *in vitro*. Neurosci Lett 80:351–355.

Roman F, Staubli U, Lynch G (1987) Evidence for synaptic potentiation in a cortical network during learning. Brain Res 418:221–226.

Ryback RS (1971) The continuum and specificity of the effects of alcohol on memory. A review. Q J Stud Alcohol 32:995–1016.
Schummers J, Bentz SD, Browning MD (1997) Ethanol's inhibition of LTP may not be mediated solely via direct effects on the NMDA receptor. Alcohol Clin Exp Res 21:404–408.
Schwartzkroin PA, Wester K (1975) Long-lasting facilitation of a synaptic potential following tetanization in the *in vitro* hippocampal slice. Brain Res 89:107–119.
Shors TJ, Seib TB, Levine S, Thompson RF (1989) Inescapable versus escapable shock modulates long-term potentiation in the rat hippocampus. Science 244:224–226.
Sinclair JG, Lo GF (1986) Ethanol blocks tetanic and calcium-induced long-term potentiation in the hippocampal slice. Gen Pharmacol 17:231–233.
Tako A, Beracochea D, Lescaudron L, Jaffard R (1991) Differential effects of chronic ethanol consumptions or thiamine deficiency on spatial working memory in Balb/c mice: A behavioral and neuroanatomical study. Neurosci Lett 123:37–40.
Teyler T, DiScenna P (1987) Long-term potentiation. Ann Rev Neurosci 10:131–161.
Weisz DJ, Clark GA, Thompson RF (1984) Increased responsivity of dentate granule cells during nictitating membrane response conditioning in rabbit. Behav Brain Res 12:145–154.
White G, Lovinger DM, Weight FF (1990) Ethanol inhibits NMDA-activated current but does not alter GABA-activated current in an isolated adult mammalian neuron. Brain Res 507:332–336.
Wright JM, Weight FF (1992) Effects of ethanol on NMDA receptor-channels in single channel recordings. Soc Neurosci Abstr 18:
Yamamoto C, Chujo T (1978) Long-term potentiation in thin hippocampal sections studied by intracellular and extracellular recordings. Exp Neurol 58:242–250.
Zimmerberg B, Sukel HL, Stekler JD (1991) Spatial learning of adult rats with fetal alcohol exposure: deficits are sex-dependent. Behav Brain Res 42:49–56.

13

OF MICE AND MINIS

Novel Forms and Analyses of Ethanol Effects on Synaptic Plasticity

Richard A. Morrisett and Mark P. Thomas

The Institute for Neuroscience and
The Division of Pharmacology and Toxicology
The College of Pharmacy
The University of Texas
Austin, Texas 78712-1074

1. INTRODUCTION

Our basic understanding of synaptic neurobiology has undergone major shifts over the past decade. These revisions in this basic field, in combination with the application of advanced techniques in electrophysiology and molecular biology, have had a great impact on the study of the synaptic basis of alcohol-related brain disorders (intoxication, tolerance, dependence, withdrawal hyperexcitability, prenatal alcohol effects and neurotoxicity). One of the more definitive shifts is relative to the original Meyer-Overton derived work. Previously many investigators focused upon lipid effects and related *intra*-neuronal signaling processes (i.e. effects of ethanol on spike generation, conduction and repolarization mechanisms) (Chin and Goldstein, 1977; Hunt, 1985). Since the demonstration of specific effects of ethanol on distinct synaptic receptors, there has been a refocusing of work on the effects of ethanol on *inter*-neuronal signaling processes. Most investigators today agree the involvement of synaptic receptors in alcohol-related disorders is paramount and likely a major site of ethanol action on neural systems.

But such shifts do not occur in a vacuum, and progress in basic neurobiology has continued to evolve in complexity and breadth. Advances in two basic areas of particular pertinence to alcohol neurobiology include the increasing numbers of novel forms of synaptic plasticity as well as new findings about active information processing in dendrites. Concerning the former, it is apparent that a variety of long-term alterations in synaptic transmission likely encompass major mechanisms for the development of neural plasticity, classical conditioning, and learning and memory. Indeed, recent evidence suggests addic-

The "Drunken" Synapse, edited by Liu and Hunt.
Kluwer Academic / Plenum Publishers, New York, 1999.

tion mechanisms exhibit many of the integral components displayed by these plasticity mechanisms (Wickelgren, 1998). Synaptic plasticity is a critical component of information processing in neural systems, and it follows that long-term changes in synaptic transmission are critically involved in the development and expression of the various alcohol-related brain disorders. In this chapter, the impact of ethanol on long-term changes in synaptic transmission will be discussed using various examples and technical approaches. The goals is to present a useful and thorough view of the complex effects of ethanol on information processing in neural systems. The initial focus will involve discussion of ethanol actions on conventional forms of N-methyl-D-aspartate (NMDA) receptor-dependent plasticity and then progress to address ethanol effects on a more recently described form of synaptic potentiation. Finally, some of the developments in basic synaptic neurobiology concerning active properties of dendrites will be reviewed, and examples of their relevance to questions in alcohol neurobiology will be discussed.

2. DISTINCT FORMS OF SYNAPTIC PLASTICITY

One of the major areas of emphasis in basic synaptic neurobiology has been to understand the cellular and molecular alterations responsible for long-term potentiation (LTP) of synaptic transmission. Emphasis has been on the form due to activation of NMDA receptors in hippocampal circuits. More recently, novel forms of synaptic plasticity have been described in various central mammalian synapses. It is now generally agreed that NMDA-dependent and independent forms of both synaptic potentiation and depression exist (Collingridge and Bliss, 1995). Further dissection has revealed multiple subtypes of synaptic plasticity utilizing non-NMDA receptor-dependent mechanisms. These induction and expression mechanisms are not necessarily distinct to specific brain regions and/or synapses. Activation of voltage-gated calcium channels (VGCCs) can result in either enhancement or depression of synaptic transmission in hippocampus through distinct mechanisms (see below). For clarity in this discussion, it seems reasonable to discriminate between synaptic potentiation and depression. Since NMDA receptors encompass a major cellular target of ethanol action, the distinction between ethanol effects on synaptic plasticity mediated via NMDA receptor-dependent and independent forms is further substantiated.

2.1. Distinct Effects of Ethanol on NMDA Receptor-Dependent Synaptic Potentiation and Depression

The major forms of hippocampal synaptic plasticity require NMDA receptor activation for their induction mechanism. Strong activation of NMDA receptors due to high frequency synaptic activation results in an elevation of postsynaptic calcium levels, induction of second messenger systems and retrograde messenger production (Collingridge and Bliss, 1995; Malenka and Nicoll, 1997). The combination of these effects ultimately results in an increased sensitivity of the (S)-α-amino-3-hydroxy-5-methyl-4-isoxazolepropionic acid (AMPA)/kainate subtype of glutamate receptors, the uncovering of previously silent synapses and an increased release of glutamate. All these effects are considered, to varying degrees by different investigators, responsible for the overall increased synaptic strength seen following LTP. Conversely, prolonged low frequency activation of NMDA receptors appears to decrease synaptic strength due to phosphatase activation and a subsequent reduction in the phosphorylation state of AMPA/kainate receptors (Oliet et al.,

1997). It is generally thought that lower levels of Ca^{++} influx activates long-term depression (LTD) expression mechanisms whereas higher levels induce LTP (Johnston, et al., 1996). These distinct plasticity mechanisms share an absolute requirement for activation of NMDA receptors for their induction mechanisms. Comparison of ethanol effects on these forms of NMDA receptor-dependent plasticity reveals some unexpected results.

Figure 1 shows a typical experiment in which NMDA receptor-dependent LTP was induced due to theta-type stimulation. Bath application of ethanol inhibited LTP induction (compare a and b), whereas LTP occurred when the same conditioning stimulus was delivered to the same preparation following ethanol washout (compare c and d). The cumulative data are shown for several slices in the right panel demonstrating the strong inhibitory effect of ethanol against this form of synaptic potentiation that had previously been shown to be dependent upon the activation of NMDA receptors. These data are in strong agreement with those of others and, in combination with slice patch recordings of pharmacologically-isolated NMDA synaptic currents, suggest that inhibition of NMDA receptors is a primary site through which ethanol inhibits NMDA receptor-dependent synaptic potentiation (but see Schummers et al., 1997; Browning et al., in this volume).

We then assessed ethanol effects on NMDA receptor-dependent long-term depression as presented in Figure 2. This experiment was performed for two reasons. First, to compare the ethanol sensitivity on these major forms of NMDA-receptor dependent plasticity and for the additional reason that long-term depression is a powerful mechanism for altering synaptic transmission in central synapses. Figure 2 depicts ethanol effects upon NMDA receptor-dependent LTD in a similarly performed experiment. Low frequency activation of synaptic transmission for 7 minutes resulted in a stable and substantial reduction in the synaptic response. This long-term depression was inhibited by bath application of

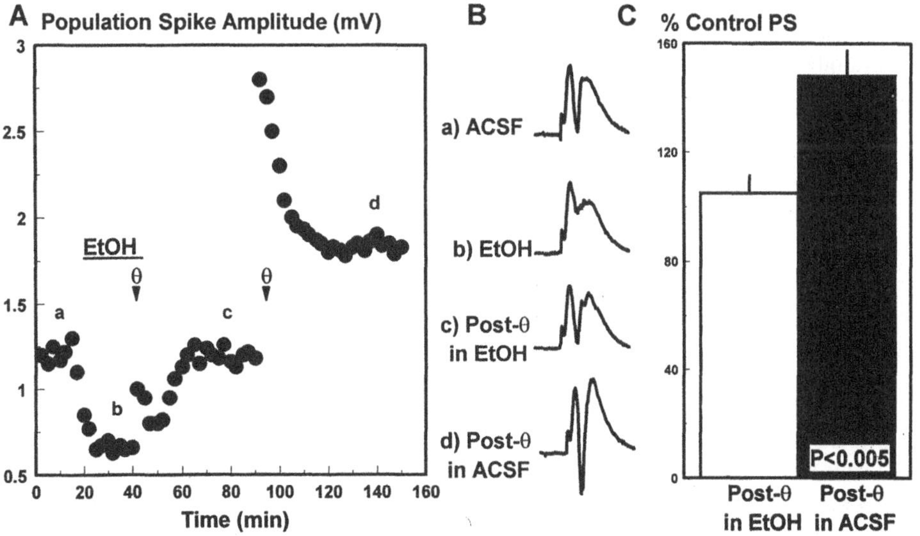

Figure 1. Ethanol inhibits NMDA receptor-dependent synaptic potentiation. Extracellular field potential recording of population spikes in dentate gyrus was used to assess potentiation of synaptic responses due to theta-like conditioning stimuli. (A) The time-course of responses to conditioning stimuli in an individual slice is shown in the absence and presence of ethanol (75 mM). (B) Typical responses as indicated from (A). (C) Cumulative data for all slices studied demonstrated a substantial and significant inhibition of NMDA receptor-dependent LTP by pre-treatment with ethanol (reprinted by permission from Morrisett and Swartzwelder, 1993).

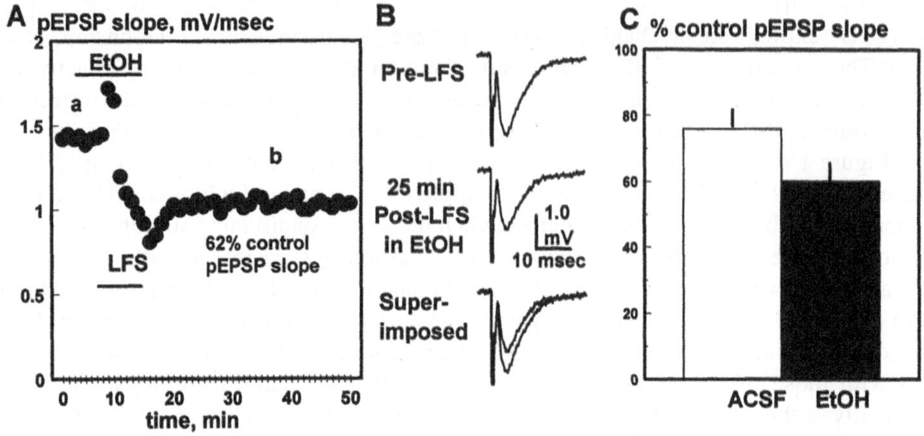

Figure 2. Lack of ethanol effect against NMDA receptor-dependent synaptic depression. (A) Field potential recordings from area CA1 demonstrate the time course of prolonged, low frequency conditioning stimuli delivered in the presence of ethanol (75 mM). (B) Typical responses from the time points indicated in (A). (C) Cumulative data from all slices tested revealed no difference in the expression of NMDA receptor-dependent LTD (D-APV sensitive) if the conditioning stimuli were delivered in artificial cerebrospinal fluid or in ethanol ($p < 0.0001$, for both groups versus pre-low frequency stimulation (LFS) excitatory postsynaptic potential (EPSP) slope, $n = 6–8$ each group indicating LTD induction; $p > 0.5$ for ethanol (EtOH) versus artificial cerebrospinal fluid (ACSF) interaction).

NMDA receptor antagonists but was completely insensitive to bath application of ethanol at the same concentration as that used for the previous LTP experiment. In light of the demonstration of ethanol sensitivity of NMDA receptor-dependent LTP, the lack of an effect of ethanol against another NMDA receptor-dependent process is unexpected.

A number of possible explanations could account for these distinct effects of ethanol on NMDA receptor-dependent LTP versus NMDA receptor-dependent LTD. One explanation involves distinct effects of ethanol on NMDA receptor subtypes differentially coupled to LTP versus LTD induction. Pharmacological differences in the sensitivity of these forms of plasticity to different NMDA receptor antagonists have been reported (Hrabetova and Sacktor, 1997). NMDA receptor-dependent LTP, LTD and depotentiation (reversal of a potentiated response with a LTD-like conditioning paradigm) all were inhibited by the classical NMDA receptor antagonist, D-APV. In contrast, another receptor antagonist, D-CPP, inhibited only LTP and not LTD or depotentiation. These authors suggest that distinct NMDA receptor subtypes might be responsible for these pharmacological differences of the different forms of NMDA receptor-dependent plasticities. Evidence for NMDA receptor subtypes has been long proposed by Monaghan and has been further substantiated pharmacologically with expressed heteromeric NMDA receptors by Buller et al., (1994). Several investigators have observed some relatively small differences in ethanol sensitivity of heteromeric NMDA receptors such that NR2A or 2B-containing subunits display a greater ethanol sensitivity than those that contain NR2C or 2D subunits (Buller, et al., 1995; Chu et al., 1995; Mirshahi and Woodward, 1995). However, the degree of differences observed does not appear to account for the complete insensitivity of LTD to ethanol. Therefore, we feel that evidence for marked differences in the ethanol sensitivity of expressed recombinant NMDA receptors is not particularly supportive of this hypothesis.

There are additional pharmacodynamic aspects that should be considered and may relate to the differences between LTP and LTD sensitivity to ethanol. It might be considered counter-intuitive that the plasticity that exhibits the most ethanol sensitivity occurs under circumstances in which the greatest degree of neuronal activation and calcium influx is induced. One might presume this form of plasticity would be less likely to be sensitive to inhibition, since it may have a greater safety factor. Since we and others have shown equivalent degrees of ethanol inhibition of NMDA receptor-mediated responses under a variety of stimulation intensities and conditions, ethanol inhibition is most likely independent of the degree of NMDA receptor activation (Morrisett and Swartzwelder, 1993; Morrisett, unpublished observations; Schummers et al., 1997).

In native systems though, ethanol is not a completely effective NMDA receptor antagonist. Most investigators report degrees of inhibition on the order of 50–75%, but rarely greater than that level. Responses remaining in the presence of ethanol may be sufficient to support NMDA receptor-dependent LTD induction, while the mechanism for LTP induction may not have a sufficient safety factor to overcome a more marginal degree of NMDA receptor inhibition. Such a consideration is promoted by the similar observation that not all the inhibitory effect of ethanol against NMDA receptor-dependent LTP can be ascribed to ethanol inhibition of NMDA receptor function itself (Schummers et al., 1997). This observation adds a further level of complexity and brings forth the possibility that ethanol may inhibit more distal components of plasticity expression mechanism(s) in addition to its actions on NMDA receptors. Distinguishing between direct effects of ethanol on induction versus expression mechanisms represents an important new direction for studying alcohol effects on plasticity systems (see below).

2.2. Chronic Ethanol Induced Alterations in NMDA Receptor Function

Changes in synaptic strength in conventional LTP and LTD are expressed due to alterations in the presynaptic and postsynaptic function of glutamatergic neurons. Postsynaptically, enhancement of glutamatergic receptor function has been linked to expression of synaptic plasticity (non-NMDA AMPA/kainate subtypes). NMDA receptors are usually quiescent at most central synapses and do not normally contribute substantially to fast synaptic transmission. Nevertheless, a prominent effect of ethanol on glutamatergic synaptic transmission is due to inhibition of plasticity induction mechanisms by virtue of ethanol actions at NMDA receptors. In this regard, ethanol appears to alter non-NMDA glutamatergic transmission via an indirect effect on plasticity induction mechanism(s).

On the other hand, a number of alcohol-related brain disorders may result from long-term alterations in synaptic transmission induced by compensatory responses to chronic ethanol exposure. Perhaps the greatest interest in this area, and clinical relevance as well, has been in the molecular and electrophysiological mechanisms of alcohol withdrawal seizures. In this section, evidence for long-term alterations in NMDA receptors themselves due to chronic ethanol exposure and subsequent withdrawal will be presented. This discussion will also include reference to other pertinent ion channels that have been implicated in ethanol withdrawal effects.

One of the more prevalent hypotheses in this area is that chronic ethanol exposure may activate neurochemical, electrophysiological or genomic mechanisms that ultimately result in long-term compensatory alterations in ion channel function. Subsequent withdrawal from chronic exposure under circumstances, where the compensatory upregulation of ion channel function was still active, may substantially enhance the neuronal excitability of individual neurons. Increased neuronal excitability is a critical component for burst firing, the cellular correlate of an epileptiform event (also known as the paroxysmal depolarizing shift (PDS)). Synchronicity of individual neural elements displaying PDSs may be sufficient to result in outright ictal events (seizures) throughout a neural network. Therefore, identifying the cellular alterations responsible for the induction of the PDS is a critical first step for elucidating the basic mechanisms underlying the induction and expression of alcohol withdrawal seizures.

Upregulation of NMDA receptor function is thought to represent one important component of the compensatory response to chronic ethanol exposure responsible for the induction of ictal events (Lovinger et al., 1989; Grant et al., 1990; Morrisett et al., 1990). Compensatory alterations following chronic exposure of VGCCs (Whittington and Little, 1989; 1991; 1993), as well as gamma-aminobutyric acid $(GABA)_A$ receptor-operated channels, have also been strongly implicated (Buck et al., 1991; Mhartre and Ticku, 1992; Morrow et al., 1988). Discrimination between these different compensatory alterations due to chronic exposure is paramount for identifying the mechanisms for induction of withdrawal seizures. One critical question involves the exact mechanism of these different ion channels in the induction phase of the cellular PDS. A model describing the role of these channels in PDS generation relative to one another is presented in Figure 3. For brevity, alterations in VGCC and $GABA_A$ receptors are grouped together under the heading of *Other Depolarizing Effects* but should be considered as independent.

In the simplest model, an alteration in a single ion channel type may be sufficient to drive the cellular PDS. For example, the increased function of NMDA receptors may be sufficient to *completely* provide the necessary enhancement for the generation of the PDS and therefore ultimately trigger the ictal event. Other alterations in neural function, which occur during chronic exposure, may not actually be involved in the generation of the PDS (Figure 3, model 1). Conversely, increased NMDA receptor function, in combination with changes in VGCC and $GABA_A$ channels, may elicit a more complex series of cellular alterations responsible for PDS generation. In that case, alteration in the activity of all these channel types following chronic exposure might be absolutely necessary for PDS generation. Each channel may therefore be considered an integral component but neither alone is sufficient for PDS generation (Figure 3, Model 2). On the other hand, the PDS generation

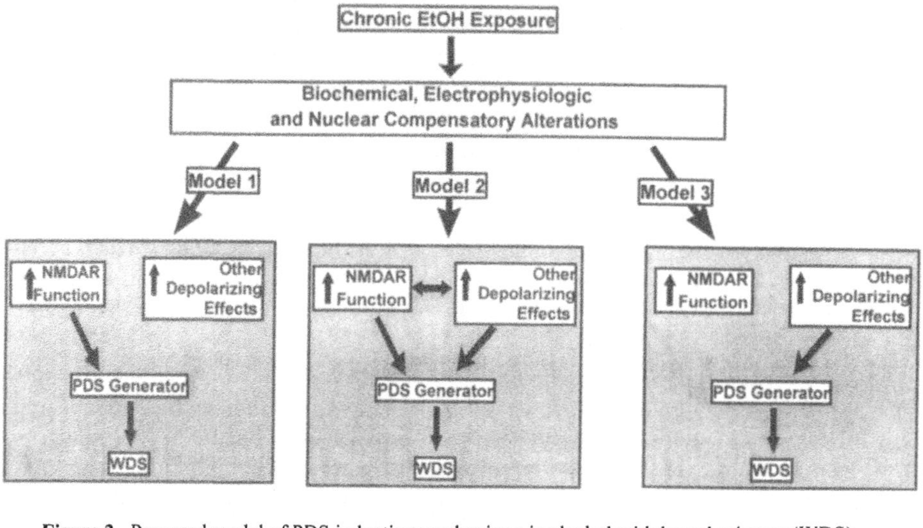

Figure 3. Proposed model of PDS induction mechanisms in alcohol withdrawal seizures (WDS).

mechanism may be due to processes completely independent of alterations in NMDA receptor function. In that case, alterations in NMDA function due to chronic ethanol exposure could be construed simply as an epiphenomenon (Figure 3, Model 3) and increased function of VGCCs or GABA channels alone or in combination would be necessary for the PDS.

The complexity and possibility of cross-talk between NMDA-, VGCC-, and $GABA_A$-dependent systems for PDS generation is daunting but represents an important future focus for understanding of the cellular basis of alcohol withdrawal seizures. A particularly important confound is the likelihood that upregulation of these various ion channels can occur in response to the ictal events themselves (i.e. as with kindling). Therefore, great care should be taken to assess the function of these channels *prior* to the expression of ictal events that occur during withdrawal from chronic exposure.

Our lab has focused upon the involvement of NMDA receptors in alcohol withdrawal hyperexcitability for a number of years (Morrisett et al., 1990; Morrisett, 1994). We have adopted the use of an *in vitro* hippocampal explant model system (Thomas et al., 1998a, b, c) to assess the role of NMDA receptors in the development and expression of withdrawal ictal events as presented in Figure 4. Population field potential recordings from area CA1 of control and chronic ethanol-exposed explants were obtained for up to nine hours. During the first two hours of recording from chronic ethanol-exposed tissue, ethanol was included in the recording solution. The NMDA and non-NMDA components of synaptic transmission were assessed electrophysiologically by virtue of their individual time courses of these different components of the synaptic response. Immediately upon washout of ethanol and therefore removal from chronic exposure, the synaptic response displayed marked changes as depicted in Figure 4, A and B. The slow component of the synaptic response (NMDA receptor-mediated) was substantially elevated relative to that

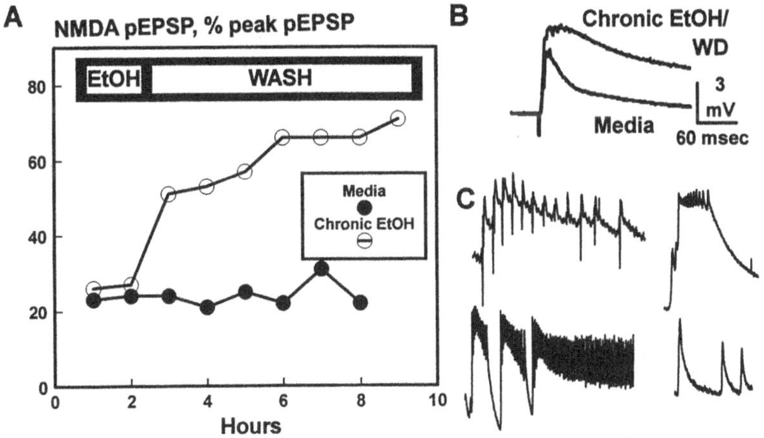

Figure 4. Enhancement of NMDA receptor-dependent synaptic potentials and hyperexcitability following withdrawal from chronic ethanol exposure in hippocampal explants. (A) Typical example of time course of changes in NMDA receptor component in synaptic transmission in area CA1 following withdrawal from chronic ethanol exposure (indicated by the bar) in comparison with an ethanol-naive control explant (media). (B) Typical synaptic responses several hours following ethanol withdrawal. Note the marked increase in the slow (NMDA receptor) component of synaptic transmission in the chronic ethanol treated explant. (C) Typical components of the electrographic seizure recorded following ethanol withdrawal and NMDA receptor-enhancement. Hyperexcitability such as this was completely inhibited by block of NMDA receptors with D-APV (25–50 μM).

seen in control explants (Figure 4B). No changes in the synaptic response elicited by non-NMDA receptors were observed. Thus, we observed a selective enhancement of the NMDA receptor-mediated component of synaptic transmission during withdrawal from chronic exposure. Immediately following ethanol washout from chronic exposure, very long-lasting and well-organized ictal events were observed and these events were inhibited by application of NMDA receptor antagonists.

These findings demonstrate that enhanced NMDA receptor function occurs upon withdrawal from chronic ethanol exposure. These changes precede the expression of alcohol withdrawal hyperexcitability and therefore are potentially related to the mechanisms involved in PDS induction rather than an enhancement due to previous seizure occurrence. Since these potential confounds were absent, we feel that the alterations in NMDA receptor function observed are, at least in part, required for the generation of the ethanol-withdrawal PDS. The results forthcoming from this *in vitro* model system are consistent with either Models 1 or 2 proposed in Figure 3. Alterations in NMDA receptor function due to chronic exposure are at least necessary and may alone be sufficient for the induction of alcohol withdrawal ictal events.

3. NOVEL FORMS OF SYNAPTIC POTENTIATION (NMDA RECEPTOR-INDEPENDENT)

The previous discussions have dealt with two different types of interactions between ethanol with NMDA receptors. The first section dealt with the induction of synaptic plasticity due to direct effects of ethanol on NMDA receptor function, which are expressed via changes in non-NMDA receptor function. The second discussion involved an alteration of NMDA receptor function that occurred following chronic exposure and was unmasked during withdrawal. Studies directed toward understanding such changes in synaptic function have been predominant in alcohol neurobiology for a number of years. Beyond the classical forms of NMDA receptor-dependent synaptic plasticity, as described above, exist the realm of novel forms of synaptic plasticity that are independent of NMDA receptor activation. These more novel forms of synaptic plasticity frequently require the activation of G-protein coupled receptors, usually those in the class of metabotropic glutamate receptors (Linden and Connor, 1995; Gereau and Conn, 1994). However, there are other novel forms of decrimental synaptic potentiation as well as sustained plasticity that are due to activation of VGCCs. Evidence is quite strong that postsynaptic VGCCs contribute to alterations in synaptic transmission underlying certain forms of synaptic plasticity. Strong activation of postsynaptic VGCCs results in synaptic plasticity independent of NMDA receptor activation (Aniksztejn and Ben-Ari, 1991; Grover and Teyler, 1990; Huang and Malenka, 1993; Kullmann et al., 1992). Furthermore, postsynaptic VGCCs may contribute significantly to Ca^{++} influx resulting from LTP-inducing high frequency trains (Miyakawa et al., 1992; Kullmann et al., 1992). Finally, postsynaptic VGCC activation has been linked to phosphorylation-dependent mechanisms enhancing AMPA channel function (Wyllie and Nicoll, 1994). We hypothesized that this form of synaptic potentiation would be sensitive to pharmacological concentrations of ethanol and results from such experiments addressing this hypothesis are described in this section.

VGCCs represent a major site of ethanol action. Acute ethanol exposure inhibits Ca^{++} influx via VGCCs and electrophysiological responses due to the activation of VGCCs (Harris and Hood, 1980; Stokes and Harris, 1982; Leslie et al., 1983; Wang et al., 1994). Whittington and Little (1989; 1993) reported that dihydropyridines block hyperexcitability in hippocampal slices prepared from chronic ethanol-treated animals, whereas control slices were insensitive to these antagonists. These data suggest that enhanced function of L-type VGCCs occurs during ethanol exposure and that this enhancement contributes to hyperexcitability seen during withdrawal. Initially, we addressed this question using standard extracellular recording techniques and determined that ethanol blocked synaptic plasticity indirectly induced by inhibition of potassium channels (performed in the presence of NMDA receptor antagonists therefore resulting in the presumed activation of VGCCs) (Zhang and Morrisett, 1993).

To assess ethanol effects on this novel form of synaptic plasticity more mechanistically, we next adopted blind slice patch whole-cell voltage clamp recording techniques for analysis of miniature synaptic currents (mEPSCs). Recordings of mEPSCs in the presence of tetrodotoxin were used to prevent the complicating factors of polysynaptic and recurrent excitation of neurons. Under these conditions the frequency of the spontaneously occurring mEPSCs can be construed as a direct measure of release events and therefore a measure of presynaptic activity. On the other hand, analysis of the amplitudes of a population of mEPSCs is used as a measure of the postsynaptic function of the synapses innervating the cell being recorded. The data are then analyzed via construction of cumulative occurrence histograms in which each individual event measure (frequency or amplitude) is ranked relative to the entire population of events recorded. Shifts in frequency and amplitude distributions, and therefore pre- and postsynaptic function can be readily observed using such an analysis paradigm.

Figure 5 depicts data from individual neurons recorded where an ATP-regenerating solution was included in the intracellular solution contained in the pipette and all recordings were performed in the presence of the NMDA receptor antagonist, D-APV. The application of depolarizing steps to the neuron through the voltage-clamp recording electrode resulted in a strong potentiation of both the amplitude and frequency of spontaneously occurring miniature EPSCs. The combined enhancement of both measures of synaptic function suggests that an increase in both the presynaptic release process as well as an increase in the function of AMPA receptors occurred due to the conditioning paradigm. In agreement with previous investigators (Miyakawa et al., 1992; Kullmann et al., 1992), this form of synaptic potentiation appeared dependent upon a rise in postsynaptic Ca^{++} due to activation of L-type VGCCs, since application of the selective antagonist, nifedipine, prevented the potentiation due to the conditioning paradigm (Figure 5).

While an increase in the function of postsynaptic receptors might be expected following strong direct activation of the cell, the presynaptic increase in frequency of mEPSCs is a more surprising finding. This suggests that a retrograde messenger may be generated due to the postsynaptic depolarizing steps which subsequently increases the activity of the presynaptic terminal. Interestingly, Wyllie et al. (1994) originally reported this presynaptic enhancement and demonstrated that the increase in mEPSC frequency was insensitive to inhibitors of nitric oxide synthesis, indicating that the retrograde messenger is likely not that chemical. Our data are in very close agreement with those originally describing this form of potentiation (Kullmann et al., 1992; Wyllie et al., 1994; Wyllie and Nicoll, 1994), and we are confident that we have reproduced this novel form of synaptic potentiation.

As described above, we have previously published evidence that ethanol inhibition of synaptic plasticity due to indirect activation of VGCCs (Zhang and Morrisett, 1993). Therefore, we assessed the ethanol sensitivity of VGCC-dependent synaptic potentiation recorded using the blind slice patch whole-cell voltage clamp configuration. Figure 5 shows the effect of ethanol on the expression of VGCC-dependent synaptic potentiation.

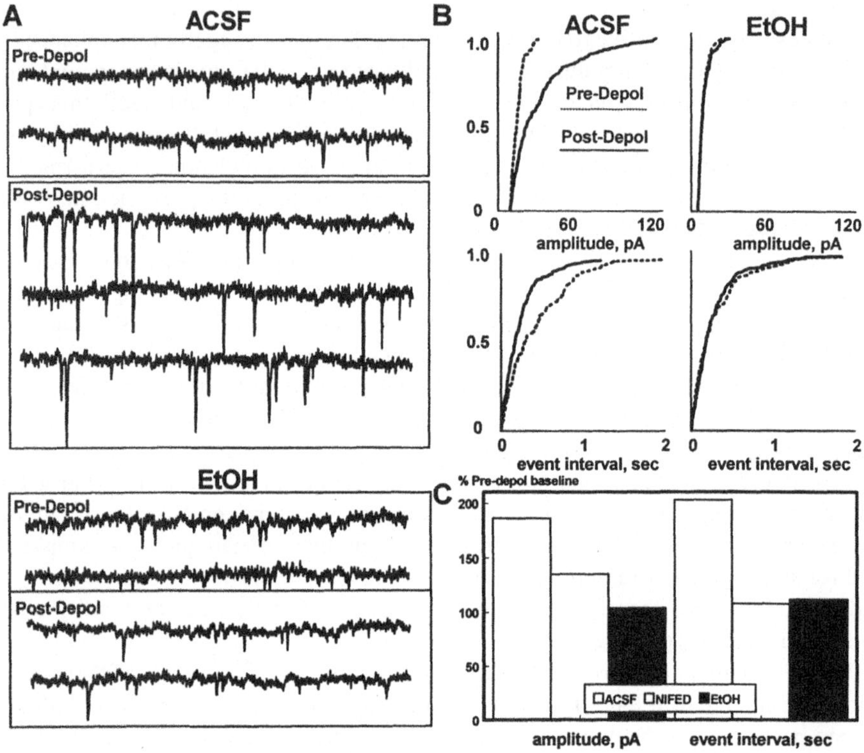

Figure 5. L-type VGCC-dependent synaptic potentiation is inhibited by ethanol. (A) Comparison of mEPSCs recorded from pyramidal cells in area CA1 before (pre-Depol) and after (post-Depol) depolarizing steps delivered post-synaptically in the absence of ACSF and presence of EtOH (75 mM). (B) Representative cumulative amplitude (top) and frequency histograms (bottom) for the cells presented in A. Note the strong shifts in both measures following the depolarizing conditioning pulses. (C) Cumulative data for changes in mEPSC parameters under normal conditions and following block of L-type VGCCs (nifedipine) as well as in the presence of ethanol ($n = 8, 4, 6$ for ACSF, nifedipine and ethanol exposed cells respectively; $p < 0.003$ and 0.05 for mEPSC amplitude measures for ACSF versus nifedipine or EtOH, respectively; $p < 0.001$ and 0.0025 for mEPSC frequency measures for ACSF versus nifedipine or EtOH, respectively).

Neither the increase in mEPSC frequency nor the increase in mEPSC amplitude was observed when the conditioning stimulation was delivered in the presence of bath-applied ethanol (75 mM). We conclude that ethanol virtually completely antagonized synaptic potentiation mediated by strong activation of L-type VGCCs and this represents a novel mechanism for ethanol to modulate information processing through effects on glutamatergic synaptic transmission.

One explanation for the inhibitory effect of ethanol against this form of potentiation is quite straightforward. Since ethanol inhibits L-type VGCCs directly, it is logical to assume that ethanol inhibits the initial step in VGCC synaptic potentiation. Ethanol action would be considered to be quite similar to the manner in which it is thought to inhibit NMDA receptor-mediated LTP, through direct inhibition of NMDA channel function.

There are other possibilities for ethanol inhibition of VGCC-dependent LTP that should not be overlooked and may indeed have some wider-reaching impact. It is conceivable that the effect of ethanol against these Ca^{++}-permeable ion channels (NMDA or VGCC-operated) is not the *sole* site for ethanol inhibition of synaptic potentiation. Some other more distal site in the conversion from the induction to the expression mechanism(s) may also be inhibited by ethanol. If the same (or a similar) site in the induction-expression mechanism is utilized by both NMDA receptor-dependent LTP and VGCC-dependent synaptic potentiation, ethanol inhibition of this site could be responsible for the observed effect against both forms of synaptic plasticity. Conversely, distinct sites could be required for both plasticity mechanisms and both could be independently inhibited by ethanol, although this more complex mechanism seems less likely. A common distal site in the expression of NMDA receptor-dependent LTP and VGCC potentiation is one that warrants consideration.

Comparing synaptic potentiation via these two different conditioning paradigms (activation of VGCCs or NMDA receptors) can be used to assess the commonality of expression mechanisms of LTP. Kullmann et al. (1992) did not observe occlusion of VGCC potentiation by prior induction of NMDA receptor-dependent LTP when assessed within the same neuron. The rundown of VGCC potentiation (the spontaneous decrease following prolonged periods of time after whole-cell break-in) did prevent a complete assessment of an interaction between these different forms of synaptic enhancement. The conversion of decrimental VGCC potentiation to a stable form has been observed due to phosphatase inhibition or by pairing synaptic activation (in the presence of NMDA receptor antagonists) with postsynaptic depolarization (Kullmann et al., 1992). The conversion to stable synaptic potentiation versus the decrimental form seen when depolarizing steps alone were delivered suggests that second messenger systems are critically involved in the stabilization process. Activation of metabotropic receptors and subsequent recruitment of protein kinases may occur, analogous to that seen in cerebellar metabotropic receptor dependent LTD (Linden and Connor, 1994). These mechanisms may represent a major site through which ethanol might inhibit expression of synaptic potentiation. Since ethanol has been demonstrated to interact with G-protein coupled systems, the possibility of ethanol interrupting a more distal site for the potentiation mechanism is substantiated. These combined effects of ethanol on potentially distinct forms of plasticity, which appear to share components of their induction mechanism at the ion channel and intracellular signaling levels, warrant more exacting study and dissection.

4. PROPAGATING BACK TO THE FUTURE: DENDRITIC ACTION POTENTIALS

Our basic concepts concerning the role of synaptic and voltage-dependent channels in information processing in neural systems have been markedly revised in the past few years. The unidirectional propagation of information from dendritic to somatic and subsequently axonal regions of the neuron has been an accepted tenet of neurobiology for most of this century (Bishop, 1956; Johnston et al., 1996). Dendrites have been generally viewed as having a largely passive role in the temporal and spatial summation of synaptic potentials. This summation of excitatory and inhibitory synaptic potentials and the propagation of this algebraically summed potential into the soma and the axon hillock was considered a critical requirement for action potential initiation, conduction and shaping (Johnston et al, 1996). Indeed, the generation of the action potential was thought to be localized solely to these latter regions.

While these basic tenets of cellular neurobiology were developed and largely accepted, other evidence for more complex processing was also forthcoming. A number of studies indicated that active conductances could be demonstrated in dendrites. Chang

(1951) demonstrated dendritic action potentials. The original description of the "fast prepotentials" were made by Spencer and Kandel (1961), who suggested the dependence of these events on dendritic action potentials. Presently, there is an abundance of evidence for the presence of dendritic voltage-dependent channels of all three major types (Na^+, Ca^{++}, and K^+; Johnston, et al., 1996). The advent of patch clamp recordings in brain slice preparations with the powerful advances in imaging techniques (high resolution infrared video microscopy and differential interference optics) has led to a remarkable increase in the ability of investigators to identify and record from dendritic sites. The combination of this technology with the simultaneous measurement of Ca^{++} signaling using fluorescent indicators has resulted in rapid leaps in our understanding of the role of dendritic voltage-dependent channels in both synaptic integration and plasticity.

The dendritic population of the various types of voltage-dependent channels has had a multi-dimensional impact on basic concepts in information processing in neural systems. There are at least three major aspects of dendritic voltage-gated channels that would appear to have distinct functional consequences. First, the sub-threshold and direct contribution of voltage-gated channels to synaptic inputs appear to result in either "boosting" (for dendritic Na^+ and Ca^{++} channels) or inhibition (dendritic K^+ channels) of synaptic responses due to the otherwise physiologic activation of ligand-gated ion channels. A related effect includes alterations in dendritic time and length constants ("shaping") and therefore involves "passive" dendritic filtering of the synaptic potential as it propagates past branch points and into soma. A second major impact relates to the role of back-propagating action potentials and therefore can be distinguished on the basis of a supra-threshold dependence. Alterations in dendritic synaptic responses, especially due to those channels clustered around the apical dendrites, are especially sensitive to back-propagating action potentials (which may either be Na^+-dependent , Ca^{++}-dependent, or a combination of both). Finally, the role of Ca^{++} in the induction phases of both synaptic potentiation and depression dictates a major potential role for dendritic VGCCs in the induction of multiple types of synaptic plasticity. Dendritic VGCCs are an obvious site for a major source for Ca^{++} required for the induction of VGCC-dependent synaptic potentiation. Any of these distinct effects of dendritic active conductances independently would be of paramount importance, together these have a great synergism of importance for the basic neuroscience research field.

4.1. Ethanol and Dendritic Voltage-Gated Channels

When we consider the complexity and sheer number of these distinct mechanisms of dendritic voltage-dependent regulation of synaptic transmission and plasticity, one is struck by the number of possible questions applicable in basic neurobiology. The combination of such questions to the field of alcohol neurobiology will almost certainly result in major revisions in our understanding of alcohol effects on information processing. One putative example concerns the differing degrees of ethanol inhibition various laboratories observe between recombinant NMDA receptors versus those studied in native systems (Buller et al., 1995). One possible mechanism for this discrepancy could be due to the effect of ethanol to inhibit subthreshold dendritic Ca^{++} channel-dependent boosting of NMDA responses. Since ethanol also inhibits VGCCs at relevant concentrations, the combination of ethanol effects on both NMDA responses as well as VGCC boosting of NMDA responses may result in a greater degree of inhibition in native tissue than that observed due to ethanol effects on NMDA channels alone. Since oocytes have no endogenous VGCCs, their absence could result in an apparent decrease in the sensitivity of the cell to ethanol. Even under voltage-clamp conditions, errors due to inadequacies of space clamp may allow for the involvement of distal dendritic voltage-gated channels elicited by synaptic or exogenous application of agonist. Under such conditions the overall electrophysiological response may be amplified by dendritic channels and therefore be a mixed response having differing ethanol sensitivity than that elicited by NMDA receptors alone.

The dependence of alcohol-related brain disorders on synaptic transmission is incontrovertible. While the technical demands for assessment of dendritic voltage-dependent channels may be great, such an analysis will be required for an accurate understanding of the cellular and electrophysiological effects of ethanol on information processing. Regenerative dendritic action potentials and their role in the generation of paroxysmal depolarizing shifts represent a potentially critical mechanism in alcohol withdrawal. Direct as well as G-protein dependent effects of ethanol on the regulation of dendritic K^+ channels in the nucleus accumbens (among other regions) will surely impact our understanding of dependence and drug seeking behavior. Alterations in the expression level of dendritic voltage-dependent channels are a likely outcome following chronic exposure and may result in alcohol-related neurotoxicity or prenatal ethanol deficits. All of these are simply examples, but they do represent the potential for important evaluation in their respective subfields of alcohol neurobiology.

5. CONCLUSIONS: DRUNKEN SYNAPSES, DENDRITES, SOMA, AND HILLOCKS

Plasticity of synapses involves processing at many cellular levels involving different induction mechanisms, expression mechanisms, different ion channels (be they ligand or voltage-gated), and the involvement of different types of second messenger systems. An efficient approach for understanding ethanol effects on synaptic plasticity would distinguish, as clearly as possible, these particular processes from one another and assess ethanol effects in isolation. Assembling an accurate model of the direct acute effects of ethanol alone is an extensive endeavor. Application of such questions to ethanol-related pathologies (chronic exposure, drug seeking behavior or prenatal effects) presents several additional levels of complexity. Many of these questions and approaches are relevant to other disorders of synaptic function or drugs of abuse. This prompts one to wonder if we are not so close to the end of the decade of the brain but at the start of a new millennium of excitement and understanding in basic and applied neurobiology.

ACKNOWLEDGMENT

The authors wish to gratefully acknowledge the support of the Alcoholic Beverage Medical Research Foundation and the National Institute of Alcohol Abuse and Alcoholism (R01 AA 9230 to RAM).

REFERENCES

Aniksztejn L, Ben-Ari Y (1991) Novel form of long-term potentiation produced by a K^+ channel blocker in the hippocampus, Nature 349:67–69.

Bishop G (1956) Natural history of the nerve impulse. Physiol Rev 36:376–99.

Buck KJ, Hahner L, Sikela J, Harris RA (1991) Chronic ethanol treatment alters brain levels of gamma-aminobutyric $acid_A$ receptor subunit mRNAs: relationship to genetic differences in ethanol withdrawal seizure severity. J Neurochem 57:1452–1455.

Buller AL, Clark HC, Schneider BJ, Morrisett RA, Monaghan DT (1994) Anatomical and pharmacological properties of NMDA receptor subtypes: Native pharmacology predicted by subunit composition. J Neurosci 14:5471–5484.

Buller AL, Morrisett RA, Monaghan DT (1995) Glycine modulates ethanol inhibition of heteromeric N-methyl-D-aspartate receptors expressed in *Xenopus* oocytes. Mol Pharmacol 48:717–723.

Chang H-T (1951) Dendritic potential of cortical neurons produced by direct electrical stimulation of the cerebral cortex. J Neurophysiol 14:1–21.

Chin JH, Goldstein DB (1977) Drug tolerance in biomembranes: a spin label study of the effects of ethanol. Science 196:684–687.

Chu B, Anantharam V, Treistman SN (1995) Ethanol inhibition of recombinant heteromeric NMDA channels in the presence and absence of modulators. J Neurochem 65:140–148.

Collingridge GC, Bliss TVP (1995) Memories of NMDA receptors and LTP. Trends Neurosci 18:54–56.

Gereau IV RW, Conn PJ (1994) A cyclic AMP-dependent form of associative synaptic plasticity induced by coactivation of β-adrenergic receptors and metabotropic glutamate receptors in rat hippocampus. J Neurosci 14:3310–3318.

Grant KA, Valverius P, Hudspith M, Tabakoff B (1990) Ethanol withdrawal seizures and the NMDA receptor complex. Eur J Pharmacol 176 289–296.

Grover LM, Teyler TJ (1990) Two components of long-term potentiation induced by different patterns of afferent activation. Nature 347:477–479.

Harris RA, Hood WF (1980) Inhibition of synaptosomal calcium uptake by ethanol. J Pharmacol Exp Therap 213:562–567.

Hrabetova S, Sacktor TC (1997) Long-term potentiation and long-term depression are induced through pharmacologically distinct NMDA receptors. Neurosci Lett 226:107–10.

Huang YY, Malenka, RC (1993) Examination of TEA-induced synaptic enhancement in area CA1 of the hippocampus: The role of voltage-dependent Ca^{++} channels in the induction of LTP. J Neurosci 13:568–576.

Hunt WA (1985) Alcohol and Biological Membranes, The Guilford Press, New York.

Johnston D, Magee JC, Colbert CM, Christie BR (1996) Active properties of neuronal dendrites. Ann Rev Neuroscience 19:165–186.

Kullmann DM, Perkel DJ, Manabe T, Nicoll RA (1992) Ca^{2+} entry via postsynaptic voltage-sensitive calcium channels can transiently potentiate excitatory synaptic transmission. Neuron 9:1175–1183.

Leslie SW, Barr EM, Chandler J, Farr RP (1983) Inhibition of fast- and slow-phase depolarization-induced synaptosomal calcium uptake by ethanol. J Pharmacol Exp Therap 225:571–575.

Linden DJ, Connor JA (1995) Long-term synaptic depression. Ann Rev Neurosci 18:319–358.

Lovinger DM, White G, Weight FF (1989) Ethanol inhibits NMDA-activated ion current in hippocampal neurons. Science 243:1721–1724.

Malenka RC, Nicoll RA (1997) Silent synapses speak up. Neuron 19:473–6.

Mhatre MC, Ticku MK (1992) Chronic ethanol administration alters gamma-aminobutyric acidA receptor gene expression. Mol Pharmacol 42:415–422.

Mirshahi T Woodward JJ (1995) Ethanol sensitivity of heteromeric NMDA receptors: Effects of subunit assembly, glycine and NMDAR1 Mg^{++}-insensitive mutants. Neuropharmacol 34:347–355.

Miyakawa H, Ross WN, Jaffe D, Callaway JC, Lasser-Ross N, Lisman, Johnston D (1992) Synaptically activated increases in Ca^{2+} concentration in hippocampal CA1 pyramidal cells are primarily due to voltage-gated Ca^{2+} channels. Neuron 9:1163–1173.

Morrisett RA, Rezvani AH, Overstreet D, Janowsky DS, Wilson WA, Swartzwelder HS (1990) MK-801 potently inhibits alcohol withdrawal seizures in rats. Eur J Pharmacol 176:103–105.

Morrisett RA, Swartzwelder HS (1993) Attenuation of hippocampal long-term potentiation by ethanol: A patch clamp analysis of glutamatergic and GABAergic mechanisms. J Neurosci 13:2264–2272.

Morrisett RA (1994) Potentiation of NMDA receptor-dependent afterdischarges in rat dentate gyrus following *in vitro* ethanol exposure. Neurosci Lett 167:175–178.

Morrow AL, Suzdak PD, Karanian JW, Paul SM (1988) Chronic ethanol administration alters gamma-aminobutyric acid, pentobarbital and ethanol-mediated $^{36}Cl^-$ uptake in cerebral cortical synaptoneurosomes. J Pharmacol Exp Therap 246:158–164.

Oliet SH, Malenka RC, Nicoll RA (1997) Two distinct forms of long-term depression coexist in CA1 hippocampal pyramidal cells. Neuron 18:969–982.

Schummers J, Bentz S, Browning MD (1997) Ethanol's inhibition of LTP may not be mediated solely via direct effects on the NMDA receptor. Alcohol: Clin Exp Res 21:404–8.

Spencer WA, Kandel E. (1961) Electrophysiology of hippocampal neurons: IV. Fast pre-potentials. J Neurophysiol 24:272–85.

Stokes JA, Harris RA (1982) Alcohols and synaptosomal calcium transport. Mol Pharmacol 22:99–104.

Thomas MP, Davis MI, Monaghan DT, Morrisett RA (1998a) Organotypic brain slice cultures for functional analysis of alcohol-related disorders: Novel versus conventional preparations. Alcoholism: Clinical Exp Res 22:51–59.

Thomas MP, Webster WW, Norgren RB, Monaghan DT, Morrisett RA (1998b) Survival and functional demonstration of interregional pathways in fore/midbrain slice explant cultures. Neuroscience 85:615–626.

Thomas MP, Monaghan DT, Morrisett RA (1998c) Evidence for a causative role of NMDA receptors in an *in vitro* model of alcohol withdrawal hyperexcitability. J Pharmacol Exp Therap (in press)

Wang X, Wang G, Lemos JR, Treistman SN (1994) Ethanol directly modulates gating of a dihydropyridine-sensitive Ca2+ channel in neurohypophysial terminals. J Neurosci 14(9):5453–5460.

Whittington MA and Little HJ (1989) Nitredipine prevents the ethanol withdrawal syndrome when administered chronically with ethanol prior to withdrawal. Brit J Pharmacol 94:385P.

Whittington MA and Little HJ (1991) Nitredipine, given during drinking, decreases the electrophysiological changes in the isolated hippocampal slice, seen during ethanol withdrawal. Brit J Pharmacol 103:1677–1684.

Whittington MA, Little HJ (1993) Changes in voltage-operated calcium channels modify ethanol withdrawal hyperexcitability in mouse hippocampal slices. Exp Physiol 78:347–370.

Wickelgren I (1998) Teaching the brain to take drugs. Science 280: 2045–2047.

Wyllie DJA, Nicoll RA (1994) A role for protein kinases and phosphatases in the Ca^{2+}-induced enhancement of hippocampal AMPA receptor-mediate synaptic responses. Neuron 13:635–643.

Wyllie DJA, Manabe T, Nicoll R (1994) A rise in postsynaptic Ca^{++} potentiates miniature excitatory postsynaptic currents and AMPA responses in hippocampal neurons. Neuron 12:127–138.

Zhang G, Morrisett RA (1993) Ethanol inhibits tetraethylammonium-induced synaptic plasticity in area CA1 of rat hippocampus. Neurosci Lett 156:27–30.

14

ETHANOL SUPPRESSION OF HIPPOCAMPAL PLASTICITY

Role of Subcortical Inputs

Scott C. Steffensen

Department of Neuropharmacology
Scripps Research Institute
10550 North Torrey Pines Road
La Jolla, California 92037

1. INTRODUCTION

In his opening remarks, Shepherd posited the following concept: the behavioral consequences of synaptic transmission are expressed in terms of circuits. In echoing this perspective, our research efforts are devoted to characterizing the functional neuronal ensembles underlying the amnestic, intoxicating, and rewarding properties of alcohol.

Acute ethanol intoxication produces deficits in learning and memory in humans that have been attributed to its effects on the acquisition of new information (Lister et al., 1987). For example, long-term potentiation (LTP) is a model of synaptic plasticity extensively studied in the dentate gyrus as well as in other regions of the rodent hippocampus whose induction correlates with the acquisition of several learning tasks (Bliss and Lomo, 1973; Morris et al., 1986). Current evidence indicates that ethanol impairs cognition and blocks hippocampal LTP primarily by its pharmacological action at the N-methyl D-aspartate (NMDA)-receptor-channel complex (Lovinger et al., 1989; Morrisett and Swartzwelder, 1993). However, altered expression of NMDA receptors and their sensitivity to ethanol have been shown to occur throughout development (Williams et al., 1993; Swartzwelder et al., 1995), but to a lessor degree in adult rats (Blitzer et al., 1990; Steffensen et al., 1993; Givens, 1995). Indeed, there are marked differences between *in vivo* and *in vitro* preparations for ethanol effects on synaptic transmission and synaptic plasticity, which may be based on the developmental age of the model system.

In our studies, we tested the effects of acute intoxicating doses of ethanol as well as locally applied ethanol on synaptic components of the hippocampal trisynaptic circuitry including, the dentate gyrus, CA3 and CA1 subfields *in vivo*. In our hands, whether it be

The "Drunken" Synapse, edited by Liu and Hunt.
Kluwer Academic / Plenum Publishers, New York, 1999.

in the anesthetized or the freely-moving preparation, ethanol appears to have two primary effects on hippocampal responses. One appears to be direct on hippocampal neurons and the other to be subregion specific and results from actions on a remote structure(s) that influences hippocampal function. The two primary effects of ethanol are: 1) whether administered systemically or locally, ethanol increases the excitability of putative gamma-aminobutyric acid (GABA)ergic interneurons and concomitantly decreases monosynaptic afferent-evoked population spike (PS) amplitudes in all subregions of the hippocampus and 2) systemic, but not local, ethanol exposure enhances recurrent inhibition and suppresses LTP in the dentate gyrus subregion of the hippocampus via actions on subcortical structure(s) that project to the hippocampus. The effects of ethanol on cellular activity, evoked synaptic activity as well as long-term plasticity in the dentate gyrus, CA3 and CA1 hippocampi will be described and the differences between subregions as well as between *in vivo* and *in vitro* preparations will be emphasized.

2. RESULTS

Figure 1 depicts a schematic view of the dentate gyrus, showing some of its microcircuitry. It will serve as a model for the other subregions of the hippocampus, with a few important differences. There are more than a score of morphologically- and neurochemically-distinct interneuronal types in the hilar region of the dentate gyrus (Amaral, 1978). However, electrophysiologically we can only identify four distinct neuron types. The dentate granule cells under halothane anesthesia are not spontaneously active, but they can be driven by perforant path monosynaptic afferent stimulation. Granule cells send mossy fibers to CA3 pyramidal cells and mossy fiber collaterals to GABAergic basket cells, GABAergic hilar interneurons and glutamatergic hilar mossy cells. The typical basket cell is located either in or subjacent to the granule cell layer (Ribak et al., 1978) and probably mediates both feed-forward as well as feedback inhibition (Knowles and Schwartzkroin, 1981). Hilar interneurons are typically situated subjacent to the granule cell layer or in the hilus and appear to project longitudinally within the ipsilateral hippocampus (Kosaka et al., 1985), and perhaps commissurally, to the contralateral hippocampus. Hilar interneurons evince similar evoked and spontaneous discharge activities as basket cells. They differ electrophysiologically by not producing multiple discharges when evoked by perforant path stimulation. They also appear to mediate feedback inhibition of dentate granule cells, but we feel that they also mediate inhibition of other putative GABAergic basket cells as reported for CA1 (Lacaille et al., 1987). Hilar mossy cells differ electrophysiologically from basket cells and hilar interneurons by their pronounced bursting activity. Hilar mossy cells are believed to be recurrent excitatory interneurons to dentate granule cells but it has also been demonstrated that they excite hilar interneurons or basket cells (Scharfman and Schwartzkroin, 1988). We consistently observe that activation of dentate granule cells by perforant path stimulation is required for activation of hilar mossy cells. Hilar mossy cells, like hilar interneurons, send projections to the contralateral hippocampus (Laurberg and Sorensen, 1981) and can be easily activated antidromically by commissural stimulation.

As mentioned previously, homologous circuitry exists in the other regions of the hippocampus with important differences. For instance, it is known that CA3 pyramidal cells synapse upon each other and also that they subserve the same arrangement that is provided by the hilar mossy cells in the dentate gyrus. In other words, they appear to be recurrent excitatory neurons for CA1 pyramidal cells (Schwartzkroin et al., 1990).

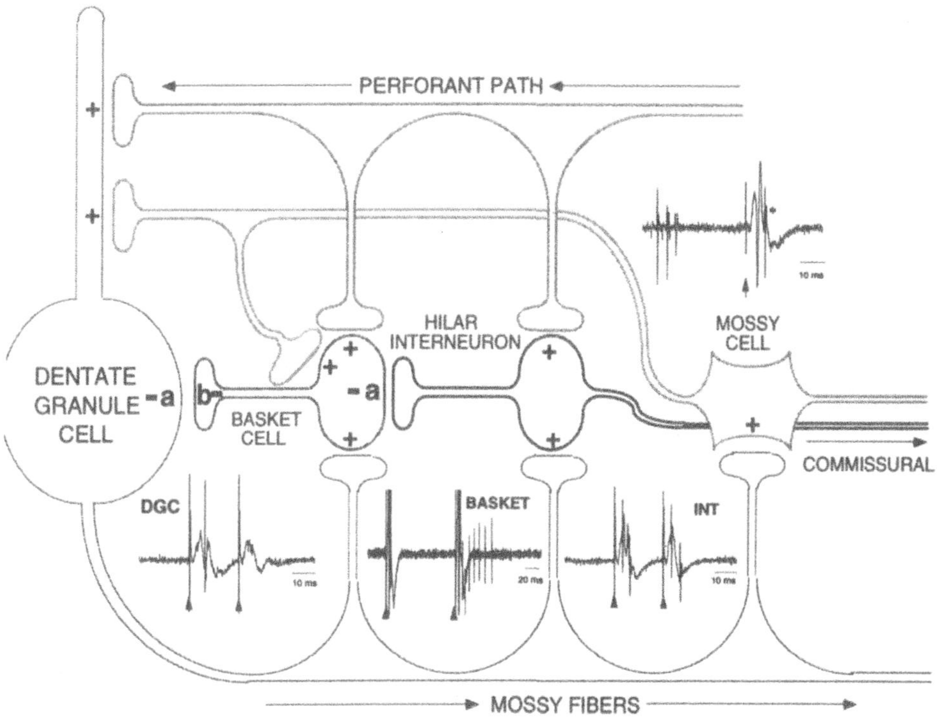

Figure 1. Electrophysiological characterization of neural circuitry in the dentate gyrus. Glutamatergic dentate granule cells, GABAergic basket cells and GABAergic hilar interneurons receive excitatory input from the entorhinal cortex via the perforant path. Dentate granule cells send mossy fibers to CA3 and mossy fiber collaterals to basket cells, hilar interneurons and hilar mossy cells. Basket cells inhibit dentate granule cells by feedforward activation via the perforant path and by feedback activation via recurrent collaterals from dentate granule cells. Hilar mossy cells excite dentate granule cells and interneurons via recurrent collaterals from dentate granule cells. Hilar interneurons inhibit basket cells by feedforward activation via the perforant path and feedback activation via recurrent collaterals from dentate granule cells. Figure insets are representative recordings demonstrating at least one distinguishing criteria for differentiation of cell types. All traces are filtered responses (1–3 kHz). Dentate granule cells were found near reversal of the population excitatory postsynaptic potential (EPSP), showed little or no spontaneous activity, discharged within the time domain of the PS, followed high frequency mossy fiber stimulation and spike discharges produced by perforant path stimulation were inhibited at inter-stimulus intervals producing inhibition of conditioned population spikes (shown at 20 ms inter-stimulus interval and threshold for the PS). Basket cells were spontaneously active non-bursting cells (mean spontaneous firing rate = 7.9 Hz ± 0.6), found approximately 50–150 µm below population EPSP (pEPSP) reversal, demonstrated marked latency fluctuations with high frequency mossy fiber stimulation, produced multiple discharges with perforant path stimulation outside the envelope of the evoked field potential (shown at 50% maximum PS stimulus level and 80 ms inter-stimulus interval, had perforant path spike latencies that typically occurred earlier than the PS and conditioned spikes were not inhibited at inhibitory inter-stimulus intervals. Basket cells were also driven orthodromically, but not antidromically, by contralateral hilar stimulation. Hilar interneurons had similar characteristics as basket cells but did not produce multiple discharges following the field potential. Hilar mossy cells were spontaneously active bursting cells (mean spontaneous firing rate = 4.8 Hz ± 0.7), encountered approximately 100–400 µm below reversal of the pEPSP, had perforant path spike latencies that were later than the PS and discharges produced by stimulation of the perforant path were not inhibited at inhibitory inter-stimulus intervals for the PS. Finally, hilar mossy cells were driven antidromically by contralateral hilar stimulation.

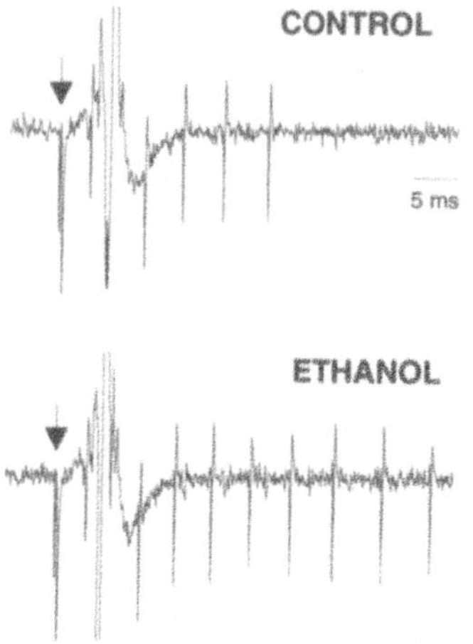

Figure 2. Systemic ethanol exposure increases interneuron discharges in CA3. Interneurons are characterized by short duration action potentials, the lack of bursting activity and the lack of spike inhibition at inter-stimulus intervals producing inhibition of the PS. The representative filtered recordings above demonstrate multiple interneuron discharges evoked in CA3 by stimulation of the commissural input. Arrow marks stimulus artifact. In the top filtered trace shown above, one discharge occurs before and four after the field potential. Following 1.2 g/kg intraperitoneal ethanol, interneuron discharges increased in number. The time to onset of effect is 3–5 min with the peak effect occurring at 20–30 min and recovery after 2 hr.

We studied the effects of systemic and local administrations of ethanol on spontaneous firing rates and evoked discharges of interneurons in the hippocampus. As a general rule, ethanol appears to decrease the spontaneous firing rate of all interneurons except hilar mossy cells, whose firing rate is consistently increased by systemic ethanol exposure. Notwithstanding that most hippocampal neurons are inhibited by ethanol, monosynaptic afferent activation of basket cells in **all** subregions appears to be increased following ethanol. Figure 2 shows the effects of acute intoxicating doses of ethanol (1.2 g/kg) on putative basket cell interneuron discharges in the CA3 region of the hippocampus. Indeed, whether ethanol is given systemically or locally, it increases the excitability of basket cell interneurons. Concomitant with this increase in interneuronal discharges is a decrease in principal cell PS amplitudes evoked in all three subregions of the hippocampus by stimulation of their monosynaptic afferent inputs, the perforant path input to dentate, the mossy fibers or commissural input to CA3 and the Schaffer collaterals or commissural input to CA1. Evoked PS amplitudes are believed to represent the synchronous firing of dentate granule cells in the dentate gyrus and pyramidal cells in CA1 and CA3. Figure 3 summarizes the effects of systemic and local administration of ethanol on PS amplitudes and interneuron discharges recorded in the dentate gyrus. The effect was similar for the other subfields which suggests that principal cell excitability in all subregions of the hippocampus is reduced by increased GABA-mediated feedforward inhibition.

Contrary to what has been reported *in vitro* wherein ethanol decreases EPSPs, we found that ethanol given systemically or locally does not decrease pEPSP slopes in the dentate gyrus or CA1 hippocampus, across stimulus levels from threshold to maximum. In fact, the tendency in both the dentate and CA1 is towards ethanol-induced increases in pEPSP slopes. However, in CA3, ethanol significantly decreases pEPSP slopes evoked by stimulation of the commissural input. Figures 4–6 summarize the effects of ethanol on pEPSP slopes and PS amplitudes in the dentate gyrus, CA1 and CA3 hippocampi, respectively.

Although systemic and local ethanol exposures appear to increase GABA-mediated feedforward inhibition in all hippocampal subregions, the effects on feedback recurrent inhibition are region-specific and result from ethanol effects on remote inputs to the hippocampus. Paired-pulse modulation of cellular PS responses is interval-dependent and results from the aftereffects of activation of principal cells in each subfield. Paired-pulse inhibition of PS amplitudes is a known measure of GABA mediated recurrent inhibition in

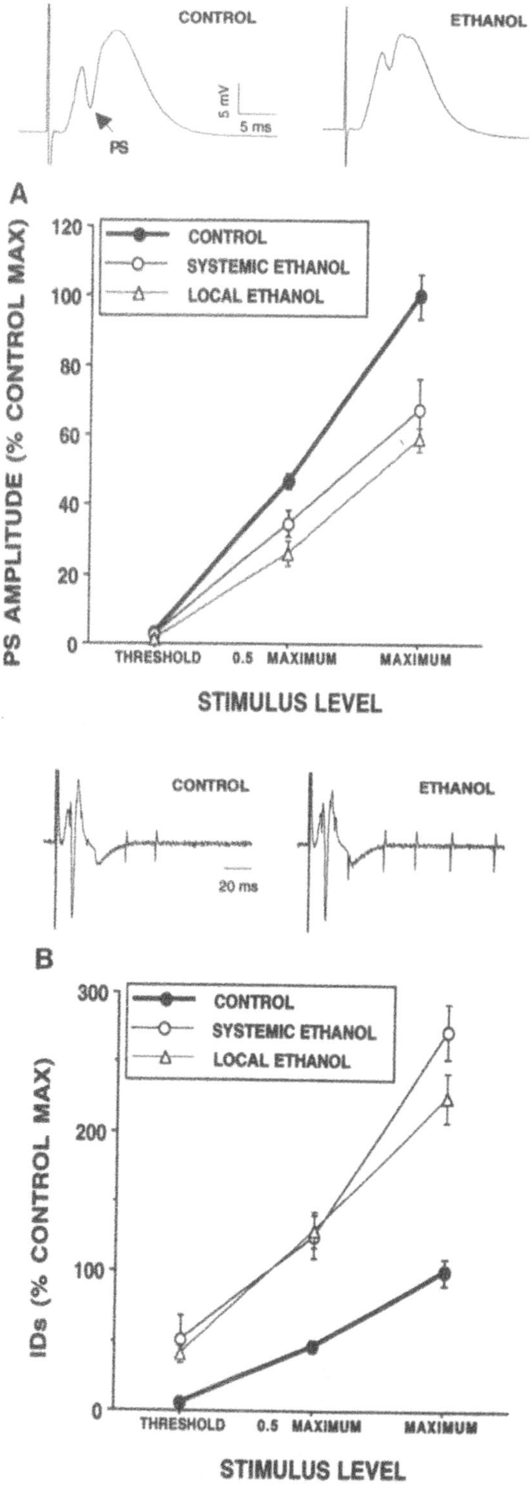

Figure 3. Systemic and local ethanol administrations decrease evoked PS amplitudes and increase interneuron discharges in the dentate gyrus. (A) Insets show unfiltered recordings of field potentials evoked in the hilar region of the dentate gyrus by stimulation of the perforant path before and after systemic administration of ethanol (1.2 g/kg). The PS is the fast-falling negative potential on the pEPSP. The graph below summarizes the effects of systemic administration and microelectroosmotic application of ethanol on PS amplitudes in the dentate gyrus across stimulus levels: threshold, 50% maximum and maximum. Ethanol significantly decreases PS amplitudes ($P < 0.001$; $N = 10$) across stimulus levels. (B) Insets show filtered recordings of an interneuron recorded in the hilar region of the dentate gyrus evoked by perforant path stimulation before and after systemic administration of ethanol. Note that the spike discharges occur at latencies beyond the time domain of the PS. Both systemic and local ethanol exposures significantly increase interneuron discharges ($P < 0.001$).

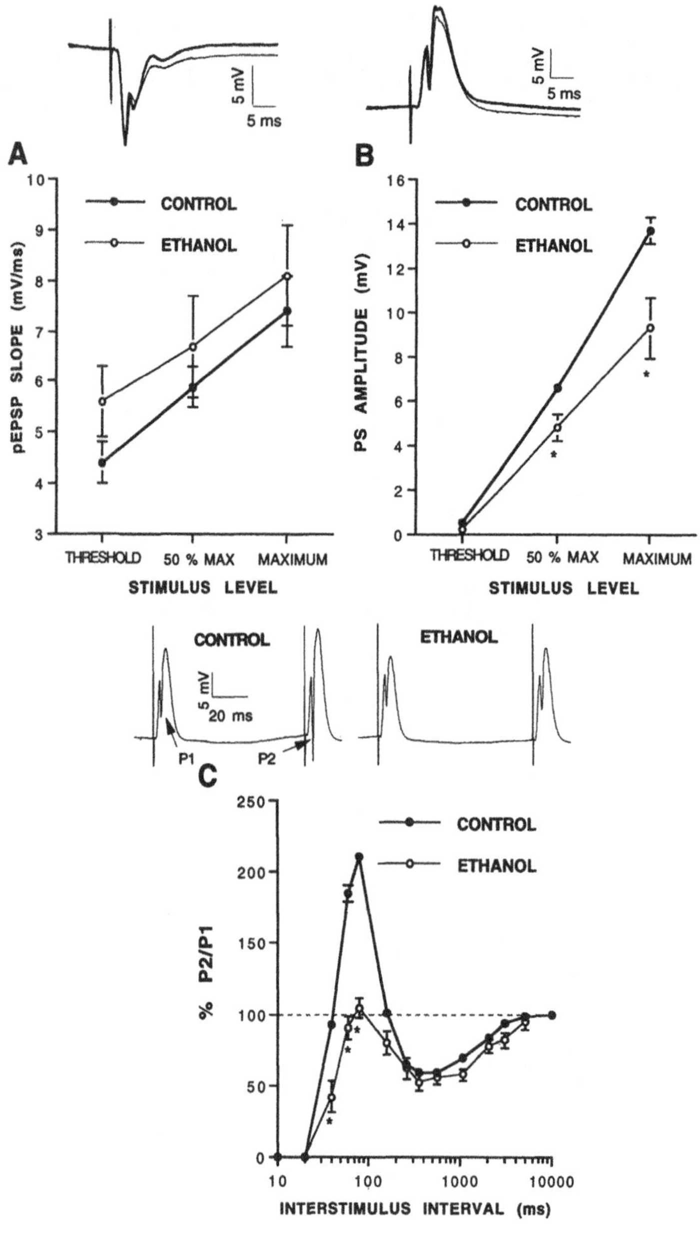

Figure 4. Effects of systemic administration of ethanol on field potential responses in the dentate gyrus subfield of the hippocampus. (A) Insets show representative superimposed pEPSPs evoked in the molecular layer of the dentate gyrus by stimulation of the perforant path before and after systemic ethanol exposure. Heavy line is control. Systemic administration of acute intoxicating doses of ethanol (1.2 g/kg; blood alcohol level (BAL)s = 140 mg%) slightly, but not significantly ($P > 0.05$), increase pEPSP slopes recorded in the molecular layer of the dentate gyrus by stimulation of the perforant path ($N = 10$). (B) Insets show representative superimposed field potentials evoked in the hilar region of the dentate by stimulation of the perforant path before and after systemic ethanol exposure. Heavy line is control. Acute ethanol significantly reduces PS amplitudes ($P < 0.001$; $N = 10$). (C) Insets show representative recordings of waveforms obtained in the dentate gyrus at 50 % maximum stimulus level by paired stimulation of the perforant path at 80 ms inter-stimulus interval. At this interval conditioned PSs are facilitated. Acute ethanol increases paired-pulse inhibition across inter-stimulus intervals from 40 to 80 ms.

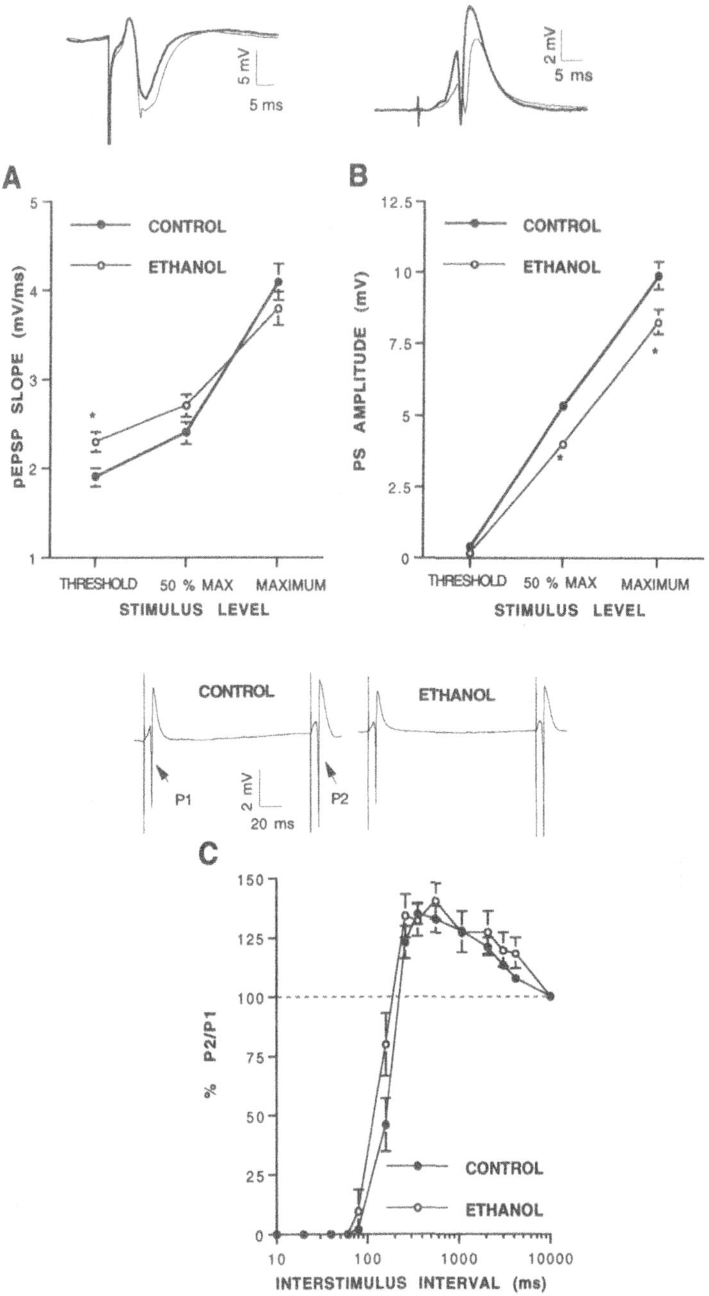

Figure 5. Effects of systemic administration of ethanol on field potential responses in the CA1 subfield of the hippocampus. (A) Insets show representative superimposed pEPSPs evoked in the molecular layer of the CA1 hippocampus by stimulation of the commissural input before and after systemic ethanol exposure. Heavy line is control. Systemic administration of acute intoxicating doses of ethanol (1.2 g/kg; BALs = 140 mg%) slightly increase pEPSP slopes ($P < 0.001$; $N = 8$). (B) Insets show representative superimposed field potentials evoked in CA1 by stimulation of the commissural input before and after systemic ethanol exposure. Heavy line is control. Acute ethanol decrease PS amplitudes ($P < 0.001$; $N = 8$). (C) Insets show representative recordings of waveforms obtained in CA1 at 50 % maximum stimulus level by paired stimulation of Schaffer collaterals at 160 ms inter-stimulus interval. At this interval conditioned PSs are slightly facilitated. Acute ethanol has no effect on paired-pulse responses ($P > 0.05$; $N = 8$).

the hippocampus. We studied the effects of systemic and local administration of ethanol on paired-pulse responses in the three subfields of the hippocampus (Figures 4–6C).

Figure 4C shows the effects of acute intoxicating ethanol on paired-pulse responses in the dentate gyrus. Dentate paired-pulse responses are characterized by a triphasic oscillation of conditioned PS amplitude, expressed as percent of unconditioned PS amplitude, consisting of a relatively short period of absolute inhibition, then facilitation, and then a long period of late relative inhibition. Systemic ethanol administration significantly increases either paired-pulse inhibition or decreases paired-pulse facilitation in the dentate gyrus. The paired-pulse curve of CA1 differs significantly from the curve of the dentate. There is considerably more inhibition in CA1 than the dentate and there is no late period of relative inhibition (Figure 5C). Interestingly, paired-pulse responses for all subregions in the anesthetized rat do not vary much from paired-pulse responses in the freely-behaving rat. This is not an effect of anesthesia. In fact, the only difference in freely-moving rats is that the paired-pulse curve is slightly state-dependent, with decreasing inhibition while the animal is undergoing hippocampal theta rhythm. Systemic ethanol exposure does not significantly alter paired-pulse responses in CA1. In CA3, a short period of absolute inhibition and then a long period of facilitation (Figure 6C) characterize paired-pulse responses. Systemic ethanol exposure does not significantly alter paired-pulse responses in CA3.

There are notable differences between paired-pulse responses *in vivo* and *in vitro*. For example, there is considerably less paired-pulse inhibition in the dentate and CA1 hippocampi *in vitro* compared to *in vivo* (Figure 7). This is a profound difference that is apparent but which is not appreciated until lined up together to graphically illustrate the differences between the two preparations. For many years our group has been trying to reconcile these differences, especially regarding differences between *in vivo* and *in vitro* preparations for the inhibitory effects of ethanol on NMDA responses and the lack of effects of ethanol on GABA inhibition seen in the slice preparation. As mentioned previously, some populations of hippocampal interneurons project and synapse at sites remote to the plane of section of the typical 300–400 μm slice and their synaptic influence would not be exerted within that slice. For example, we know that the two type of neurons whose synaptic influence are missing in the slice preparation are the hilar mossy cells and hilar interneurons that project 1–2 mm longitudinally and 12–20 mm contralaterally in the rat hippocampus. As mentioned previously, hilar mossy cells are the one population of hippocampal neurons that are consistently excited by acute ethanol. In the absence of their influence, hilar interneurons would not undergo an increase in synaptic transmission from hilar mossy cells; hence, their feedback inhibition would not be felt. Indeed, while systemic application of ethanol increases GABA-mediated feedback or recurrent inhibition in the dentate gyrus, local application of ethanol is without effect (Figure 8).

Numerous reports have shown that ethanol decreases NMDA responses in the hippocampus. However, in the *in vivo* preparation it is difficult to obtain pharmacologically isolated excitatory synaptic responses. Therefore, studying the effects of ethanol on NMDA responses *in vivo* is somewhat problematical. One *in vitro* study reported that LTP (2–3 fold) of conditioned dentate pEPSPs (100–400 ms inter-stimulus intervals) produced by 5 Hz stimulation did not require pharmacological isolation and showed clear ethanol inhibition of the potentiated NMDA receptor-mediated pEPSPs with an IC_{50} of 58 mM (Morrisett and Swartzwelder, 1993). Surely, many variables could not be replicated exactly *in vivo*, such as the frequency of stimulation needed to produce LTP (*e.g.*, 5 Hz stimulation for 2 s produces epileptiform responses and subsequent long-term depression *in vivo*). Nonetheless, we sought to replicate LTP qualitatively with those stimulation patterns that

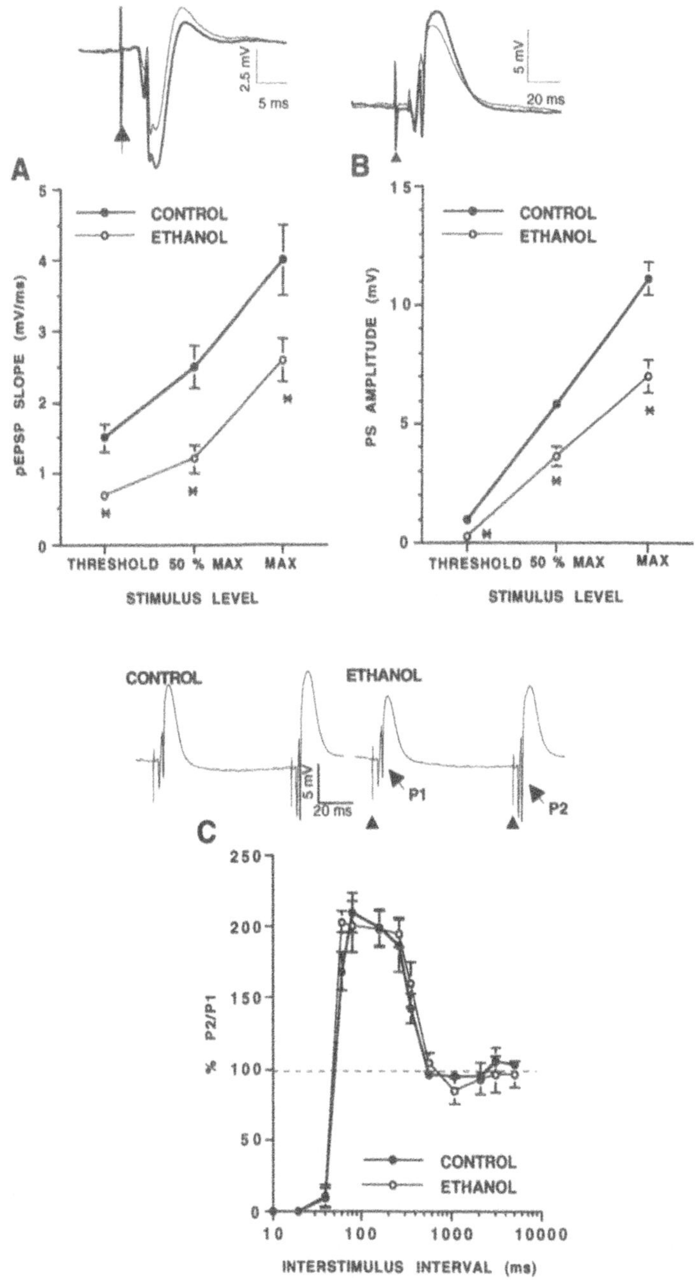

Figure 6. Effects of systemic administration of ethanol on field potential responses in the CA3 subfield of hippocampus. (A) Insets show representative superimposed pEPSPs evoked in the molecular layer of the CA3 hippocampus by stimulation of the commissural input before and after systemic ethanol exposure. Heavy line is control. Systemic administration of acute intoxicating doses of ethanol (1.2 g/kg; BALs = 140 mg%) significantly decrease pEPSP slopes recorded in the molecular layer of the CA3 by activation of the commissural input activated by stimulation of the contralateral hilus ($P < 0.001$; $N = 7$). (B) Insets show representative superimposed field potentials evoked in CA3 by stimulation of the commissural input before and after systemic ethanol exposure. Heavy line is control. Acute ethanol significantly decreased PS amplitudes ($P < 0.001$; $N = 7$). (C) Insets show representative recordings of waveforms obtained in CA3 at 50 % maximum stimulus level by paired commissural stimulation at 160 ms inter-stimulus interval. At this interval conditioned PSs are slightly facilitated. Acute ethanol has no effect on paired-pulse responses ($P > 0.05$).

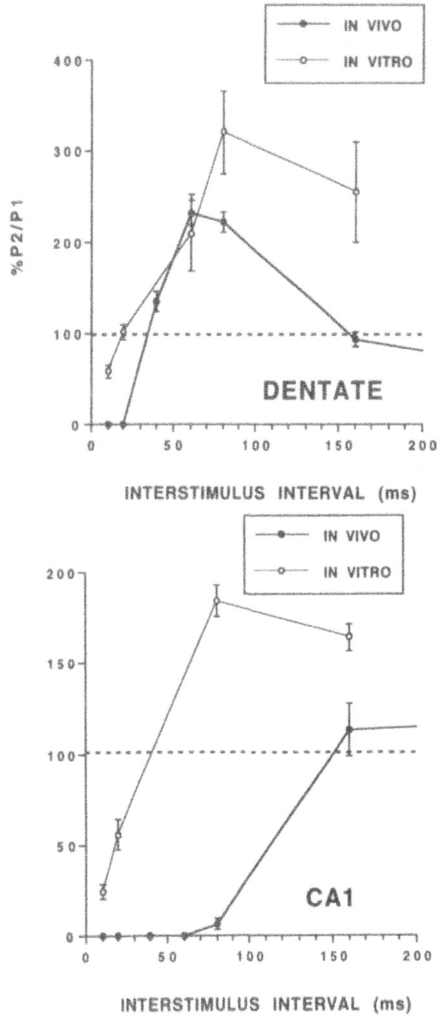

Figure 7. Differences between *in vivo* and *in vitro* paired-pulse responses recorded in the dentate gyrus and CA1 hippocampus. (A) Paired-pulse responses in the dentate gyrus *in vivo* are characterized by significantly more inhibition than corresponding paired-pulse intervals in the slice preparation. (B) Paired-pulse responses in the CA1 hippocampus *in vivo* show markedly increased paired-pulse inhibition relative to similar conditioning intervals recorded in the slice preparation.

produce robust and reproducible LTP of PSs and pEPSPs in the dentate gyrus. Although we could produce robust LTP of PSs and pEPSPs in the eight young rats studied (15–30 day old), no significant increase in pEPSP duration (at 200 ms inter-stimulus interval as performed in the *in vitro* study) across inter-stimulus intervals was evident (Figure 9).

Consistent with *in vitro* studies, ethanol decreases PS amplitudes in all hippocampal subfields. However, contrary to *in vitro* studies, ethanol tends to **increase** pEPSP slopes in both the dentate gyrus and CA1 hippocampus *in vivo*. We studied the effects of ethanol on LTP of PS amplitudes and pEPSP slopes in the dentate gyrus *in vivo*, two well-studied, reproducible and robust phenomena. We found that systemic ethanol exposure suppresses LTP of PSs (Figure 10A), but not LTP of pEPSPs, and that local administration of ethanol has no effect on LTP of the dentate gyrus. These findings underscore the marked differences between *in vivo* and *in vitro* preparations regarding the effects of ethanol on synaptic transmission and plasticity.

As systemic ethanol exposure enhances paired-pulse inhibition and suppresses LTP, but local ethanol exposure had no effect on either of these measures, we postulated that

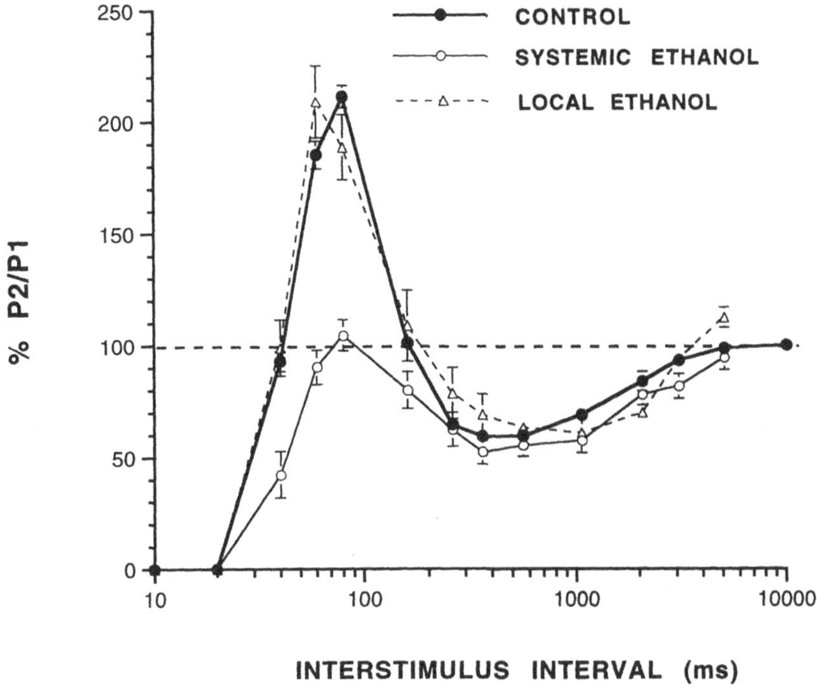

Figure 8. Lack of effect of local ethanol on recurrent inhibition in the dentate gyrus. Ethanol is microelectroosmotically applied (0.3 M; +200 nA) from a micropipette located 200 μm from the recording electrode. Though *in situ* (local) ethanol effectively decreases PSs and increases interneuron discharges (see Figure 3), it has no effect on paired-pulse responses in the dentate gyrus. The enhancement of paired-pulse inhibition by systemic administration of ethanol is shown in the graph for comparison.

the effects of ethanol on short- and long-term plasticity in the dentate gyrus subfield of the hippocampus result from actions on a remote input to the hippocampus. Our first attempt to prove or disprove this hypothesis was to study the effects of systemic administration of ethanol following deafferentation of the hippocampus. This was accomplished by electrolytic lesioning of the septo-hippocampal nucleus and fimbria-fornix, which have been shown to lesion subcortical inputs from the locus coeruleus, the raphe nucleus, the ventral tegmental area, the medial septal region, as well as the cholinergic and GABAergic inputs from the medial septal region. We found that ethanol enhancement of paired-pulse inhibition and lesions of the septo-hippocampal nucleus and fimbria-fornix effectively block suppression of LTP (Figure 10B).

Of the subcortical inputs mentioned above, we wanted to determine which structure(s) was responsible for ethanol actions on paired-pulse responses or recurrent inhibition and LTP of dentate gyrus. The first area of interest was the medial septum, as we knew that the septo-hippocampal input was a strong one and has been known to trigger or to generate theta rhythm in the hippocampus. We soon determined that septo-hippocampal modulation, in particular inhibition, facilitation and disinhibition were unaffected by systemic ethanol exposure. Some investigators (Brodie et al., 1990; Mereu et al., 1984) had shown that dopamine (DA) neurons were sensitive to ethanol, and another group (Yanagihashi & Ishikawa, 1992) had shown that D_1 receptor agonists blocked LTP in the dentate gyrus *in vivo*. Based on this evidence, we decided to pursue the ventral tegmental area

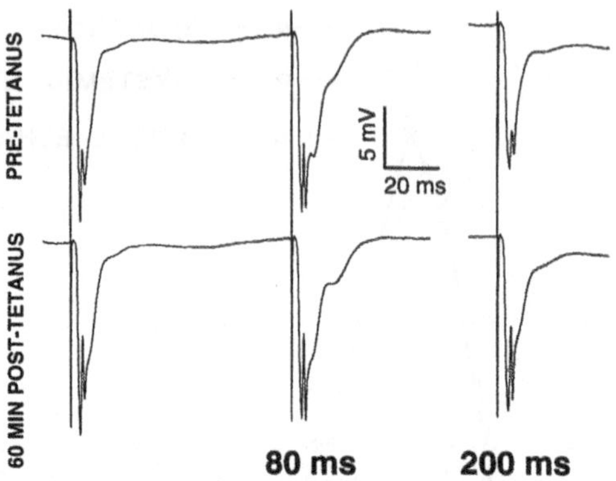

Figure 9. Lack of effects of tetanization on pEPSP duration in the dentate gyrus *in vivo*. These studies were performed in attempts to replicate an earlier study demonstrating LTP of NMDA receptor-mediated EPSP components in young rats (Morrisett and Swartzwelder, 1993). Stimulation of the perforant path evoked pEPSPs recorded from the molecular layer of the dentate gyrus. The pEPSPs depicted here were recorded at maximum EPSP amplitude from a 25 day old rat. In pre-tetanus responses, though little paired-pulse facilitation of EPSP slopes is evident at 80 ms, a slight paired-pulse inhibition is apparent at 200 ms inter-stimulus interval. Following tetanization of the perforant path that produced robust potentiation of PS amplitudes (not shown here), maximal EPSP slopes were only slightly potentiated and paired-pulse responses either at 80 ms or 200 ms were not significantly affected by tetanization. Although 200 ms inter-stimulus interval is facilitatory *in vitro*, it is inhibitory to both conditioned PSs and pEPSPs *in vivo*, but even at the 80 ms inter-stimulus interval that is facilitatory *in vivo*, there was no increase in pEPSP duration across stimulus levels: threshold, 50% maximum and maximum. Only maximum stimulus level is shown in this figure.

(VTA), which contains DA neurons that project to limbic structures, as a possible site that may be mediating the effects of ethanol on the dentate gyrus. We knew there was a sparse DA input from the VTA to the hippocampus and an abundant DA input to the lateral septal nucleus. We felt this was the most reasonable approach because of the well studied septo-hippocampal modulation of hippocampal responses.

We were the first to demonstrate that stimulation of the VTA effectively produces a small field potential in the hippocampus of about 1–2 mV in amplitude. If the perforant path is stimulated during the rising edge of this field potential, facilitation of dentate PSs occurs (Figure 11). We have termed this phenomenon, VTA facilitation. VTA conditioning only facilitates PS amplitudes but not pEPSP slopes or amplitudes (Figure 12), suggesting that it has a modulatory influence on inhibitory processes in the dentate gyrus. The field potential is centered around the hilar region, and may be the result of activation of inhibitory interneurons in that region. VTA facilitation is a robust phenomenon. We have observed facilitation of PSs amplitudes as great as 28 mV. VTA facilitation can be blocked by the same septo-hippocampal lesions that block the effects of ethanol on paired-pulse responses or LTP in the dentate gyrus. VTA facilitation is also blocked by systemic administration of ethanol (Figure 13), and VTA lesions block the effects of systemic ethanol exposure to enhance paired-pulse inhibition in the dentate gyrus. But perhaps the most supportive evidence for our hypothesis that ethanol acts on a remote structure to influence hippocampal physiology is that ethanol applied locally to the VTA increases dentate paired-pulse inhibition similar to systemic ethanol exposure (Figure 14).

We sought to determine if DA was involved in ethanol enhancement of recurrent inhibition and found that D_1 subtype receptor antagonists block the effects of ethanol on paired-pulse responses in the dentate gyrus. In addition, local applications of D_1 antago-

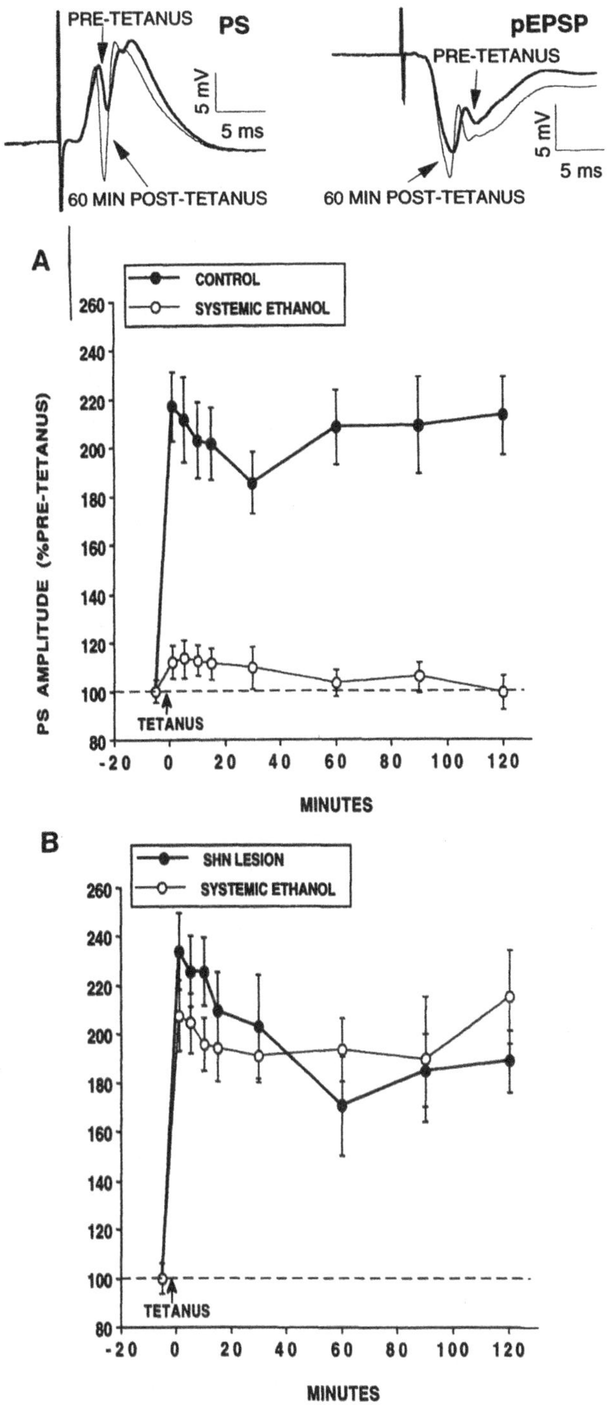

Figure 10. Septal lesions block ethanol suppression of LTP in the dentate gyrus. (A) Insets show representative superimposed field potential recordings before and 60 min after tetanization of the perforant path. The waveforms demonstrate PSs and pEPSPs recorded simultaneously from the hilar and molecular regions of the dentate gyrus with staggered electrodes. Acute intoxicating doses of ethanol producing BALs of 140 mg% suppress LTP of PSs ($P < 0.001$; $N = 10$), but not pEPSPs (not shown in graph; $P > 0.05$; $N = 10$). (B) Electrolytic lesions of the septo-hippocampal nucleus block ethanol suppression of LTP ($N = 7$).

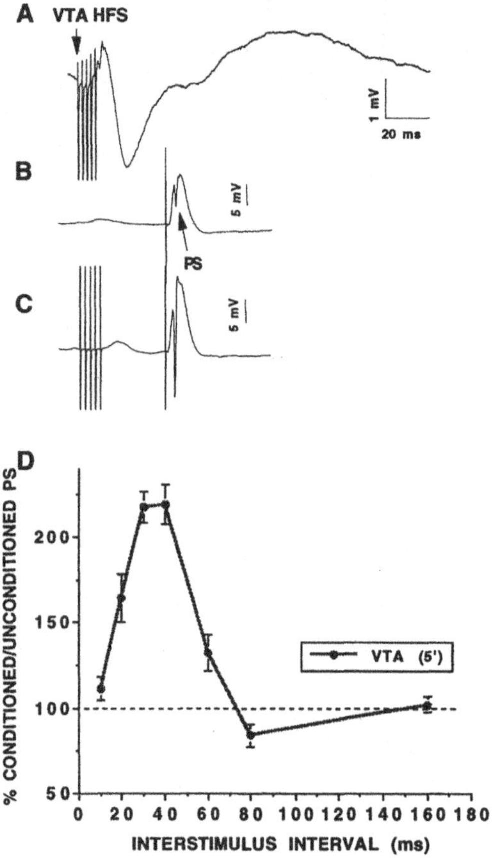

Figure 11. VTA conditioning modulates evoked responses in the dentate gyrus. High frequency stimulation (HFS; 5 pulses, 2.5 ms intervals) of the ventral tegmental area (VTA) elicited a small field-potential recorded in the hilar region of the dentate gyrus with a predominant negative-going component at a peak latency of approximately 25 ms (A). Stimulation of the perforant path evoked a positive-going field potential recorded in the hilar region (B) whose waveforms consisted of a prominent relatively fast negative-going PS superimposed on the field EPSP/inhibitory postsynaptic potential (IPSP) (shown here at half-maximal PS amplitude). Conditioning the perforant path to dentate response with VTA stimulation (shown in C at 40 ms interval) significantly increased the PS amplitude. VTA HFS markedly increased PS amplitudes at conditioning intervals 20–60 ms and slightly decreased PS amplitudes at 80 ms conditioning interval (D: ** = $P < 0.001$, * = $P < 0.05$; N = 17). Recordings A-C are shown with the same time scale as graph D to demonstrate the relationships between the VTA-evoked field potential and the corresponding alterations in conditioned responses in the dentate. Note that VTA facilitation and inhibition correspond to the negative and positive phases of the VTA-evoked field potential, respectively.

nists are more efficacious in the lateral septal nucleus (Figure 15), suggesting that VTA DA modulation of dentate recurrent inhibition is acting through the lateral septal nucleus. Furthermore, D_1 receptor antagonists also block the suppressive effects of ethanol on LTP, whether given systemically or locally into the lateral septal nucleus (Figure 16).

While passing ethanol into the VTA, we observed that a homogeneous population of neurons is especially sensitive to the effects of either systemic or local administration of ethanol. At first we wanted to study VTA DA neurons, hoping that we could replicate the findings that others have seen *in vitro* and be able to find that alcohol increases the discharging or spontaneous firing rate of DA neurons. It was somewhat problematical, since under halothane anesthesia, DA neurons are difficult to find because they fire very slowly or are silent. We have recently characterized the VTA non-DA neurons electrophysiologically utilizing intracellular and extracellular recording techniques *in vivo* and have subsequently labeled them with neurobiotin to determine their neurochemical, morphological and ultrastructural signature. Indeed, they are GABA containing neurons that project to the cortex and receive inputs from corticolimbic structures (Steffensen et al., 1998). Some of their extracellular properties are shown in (Figure 17). We have studied these neurons in both anesthetized and freely-behaving rats, and we have recently found that ethanol (0.2–0.4 g/kg) markedly inhibits the firing rate of these VTA non-DA neurons (Figure 18).

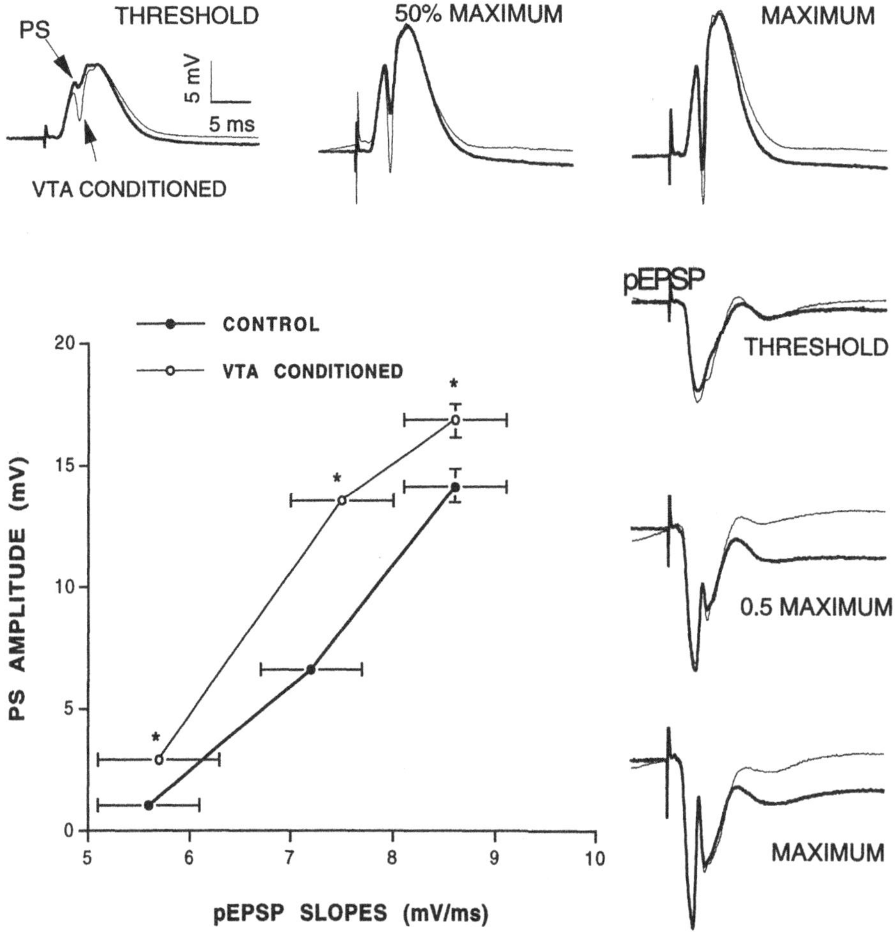

Figure 12. VTA stimulation facilitates dentate responses without altering excitatory monosynaptic transmission. Insets show representative field potential recordings evoked in the hilar region (top three traces) and molecular layer (bottom three traces at right) of the dentate gyrus demonstrating VTA facilitation of perforant path to dentate PS amplitudes, but not pEPSP slopes, across stimulus levels: threshold, 50% maximum and maximum PS amplitude. The graph summarizes the effects of VTA conditioning on the EPSP-PS coupling curve. Conditioning the perforant path to dentate response with VTA stimulation significantly increased PS amplitudes, but not pEPSP slopes simultaneously recorded in the molecular layer of the dentate with staggered electrode pairs ($P < 0.001$; $N = 12$).

3. SUMMARY AND CONCLUSIONS

The mechanisms responsible for the effects of ethanol intoxication on short- and long-term plasticity in the dentate gyrus of adult rats are different from those reported in the dentate gyrus of immature rats (Morrisett and Swartzwelder, 1993; Swartzwelder et al., 1995). Whereas ethanol may block LTP in the immature dentate gyrus by directly reducing NMDA receptor function, its suppression of LTP in the adult dentate gyrus is primarily indirect via its actions on subcortical afferents (Steffensen et al., 1993). In fact, several subcortical structures are known to modulate short- and long-term plasticity in the hippocampus via their actions on inhibitory interneurons (Robinson and Racine, 1982;

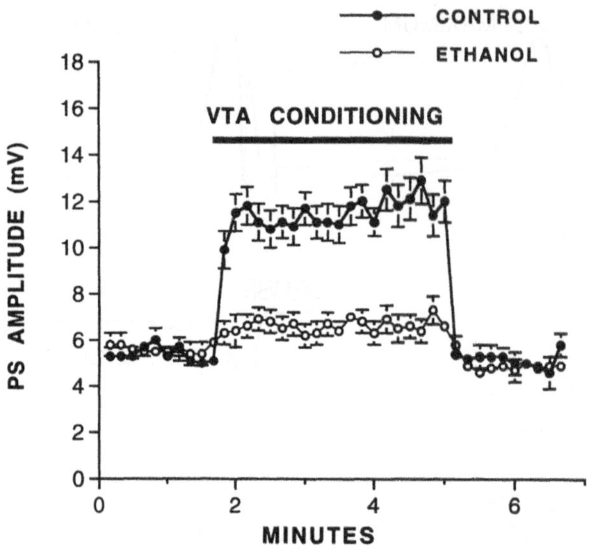

Figure 13. Systemic administration of ethanol suppresses VTA facilitation of dentate PSs. Stimulation of the VTA facilitates perforant path to dentate PSs more than two-fold. Systemic administration of ethanol (1.2 g/kg) suppressed VTA facilitation ($P < 0.001$; $N = 12$).

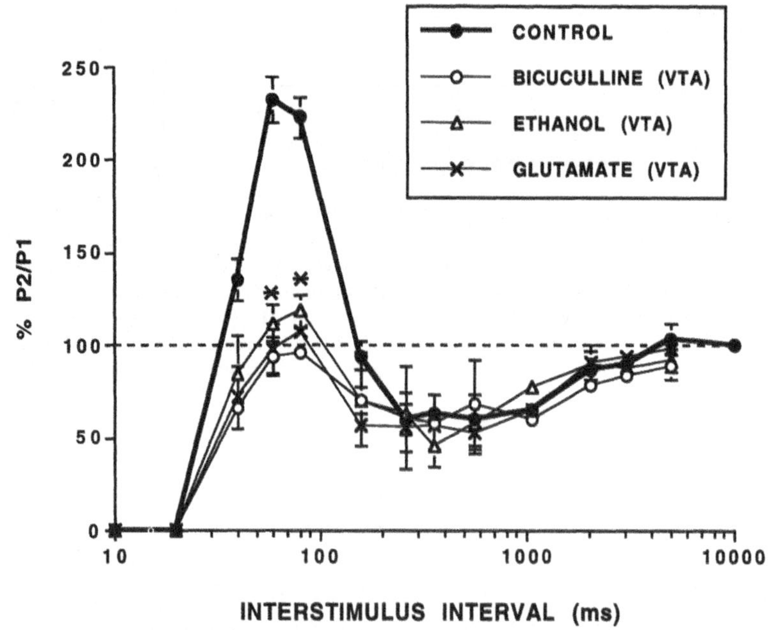

Figure 14. Local application of bicuculline, ethanol, and glutamate to the VTA increases dentate recurrent inhibition. Microelectrophoretic application of bicuculline (25 nA) and glutamate (25 nA) markedly increased spontaneous firing rates of VTA non-DA neurons (data not shown) and simultaneously increased recurrent inhibition in the dentate gyrus ($N = 3$ each; $P < 0.001$ at inter-stimulus intervals 60 and 80 ms). In addition, ethanol microelectrophoresis (200 nA) moderately increased VTA neuronal activity (data not shown) and, similar to bicuculline and glutamate, increased dentate recurrent inhibition ($N = 3$).

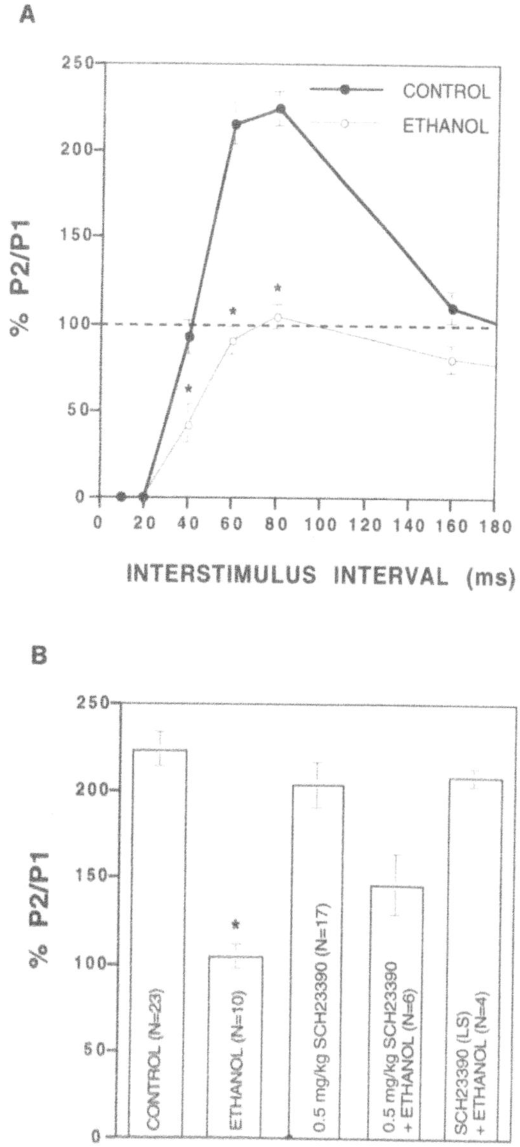

Figure 15. DA antagonists attenuate ethanol-induced enhancement of recurrent inhibition in the dentate gyrus. (A) Systemic administration of (1.2 g/kg) ethanol producing BALs at 15 min of 151.5 ± 12.4 significantly increased the early phase of paired-pulse inhibition in the dentate gyrus ($P < 0.001$; $N = 11$). (B) Intraperitoneal administration of the D1 receptor subtype-selective DA antagonist SCH23390 (0.5 mg/kg) had no effect on paired-pulse responses ($P > 0.05$; $N = 9$) but moderately reduced ethanol's ability to increase recurrent inhibition ($P < 0.05$ vs. $P < 0.001$; $N = 6$). Microelectrophoretic application of SCH23390 into the lateral septal nucleus, however, markedly reduced ethanol enhancement of paired-pulse inhibition.

Bilkey and Goddard, 1985; Buzsaki et al., 1989; Mizumori et al., 1989). Developmental differences in the induction and expression of hippocampal LTP (Bekenstein and Lothman, 1991; Bronzino et al., 1994; Izumi and Zorumski, 1995) could be influenced by a number of factors, including a delay in the functional maturation of GABAergic interneurons (Michelson and Lothman, 1989), changes in glutamate binding (Baudry et al., 1981), or a reduction in the sensitivity of hippocampal NMDA receptors (Morrisett et al., 1990). This may be reflected in the disparities between *in vitro* and *in vivo* preparations for ethanol effects on synaptic plasticity. Moreover, since the synaptic influence of some hippocampal neurons is exerted beyond the plane of the slice, any interpretation for the effects of ethanol on excitatory or inhibitory synaptic transmission must be considered in

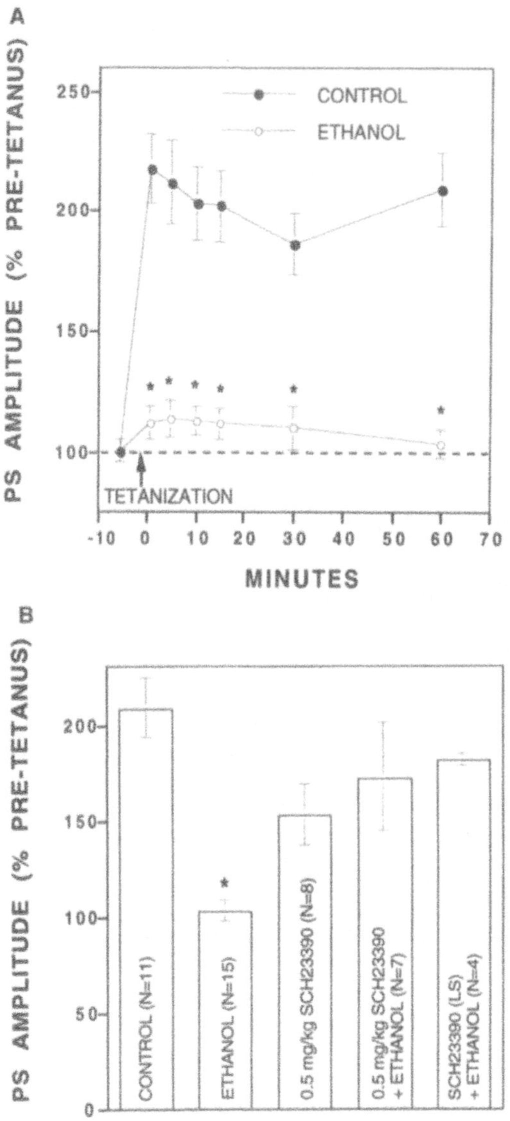

Figure 16. DA antagonists attenuate ethanol-induced suppression of LTP in the dentate gyrus. Tetanization of the perforant path produced a marked and prolonged increase in dentate PS amplitudes. Administration of ethanol (1.2 g/kg, i.p.) produced acute intoxicating levels of ethanol (BALs = 151.5 ± 12.4) and suppressed LTP of dentate PS amplitudes when tetanized within 20 min following injection. (B) While the DA D_1 receptor subtype-selective antagonist SCH-23390 moderately decreased LTP amplitude and reduced ethanol suppression of LTP, *in situ* microelectrophoretic application of SCH-23390 into the lateral septal nucleus markedly reduced ethanol suppression of dentate LTP. Asterisks represent significance levels of $P < 0.001$ (two-tailed t-test at each point).

ight of the fact that inhibition is severely compromised in the slice preparation, thus pharnacological effects would be effectively biased against inhibition. Nonetheless, these diferences may be effectively capitalized on in formulations of the mechanistic aspects of thanol actions on synaptic transmission in the hippocampus.

We have previously demonstrated that local application of ethanol into the dentate yrus reduces PS amplitudes, but produces no effect on dentate LTP (Steffensen et al., 993). These findings suggest that extra-hippocampal inputs are likely responsible for thanol suppression of dentate LTP. The fact that blockade of D_1 receptors in the lateral eptal nucleus antagonized ethanol actions on dentate synaptic plasticity suggests that, in dult rats, ethanol intoxication alters the activity of the SH pathway by increasing DA inuts innervating the lateral septal nucleus. Ethanol may produce these effects by increas-

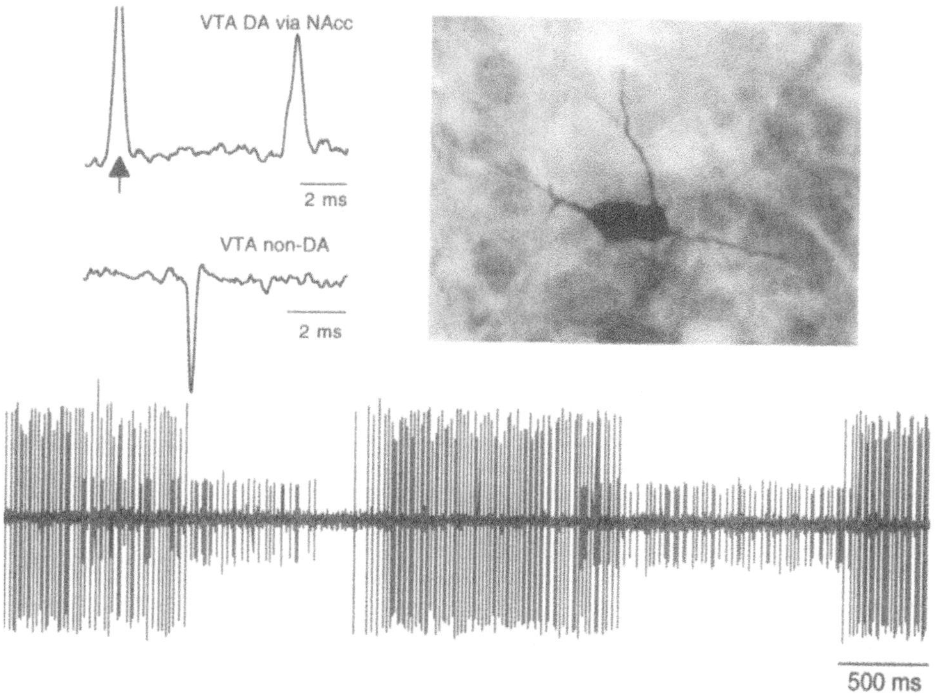

Figure 17. Extracellular electrophysiological characterization of DA and non-DA neurons in the VTA. Insets show unfiltered recordings of a VTA DA neuron evoked by stimulation of the nucleus accumbens (top) and a spontaneous non-DA neuron (bottom). VTA DA neurons were slow firing (< 1 Hz), bursting neurons that were driven by nucleus accumbens stimulation with spike durations greater than 500 μs. VTA non-DA neurons were relatively fast-firing, non-bursting cells that evinced negative-going spikes and were characterized by spike durations less than 500 μs. VTA non-DA neurons were not driven by nucleus accumbens stimulation. Under halothane anesthesia, VTA non-DA neurons evinced pronounced and persistent phasic activity as demonstrated by the two simultaneously recorded VTA non-DA neurons in the filtered trace below. The light micrograph at right shows a neurobiotin-labeled non-DA neuron in the VTA. This neuron was characteristic of all neurons identified electrophysiologically as VTA non-DA neurons and was multi-polar in shape with few dendritic processes branching from its soma.

ing the firing rate of VTA DA neurons (Gessa et al., 1985; Gessa et al., 1985; Brodie et al., 1990), which project afferents to the lateral septal nucleus and have been implicated in both ethanol reinforcement and learning processes (Assaf and Miller, 1977; Simon et al., 1980; Swanson, 1982; Koob, 1992).

Previous studies have demonstrated that subcortical pathways modulate the induction of LTP in the adult hippocampus (Robinson and Racine, 1982; Buzsaki and Gage, 1987). We have demonstrated that electrolytic lesions of the septo-hippocampal nucleus and of presumed DA neurons in the VTA attenuate ethanol-induced alterations of paired-pulse inhibition and LTP in the dentate gyrus (Steffensen et al., 1993; Criado et al., 1994; Criado et al., 1994; Criado et al., 1996). Substantial evidence indicates that DA fibers from A10 neurons in the VTA project to both the lateral septal nucleus and the hippocampus (Hokfelt et al., 1974; Scatton et al., 1980; Swanson, 1982). Consistent with these findings, mesolimbic DA neurons have been shown to play a role in the regulation of SH cholinergic transmission and in cognitive processes (Assaf and Miller, 1977; Robinson et al., 1979; Simon et al., 1980). In fact, D_1 DA receptors have been shown to mediate the re-

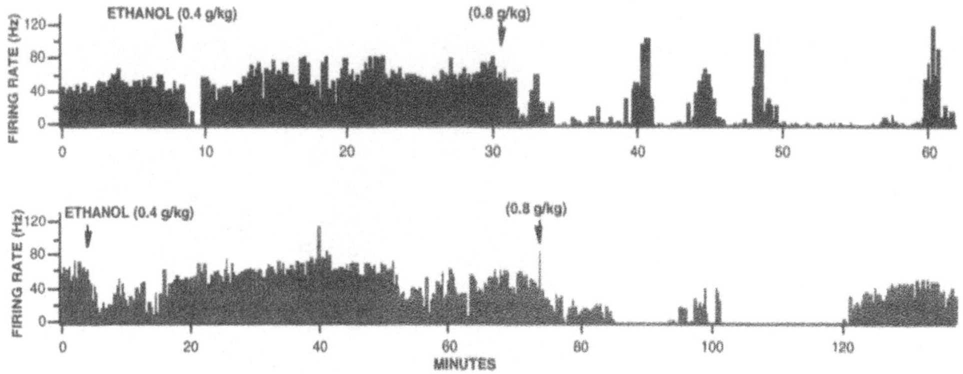

Figure 18. Effects of passive injections of intraperitoneal ethanol on the spontaneous firing rate of VTA non-DA neurons in the freely-behaving rat. The ratemeter records show two non-DA (DA) neurons recorded in the VTA of freely-behaving rats with 40 Hz (top) and 61 Hz (bottom) spontaneous firing rates, respectively. Intraperitoneal administration of ethanol (0.4 g/kg) markedly but transiently inhibited VTA non-DA firing rates. Subsequent administration of ethanol (0.8 g/kg) also markedly inhibited VTA non-DA firing rates but the duration of inhibition was more prolonged.

inforcing properties associated with voluntary ethanol self-administration (Koob et al., 1980) and D_1 agonists block the induction of LTP in the dentate gyrus (Yanagihashi and Ishikawa, 1992). Indeed, systemic administration or local application of D_1 receptor antagonists blocks ethanol enhancement of recurrent inhibition and suppression of LTP in the dentate gyrus (Criado et al., 1996).

VTA non-DA neurons are inhibited by ethanol, which may effectively disinhibit DA neurons and increase DA mesolimbic neurotransmission. In future studies we will determine the role of these neurons in mediating ethanol effects on hippocampal plasticity. We hypothesize that these neurons are critical substrates mediating ethanol effects on mesolimbic neurotransmission and that this homogenous population of GABA neurons may play a crucial role in the acquisition of ethanol motivational and reinforcing behaviors.

ACKNOWLEDGMENT

I wish to acknowledge the generous support of National Institute on Alcohol Abuse and Alcoholism (NIH grant AA10075).

REFERENCES

Amaral DG (1978) A Golgi study of cell types in the hilar region of the hippocampus in the rat. J Comp Neurol 182:851–914.

Assaf SY, Miller JJ (1977) Excitatory action of the mesolimbic dopamine system on septal neurones. Brain Res 129:353–360.

Baudry M, Arst D, Oliver M, Lynch G (1981) Development of glutamate binding sites and their regulation by calcium in rat hippocampus. Dev Brain Res 1:37–48.

Bekenstein JW, Lothman EW (1991) An *in vivo* study of the ontogeny of long-term potentiation (LTP) in the CA1 region and in the dentate gyrus of the rat hippocampal formation. Dev Brain Res 63:245–251.

Bilkey DK, Goddard GV (1985) Medial septal facilitation of hippocampal granule cell activity is mediated by inhibition of inhibitory interneurons. Brain Res 361:99–106.

Bliss TVP, Lomo T (1973) Long-lasting potentiation of synaptic transmission in the dentate area of the anaesthetized rabbit following stimulation of the perforant path. J Physiol (Lond) 232:331–356.

Blitzer RD, Gil O, Landau EM (1990) Long-term potentiation in rat hippocampus is inhibited by low concentrations of ethanol. Brain Res 537:203–208.

Brodie MS, Shefner SA, Dunwiddie TV (1990) Ethanol increases the firing rate of dopamine neurons of the rat ventral tegmental area *in vitro*. Brain Res 508:65–69.

Bronzino JD, Abu-Hasaballah RJ, Austin-LaFrance RJ, Morgane PJ (1994) Maturation of long-term potentiation in the hippocampal dentate gyrus of the freely moving rat. Hippocampus 4:439–446.

Buzsaki G, Gage F (1987) Absence of long-term potentiation in the subcortically deafferented dentate gyrus. Brain Res 484:94–101.

Buzsaki G, Ponomareff GL, Bayardo F, Ruiz R, Gage FH (1989) Neuronal activity in the subcortically denervated hippocampus: A chronic model for epilepsy. Neuroscience 28:527–538.

Criado JR, Steffensen SC, Henriksen SJ (1994) Ethanol acts via the ventral tegmental area to influence hippocampal physiology. Synapse 17:84–91.

Criado JR, Steffensen SC, Henriksen SJ (1994) Mesolimbic Dopaminergic modulation of synaptic transmission in the dentate gyrus is mediated through the septal nucleus. Soc Neurosci Abst 20:530.

Criado JR, Steffensen SC, Henriksen SJ (1996) Microelectrophoretic application of SCH-23390 into the lateral septum nucleus blocks ethanol-induced suppression of LTP, *in vivo*, in the adult rodent hippocampus. Brain Res 716:192–196.

Gessa G, Muntoni F, Boi V, Mereu G (1985) Effects of Ethanol on Mid-Brain Dopamine and NON-Dopamine Neurons. Soc Neurosci Abst 11:87.14.

Gessa GL, Muntoni F, Collu M, Vargiu L, Mereu G (1985) Low doses of ethanol activate dopaminergic neurons of the ventral tegmental area. Brain Res 348:201–204.

Givens B (1995) Low doses of ethanol impair spatial working memory and reduce hippocampal theta activity. Alcoholism Clin Exp Res 19:763–767.

Hokfelt T, Fuxe K, Johansson O, Ljungdahl A (1974) Pharmaco-histochemical evidence of the existence of dopamine nerve terminals in the limbic cortex. Eur J Pharmacol 25:108–112.

Izumi Y, Zorumski CF (1995) Developmental changes in long-term potentiation in CA1 of rat hippocampal slices. Synapse 20:19–23.

Knowles WD, Schwartzkroin PA (1981) Local circuit interactions in hippocampal brain slices. J Neurosci 1:318–322.

Koob G, Strecker R, Bloom FE (1980) Substance Alcohol Actions/Misuse 1:447–457.

Koob GF (1992) dopamine, addiction and reward. Semin Neurosci 4:139–148.

Kosaka T, Kosaka K, Tateishi K, Hamaoka Y, Yanaihara N, Wu J-Y, Hama K (1985) GABAergic neurons containing CCK-8-like and/or VIP-like immunoreactivities in the rat hippocampus and dentate gyrus. J Comp Neurol 239:420–430.

Lacaille J-C, Mueller AL, Kunkel DD, Schwartzkroin PA (1987) Local circuit interactions between oriens/alveus interneurons and CA1 pyramidal cells in hippocampal slices: electrophysiology and morphology. J Neurosci 7:1979–1993.

Laurberg S, Sorensen JE (1981) Associational and commissural collaterals of neurons in the hippocampal formation (hilus fasciae dentate and subfield CA3). Brain Res 212:287–300.

Lister RG, Eckardt MJ, Weingartner H (1987) Ethanol intoxication and memory: Recent developments and new directions. In: Recent Developments in Alcoholism, (Galanter M, ed), pp 111–126. New York: Plenum Press.

Lovinger DM, White G, Weight FF (1989) Ethanol inhibits NMDA-activated ion currents in hippocampal neurons. Science 243:1721–1724.

Mereu G, Fadda F, Gessa GL (1984) Ethanol stimulates the firing rate of nigral dopaminergic neurons in unanesthetized rats. Brain Res 292:63–69.

Michelson HB, Lothman EW (1989) An in vivo electrophysiological study of the ontogeny of excitatory and inhibitory processes in the rat hippocampus. Dev Brain Res 47:113–122.

Mizumori SJY, McNaughton BL, Barnes CA (1989) A comparison of supramammillary and medial septal influences on hippocampal field potentials and single-unit activity. J Neurophysiol 61:15–31.

Morris RGM, Anderson E, Lynch GS, Baudry M (1986) Selective impairment of learning and blockade of long-term potentiation by an N-methyl-D-aspartate receptor antagonist, AP5. Nature 319:774–776.

Morrisett RA, Mott DD, Lewis DV, Wilson WA, Swartzwelder HS (1990) Reduced sensitivity of the N-methyl-D-aspartate component of synaptic transmission to magnesium in hippocampal slices from immature rats. Dev Brain Res 56:257–262.

Morrisett RA, Swartzwelder HS (1993) Attenuation of hippocampal long-term potentiation by ethanol: a patch-clamp analysis of glutamatergic and GABAergic mechanisms. J Neurosci 13:2264–2272.

Ribak C, Vaughn J, Saito K (1978) Immunocytological localization of glutamic acid decarboxylase in neuronal somata following colchicine inhibition of axonal transport. Brain Res 140:315–322.

Robinson GB, Racine RJ (1982) Heterosynaptic interactions between septal and entorhinal inputs to the dentate gyrus: long-term potentiation effects. Brain Res 249:162–166.

Robinson SE, Malthe-Sorenssen D, Wood PL, Commissiong J (1979) dopaminergic control of the septal-hippocampal cholinergic pathway. J Pharm Exp Ther 208:476–479.

Scatton B, Simon H, Le Moal M, Bischoff S (1980) Origin of dopaminergic innervation of the rat hippocampal formation. Neurosci Lett 18:125–131.

Scharfman HE, Schwartzkroin PA (1988) Electrophysiology of morphologically identified mossy cells of the dentate hilus recorded in guinea pig hippocampal slices. J Neurosci 8:3812–3821.

Schwartzkroin PA, Scharfman HE, Sloviter RS (1990) Similarities in circuitry between Ammon's horn and dentate gyrus: local interactions and parallel processing. In: Progress in Brain Research, ed), pp 269–286.

Simon H, Scatton B, Le Moal M (1980) dopaminergic A10 neurones are involved in cognitive functions. Nature 286:150–151.

Steffensen SC, Svingos AL, Pickel VM, Henriksen SJ (1998) Electrophysiological characterization GABAergic neurons in the ventral tegmental area. J. Neurosci 18:8803–8815.

Steffensen SC, Yeckel M, Miller DR, Henriksen SJ (1993) Ethanol-induced suppression of hippocampal long-term potentiation is blocked by lesions of the septohippocampal nucleus. Alcohol Clin Exp Res 17:655–659.

Swanson LW (1982) The projections of the ventral tegmental area and adjacent regions: a combined fluorescent retrograde tracer and immunofluorescence study in the rat. Brain Res Bull 9:321–353.

Swartzwelder HS, Wilson WA, Tayyeb MI (1995) Differential sensitivity of NMDA receptor-mediated synaptic potentials to ethanol in immature versus mature hippocampus. Alcohol Clin Exp Res 19:320–323.

Williams K, Russell S, Shen Y, Molinoff P (1993) Developmental switch in expression of NMDA receptors occurs in vivo and *in vitro*. Neuron 10:267–278.

Yanagihashi R, Ishikawa T (1992) Studies on long-term potentiation of the population spike component of hippocampal field potential by the tetanic stimulation of the perforant path in rats: effects of a dopamine agonist, SKF-38393. Brain Res 579:79–86.

15

QUESTIONS AND ANSWERS OF SESSION III

Synaptic Plasticity

1. Q&As BETWEEN AUDIENCE AND INDIVIDUAL SPEAKERS

1.1. Q&As between Audience and Dr. Stevens

1.1.1. The Readily-Releasable Pool

AUDIENCE MEMBER: If you put an electrode down into the brain of a rat, say, the medial septum and measure the firing rate of neurons, the cells are spontaneously firing maybe 500–600 times over a ten-second period. Assuming those action potentials get out to their presynaptic terminals, shouldn't that deplete the synapses of their transmitter fairly rapidly?

DR. STEVENS: Yes, they would get depleted. What happens is that as you stimulate over and over again, you use the readily releasable pool. It gets smaller, but the release probability goes down. And you finally find some size of the pool where the average rate at which you're stimulating and the average rate at which it's releasing are equal, keeping that size indefinitely.

AUDIENCE MEMBER: This goes back to some very, very old work. If you continually stimulate a nerve in the periphery, the response goes down...

DR. STEVENS: Declines, yes.

AUDIENCE MEMBER: ...and release goes down—and as I've looked at all these other elaborate pieces of work about synaptic vesicles, they're supposed to go back through this cycle.

DR. STEVENS: Yes.

AUDIENCE MEMBER: What happens in this stimulus experiment, as far as what comes out and what's in the tissue itself. Is it that what starts coming out is the newly synthesized transmitter, and has it a different radioactivity component than the pool in the tissue? In other words, as you deplete the pool, you would assume that these vesicles wouldn't have very much transmitter left in them. But it appears that instead of going through this recycling cycle, they just come right back around, and they synthesize the transmitter, and it gets released. And is there any evidence for that in the brain?

DR. STEVENS: I think that this readily releasable pool is just the docked vesicle pool. There may be ten docked vesicles, although there may be 50 vesicles that are waiting behind, that are all ready to go. When you deplete the readily releasable pool, you just motor down one of those new vesicles and dock it. That takes maybe five seconds. And then this may take 45 seconds to endocytosis the vesicle over here, pump it full of transmitters again, and put it back in line. So I think that's it. I didn't mean to give the impression that there are only ten vesicles total in the synapse. There is five times that many in the real pool.

1.1.2. Paired-Pulse Facilitation and LTP: Pre- vs. Postsynaptic

AUDIENCE MEMBER: One of the things that bothered a lot of investigators who've looked at LTP over the years in terms of a release probability change is the lack of change in paired-pulse facilitation. In your opinion, is that because that's just not a good measurement of the release probability or should that change?

DR. STEVENS: The question is: Does paired-pulse facilitation change with LTP? If it didn't change, would that be bad for people who said there's a presynaptic mechanism of LTP? Dr. Roger Nicoll feels very strongly that LTP is a postsynaptic mechanism (Isaac et al., 1996). One of the reasons that he thinks so is because he has done experiments and finds no change with paired-pulse facilitation. But other people have found changes of paired-pulse facilitation, so it's not true that nobody finds it. Some people find it and some don't. But whether you find it or not is quite complicated. The amount of paired-pulse facilitation that you get depends on the initial probability of release of the synapse. If you have a synapse that has a release probability of, say, 0.4 or 0.5, above that, it'll show no paired-pulse facilitation essentially. If you have synapses with real low release probability, they'll show a lot of paired-pulse facilitation. So what you get in a macroscopic experiment, like the kinds you're talking about, is a very complicated thing. It has to do with what the initial distribution of release probabilities for all the synapses were, which ones changed by how much, and whether they were ever getting close to the place that they would be bumping into a region where they wouldn't be releasing. What Roger and other people say, they change release probability by changing calcium. That affects on paired-pulse facilitation. That's true. But that's because you changed all the synapses, not just some special subset of them.

AUDIENCE MEMBER: Do you think that's because of the initial conditions that are used in different paradigms, or do you think it's just a matter of the preparations?

DR. STEVENS: I think it's just the way the experiments are done. I think if you had control of all the variables, you probably could do an experiment that would show whichever thing you wanted to show.

AUDIENCE MEMBER: There's no easy way to standardize that?

DR. STEVENS: There's no easy way.

1.1.3. Transcription-Dependent Phase of LTP

AUDIENCE MEMBER: Have you made any of these measurements in the late phase, the sort of transcription-dependent phase of LTP?

DR. STEVENS: No. That's something I think we have the tools for doing it now, but we haven't done it yet.

1.1.4. The Structural Model of LTP

DR. MORRISETT: Along those terms, when you were talking about long-term changes in synaptic structure, Frances Edwards has a really nice model for synapses budding (Edwards, 1995). Would you care to comment on that?

DR. STEVENS: It may be true. I don't know any evidence for it.

DR. MORRISETT: I think the time course she applies is a bit sooner than a day.

DR. STEVENS: From my point of view, all of the morphological models for synaptic plasticity are just fantasy. I'm not saying that they're not true, but you just can't know if they're true or not. The reason is that you can't correlate the changes in release probability or whatever changes you make with the very same synapses. Synapses are very different. You see all sorts of things. There is a lot of sampling problems. There are a billion synapses per cubic millimeter in there. If you change two percent of them in a physiological experiment, you're not going to find them in morphological studies.

DR. MORRISETT: You don't hold back any punches. You don't think that the structural changes could account for presynaptic and postsynaptic alterations?

DR. STEVENS: I firmly believe that there are structural changes that are responsible for long-term plasticity. But I just don't know any evidence for it. I believe it.

DR. MORRISETT: But that doesn't account for the dichotomy.

DR. STEVENS: I don't think so.

1.2. Q&As between Audience and Dr. Morrisett

1.2.1. NMDA-Dependent vs. L-Type Calcium Channel-Dependent Long-Term Depression (LTD)

DR. TSIEN: There are forms of LTD that do involve L-type calcium channels. Since you have this really intriguing contrast between LTP and LTD induced by NMDA receptors, I wonder if you've gone to great lengths to see that form of L-type calcium channel-dependent LTD, to see whether ethanol differential still holds there, too.

DR. MORRISETT: In hippocampus?

DR. TSIEN: Yes. The kinds that I know best are the ones that are worked on in culture or in young animals. There's a paper by Bolshakov and Siegelbaum that is not in culture but in slices from young animals. In their work, the LTD can be almost completely abolished by nifedipine or nimodipine (Bolshakov & Siegelbaum 1994). Can you go to that circumstance and show us what ethanol does, so as to complete the pattern?

DR. MORRISETT: I didn't spend the time to characterize it. In this system, it is young, we virtually exactly reproduced Malenka's NMDA receptor-dependent LTD (Oliet et al., 1997; Nicoll & Malenka, 1995). And it was age-dependent and APV-sensitive.

DR. TSIEN: I'm not criticizing that. I'm just saying that there's a Bolshakov and Siegelbaum paradigm, and it's very reproducible. Dr. Chuck Stevens and we have seen similar types of LTD that's dependent on L-type channels. Wouldn't it be nice to know how that form of plasticity responds to ethanol?

DR. MORRISETT: It would be critical. I think it would be paramount. I'm making the assumption that in our system the complete block of the low-frequency induced LTD with APV means that that form is NMDA receptor-dependent.

DR. TSIEN: No problem.

DR. MORRISETT: What's the induction mechanism of the L-type channel or the induction paradigm that Siegelbaum used?

DR. TSIEN: It's something like 5 Hz up to a certain amount of time, and it's dependent on intracellular calcium. It's postsynaptically induced. It's all the things that you would be looking for. It seems a big worry that you can have a differential effect. You're focusing on the NMDA receptors. I agree with you that if there were only one kind of NMDA receptor in one place, you'd have a hard time. But suppose what you're generally doing is decreasing the efficiency of the NMDA receptor. You have less calcium coming in. If the LTP is very strictly dependent on achieving very high levels of free calcium, that might be more susceptible, whereas if the LTD depends upon getting calcium into some sort of intermediate range of intracellular calcium, you might still get LTD. We don't know enough to rigorously exclude the possibility that ethanol is mainly working on the NMDA receptor. There isn't enough evidence yet to rule out that parsimonious and rather boring explanation.

DR. MORRISETT: Your point is extremely well taken. From my point of view, it all depends on how you look at the degree of ethanol inhibition of the NMDA response relative to exactly the requirement for the induction mechanism. And actually I think you're probably more right than I am, because ethanol is an incomplete antagonist. I think most everybody in this room who has ever looked at ethanol and NMDA effects see a residual ethanol-insensitive NMDA response.

1.2.2. Chronic Alcohol Exposure—Paroxysmal Depolarizing Shift (PDS)—Withdrawal Seizure

AUDIENCE MEMBER: Dr. Morrisett, in your final slide you mentioned something about a cellular PDS mechanism underlying the withdrawal seizure. Could you just explain that for me? I don't know what that is.

DR. MORRISETT: Basically, prolonged exposure to ethanol will cause an up-regulation of NMDA receptors, either their regulation, their activity, or the number of channels themselves. In terms of the cellular mechanism for expression or the increase in excitability, you could explain that by an increase in NMDA receptor function—as opposed to the classical concept of the role of NMDA receptors in plasticity. So in terms of the expression of a withdrawal seizure and the cellular PDS or paroxysmal depolarizing shift or the synaptic drive required to get that ictal event kicked off on a cellular level, I didn't get a chance to really give you the total argument. But in terms of the direct correlation that we see between the acute withdrawal after chronic exposure—that is, while we're doing the recording—we can see the up-regulation of the NMDA component of synaptic transmission that immediately precedes the expression of withdrawal seizures in a hippocampal explant preparation. This observation suggests to us that the excitatory synaptic drive that may underlie the cellular PDS could be, to a large extent, NMDA receptor mediated.

AUDIENCE MEMBER: Do you know if anyone has ever seen that in a slice—not cultured—a slice taken from a treated animal?

DR. MORRISETT: Hillary Little has seen increased NMDA receptor function after chronic ethanol exposure (Whittington et al., 1995; Ripley & Little 1995), and John Littleton, too.

AUDIENCE MEMBER: Right. But not ictal events like that one.

DR. MORRISETT: Not ictal events. That's the value of the explant preparation. I'm not even talking at all about L-type channel function here.

1.3. Q&As between Audience and Dr. Steffensen

1.3.1. Effects of Anesthesia vs. Effects of Ethanol

AUDIENCE MEMBER: Have you considered the possibility that the differences between your *in vitro* and *in vivo* physiology might have something to do with your halothane anesthesia?

DR. STEFFENSEN: Definitely. I think I mentioned all of these studies were done in parallel in freely-behaving animals. The reason why we choose halothane anesthesia is because it has the least amount of effects on ethanol effects or at least interactive effects on ethanol effects on hippocampal responses. But the effects are identical in the freely-behaving animal, so we're assured that it's not the effects of anesthesia.

1.3.2. Properties of GABAergic Neuron in the Ventral Tegmental Area (VTA)

DR. GONZALES: (Rueben Gonzales from the University of Texas). I had several questions for Scott Steffensen. I'm really interested in this GABAergic cell that you've isolated in the VTA. Is it in the VTA? Where exactly is it? Do you have any idea what could be the actual mechanism for this dramatic inhibition? What is driving that cell? You said it had a high firing rate. Any information on this cell would be appreciated.

DR. STEFFENSEN: The cells are located slightly dorsal to VTA dopamine neurons. We do find them interspersed amongst dopamine neurons. For the most part, they're slightly dorsal, which I believe is just the opposite from the situation in substantia nigra where the reticulata neurons are actually ventral to the dopamine neurons and slightly lateral from what I remember. So, in some ways, it's a little odd. But we have, indeed, shown that they are forming contacts to dopamine neurons. They're not just local circuit neurons. They project to the thalamus. We have a little bit of evidence that they project to the basolateral nucleus of the amygdala. They receive an NMDA receptor mediated-input from the thalamus that we believe is actually contributing to their spontaneous firing rate. They have an intrinsic firing rate, but if you give local application of APV or systemic administration of MK801, the spontaneous firing rate decreases about 30 or 40 %. So we know that there's an NMDA component that modulates or somehow drives their spontaneous activity. The input from the ventral lateral thalamus to the VTA is also NMDA input, as driven activity can be blocked by NMDA receptor blockers as well. But we know their spontaneous activity is partly intrinsic, because APV or NMDA antagonists do not block the spontaneous firing rate completely. I think I mentioned that there was spontaneous firing in the anesthetized animals around 20 Hz. In the freely-behaving animal, the mean firing rate is somewhere around 60 or 70 Hz, and it's modulated by movement. We're not sure what type of movement, but we feel it's during the onset of movement, and may be associated with goal-directed behaviors, but we have not parsed that out. Their activity sometimes can exceed 160 Hz during movement.

1.4. Q&As between Audience and Dr. Browning

1.4.1. Effects of Ethanol on NMDA Receptor-Mediated LTP

DR. SIGGINS: Am I right in assuming that you're looking at ethanol effects on NMDA EPSPs recorded extracellularly?

DR. BROWNING: Exactly. Fields EPSPs.

DR. SIGGINS: And it seemed like the ethanol concentrations you were using were fairly high. Do you have any idea of what the IC_{50}s might be in that preparation? I know that David Lovinger and co-workers did some similar studies in hippocampus using the same kind of extracellular methodology and got a fairly low IC_{50} at about 35 mM or so (Lovinger et al., 1989).

DR. BROWNING: At 50 mM, the inhibition is modest.

DR. SIGGINS: Right. That's what I was thinking. My question is if you were looking at intracellular recording of ethanol effects on locally applied NMDA or on NMDA EPSPs, would you see a more pronounced effect of ethanol?

DR. BROWNING: I don't have any way of telling.

AUDIENCE MEMBER: And then would it correlate better with the LTP suppression?

DR. BROWNING: If I had to guess, in my preparation, I would say no. But I don't have that data, so I really can't say.

DR. MORRISETT: Before we get off that point, just to address that, we've done that in explants. Mark Thomas in our lab did that with pharmacologically isolated NMDA EPSCs (Thomas et al., 1998). There has been a little bit of variability through the years between different labs in terms of ethanol sensitivity, especially comparing native and recombinant systems. For pharmacologically isolated NMDA-evoked responses, we've actually seen fairly stable inhibitory effects through these different preparations that we've used, usually on the order of 50 to 70 % inhibition at 70 mM, and that's similar to what you're going to be seeing at 100 mM. The level of inhibition that we've seen just slightly greater degree of inhibition than what you have seen.

DR. SIGGINS: I thought he was seeing 20 %, 17 to 20 %.

DR. MORRISETT: Against the field EPSP? I was under the impression...

DR. SIGGINS: Where did Dave Lovinger go? What was his percentage?

DR. LIU: He's left.

DR. BROWNING: Our experiments were done on 8–12 week old animals. I think if you use real young animals, you'll see quite a bit bigger NMDA inhibition by ethanol as shown by Swartzwelder (Swartzwelder et al., 1995).

DR. SIGGINS: I think, in talking to Forest Weight, they noticed what you're referring to, which they call an upward drift in the IC_{50}s.

DR. MORRISETT: Right. Exactly.

DR. SIGGINS: Especially in the expression systems, they get IC_{50}s into the hundreds of mM of ethanol.

DR. BROWNING: I think it's important to look at the age of animal you are using and other conditions as well (e.g., age, preparation, incubation temperature, perfusion conditions, synaptic vs. NMDA application etc.). In our hand, with adult animals we see a 20 % NMDA receptor inhibition by 100 mM ethanol.

DR. SIGGINS: And that's NMDA EPSP amplitude or slope?

DR. BROWNING: Slope.

DR. MORRISETT: I think it's hard to correlate in my mind: How does slope reflect on percentage inhibition of a current, what's essentially underlying the slope?

DR. BROWNING: I've always liked the slope measurement because it minimizes polysynaptic effects. And remember when we used the other inhibitors, we were using the same slope measurement. So, in other words, when you inhibited that slope the same amount with those other inhibitors, they didn't block LTP as well as ethanol did.

2. DISCUSSION BETWEEN AUDIENCE AND SPEAKERS OF SESSION III

2.1. "Equal Opportunity" for Action Potential to Reach the Terminal

DR. ALGER: Dr. stevens, in interpreting your work on the organotypic slice, did you assume that the action potential has the same probability of invading each synaptic terminal of a given cell on each trial?

DR. STEVENS: We have recorded calcium transients presynaptically. As far as we can tell, the action potential gets every place. It doesn't fail. The calcium transients are always there presynaptically, and wherever you look, you see them.

DR. STEFFENSEN: And the corollary to that is that you can assume all synapses were potentiated there?

DR. STEVENS: No. Different synapses presumably potentiate different amounts, but they all have an equal chance.

DR. STEFFENSEN: I thought you were implying that they were all potentiated.

DR. STEVENS: I wouldn't think so.

2.2. Effects of Ethanol on Action Potential in Fine Branches

DR. ALGER: There is some evidence from acute slice preparations that, as measured by the synaptic terminal calcium transients, action potential invasion of all synapses of a given cell is not uniform, perhaps because of branch point conduction block or other factors. The question is: Does ethanol have effects on action potential conduction in fine terminal branches that could alter its probability of invading all synaptic terminal?

DR. ?: At very high concentrations, there's a good deal of evidence that alterations in potassium channel function can cause alterations in action potentials. Dr. Sarah Appel might want to address this more. She's the potassium channel person. But I'm not aware of any changes that are substantial in voltage-gated channels.

2.3. Ethanol Sensitivity of LTP: *in Vivo* vs. *in Vitro*

DR. BROWNING: There's one other point that I wanted to ask Scott Steffensen. There are a number of labs besides our own that have shown high frequency-induced LTP to be ethanol-sensitive, whereas what you are saying is that if you lesion the hippocampus *in vivo* to essentially make a disconnected hippocampal preparation, you don't see an ethanol effect on high frequency LTP.

DR. STEFFENSEN: That's correct.

DR. BROWNING: Okay. Slice labs that have seen inhibition there. What's your latest explanation for the discrepancy between your result and that from other groups?

DR. STEFFENSEN: The only way I could possibly reconcile the differences is—and I've given considerable thought to this—is the fact that inhibition is somewhat compromised in the slice preparation. So when it is compromised, things are going to be biased away from inhibition and towards either excitatory synaptic transmission, whether it be NMDA receptor-mediated or not. Perhaps if you're having alcohol effects on both NMDA receptors and GABA receptors and it's biased in this case towards—or away from inhibitory processes, you're not going to see the effects on inhibitory neurotransmission as much as we see it *in vivo* with a full complement of inhibition.

DR. BROWNING: Has anybody else looked at plasticity *in vivo*?

DR. STEFFENSEN: Not too many people, no.

DR.SIGGINS: I remember a paper by Ben-Ari and co-workers, showing that the interneurons in a slice didn't actually die after anoxia, i.e., they weren't actually lost, but they were sort of dysfunctional. They were more difficult to activate by feed-forward kinds of excitatory input from pyramidal neurons. It's a fairly recent paper by Ben-Ari's group (Khazipov et al., 1995).

DR. BROWNING: That could account for your findings of the differences.

2.4. Future Direction of Research on Synaptic Plasticity

DR. LIU: I have a question for Dr. Stevens. Sitting here for the whole afternoon as a pioneer in the synaptic plasticity field, what's your opinion about the plasticity studies in the alcohol research field?

DR. STEVENS: Well, plasticity's complicated business, and I don't have anything very insightful or wonderful to say about it. But in general, I think that alcohol, like any other potent effect, is very important. It's important to understand what the mechanisms are, and so for people who work on LTP and don't take a primarily pharmacological approach, I think that we have something to offer to the people who study alcohol effects, because anything that we turn up they can fit into their scheme. But on the other hand, any alcohol effects are very important for understanding LTP mechanisms. For example, Mike Browning can knock out the NMDA receptors with his 100 mM alcohol and have a bigger effect than you would have with blocking them to the same extent with AP-5, tells you there's something really more complicated going on. So I think it's important to go both ways from the drug manipulations and from the basic science part.

DR. LIU: Any more questions? I guess everybody's as worn out as I am. I'm very touched that all of you are still here after such a long day. I can see a lot of familiar faces who are already devoted to alcohol research, and I'm also very happy to see a lot of new faces here, and hopefully you will join our research field in the future. I'd like to thank all of our speakers and all of the people who have contributed to this very successful symposium.

REFERENCES

Bolshakov VY, Siegelbaum SA (1994) Postsynaptic induction and presynaptic expression of hippocampal long-term depression. *Science* **264(5162):**1148–1152

Edwards FA (1995) LTP--a structural model to explain the inconsistencies. *Trends Neurosci* **18(6):**250–255

Isaac JT, Oliet SH, Hjelmstad GO, Nicoll RA, Malenka RC (1996) Expression mechanisms of long-term potentiation in the hippocampus. *J Physiol Paris* **90(5–6):**299–303

Khazipov R, Congar P, Ben-Ari Y (1995) Hippocampal CA1 lacunosum-moleculare interneurons: comparison of effects of anoxia on excitatory and inhibitory postsynaptic currents. *J Neurophysiol* **74(5):**2138–2149

Lovinger DM, White G, Weight FF (1989) Ethanol inhibits NMDA-activated ion current in hippocampal neurons. *Science* **243:**1721–1724

Nicoll RA, Malenka RC (1995) Contrasting properties of two forms of long-term potentiation in the hippocampus. *Nature* **377**:115–118

Oliet SH, Malenka RC, Nicoll RA (1997) Two distinct forms of long-term depression coexist in CA1 hippocampal pyramidal cells. *Neuron* **18(6)**:969–982

Ripley TJ, Little HJ (1995) Ethanol withdrawal hyperexcitability *in vitro* is selectively decreased by a competitive NMDA receptor antagonist. Brain Res 699:1–11

Swartzwelder HS, Wilson WA, Tayyeb MI (1995) Differential sensitivity of NMDA receptor-mediated synaptic potentials to ethanol in immature versus mature hippocampus. *Alcohol Clin Exp Res* **19(2)**:320–323

Thomas MP, Davis MI, Monaghan DT, Morrisett RA (1998) Organotypic brain slice cultures for functional analysis of alcohol-related disorders: novel versus conventional preparations. *Alcohol Clin Exp Res* **22(1)**:51–59

Whittington MA, Lambert JD, Little HJ (1995) Increased NMDA receptor and calcium channel activity underlying ethanol withdrawal hyperexcitability. Alcohol Alcohol. 30:105–114

CONTRIBUTORS

Gary L. Aistrup
Northwestern University
Chicago, IL 60611-3008

Bradley E. Alger
University of Maryland
Baltimore, MD 21201

Scott Bentz
Wright State University
Dayton, OH 45435

Michael Browning
University of Colorado
Denver, CO 80262

Benson Chu
University of Massachusetts
Worcester, MA 01655

Alejandro M. Dopico
University of Massachusetts
Worcester, MA 01655

Thomas V. Dunwiddie
University of Colorado
Denver, CO 80262

Walter A. Hunt
National Institute on Alcohol Abuse and Alcoholism
Bethesda, MD 20892-7003

Jon M. Lindstrom
University of Pennsylvania
Philadelphia, PA 19104-6074

Yuan Liu
National Institute on Alcohol Abuse and Alcoholism
Bethesda, MD 20892-7003

David M. Lovinger
Vanderbilt University
Nashville, TN 37232-0615

Samuel G. Madamba
The Scripps Research Institute
La Jolla, CA 92037

William Marszalec
Northwestern University
Chicago, IL 60611-3008

Richard A. Morrisett
University of Texas
Austin, TX 78712-1074

Haruhiko Motomura
Northwestern University
Chicago, IL 60611-3008

Keiichi Nagata
Northwestern University
Chicago, IL 60611-3008

Toshio Narahashi
Northwestern University
Chicago, IL 60611-3008

Zhiguo Nie
The Scripps Research Institute
La Jolla, CA 92037

Douglas W. Sapp
University of Connecticut
Farmington, CT 06030-6125

James Schummers
Massachusetts Institute of Technology
Cambridge, MA 02139

Gordon M. Shepherd
Yale University
New Haven, CT 06510

George R. Siggins
The Scripps Research Institute
La Jolla, CA 92037

Scott C. Steffensen
The Scripps Research Institute
La Jolla, CA 92037

Charles F. Stevens
Salk Institute
La Jolla, CA 92037

Hideharu Tatebayashi
Northwestern University
Chicago, IL 60611-3008

Mark P. Thomas
University of Texas
Austin, TX 78712-1074

Steven N. Treistman
University of Massachusetts
Worcester, MA 01655

Richard W. Tsien
Stanford University
Stanford, CA 94305-5426

Fan Wang
University of Pennsylvania
Philadelphia, PA 19104-6074

Hermes H. Yeh
University of Connecticut
Farmington, CT 06030-6125

Jay Z. Yeh
Northwestern University
Chicago, IL 60611-3008

Qing Zhou
Vanderbilt University
Nashville, TN 37232-0615

INDEX

The manufacturer's authorised representative in the EU is Springer Nature Customer Service Centre GmbH, Europaplatz 3, 69115 Heidelberg, Germany. If you have any concerns regarding our products, please contact ProductSafety@springernature.com

Printed and bound by CPI Group (UK) Ltd, Croydon, CR0 4YY
23/04/2026
02095622-0004